Cryopreservation of Fish Gametes

Judith Betsy • Stephen Kumar
Editors

Cryopreservation of Fish Gametes

Springer

Editors
Judith Betsy
Department of Aquaculture, Fisheries
College and Research Institute
Tamil Nadu Dr. J. Jayalalithaa
Fisheries University
Tuticorin, Tamil Nadu, India

Stephen Kumar
Directorate of Sustainable Aquaculture
Tamil Nadu Dr. J. Jayalalithaa
Fisheries University
Thanjavur, Tamil Nadu, India

ISBN 978-981-15-4024-0 ISBN 978-981-15-4025-7 (eBook)
https://doi.org/10.1007/978-981-15-4025-7

This Springer imprint is published by the registered company Springer Nature Singapore Pte Ltd.
The registered company address is: 152 Beach Road, #21-01/04 Gateway East, Singapore 189721, Singapore

Preface

Global fish production has been growing at a pace of around 10% annually during the past one decade. The pushing factor behind this growth is the aquaculture production. Many countries could adopt species specific and region-specific aquaculture practices because of the availability of seeds for stocking, which is the result of success in induced breeding of fishes. It is not an underestimation that induced breeding of fishes brought in a major thrust for the fish culture in the world. However, the intricacies in the maturation and induced breeding process demand a better understanding of the basic aspects related to fishes. It is a natural wonder to note that when feed and water quality alone can be managed for the captive maturation of few fishes, some other fishes need a lot more than that. These complexities demand more focus on breeding research of fishes.

Although there is a demand for many fishes and shellfishes, still naturally exploited stock is meeting the consumers' demand owing to the non-availability of a technology for their culture and seed production. Hybridization and development of new strains are the two imminent stages in the captive production of fish seeds. But there is not much outcome in the above two aspects in aquaculture, excepting GIFT and common carp aquaculture. Similar developments are felt essential in other important species aquaculture. At this point, gametology and knowledge on improvement of quality gametes assume importance. This can further lead to the advanced era of preservation and use of quality germplasm as it was done in animal reproduction. In aquaculture, it is yet to make an in-road. Nevertheless, research in this line is assuring new avenues for quality seed production.

Making of a referral book in the cryopreservation of fish gametes has the genesis when we were in our fourth year of research in spermatology and cryopreservation in carps. The contributions that were referred from the animal semen cryopreservation can no way be ignored. But in all these things, it was realized that a detailed collection of important aspects of cryopreservation of fish gametes with due acknowledgements of contributions of Lazzaro Spallanzani (1776) is needed as a one-stop reference for the beginners and early stage researchers in this line. Therefore, the concept of this book was born.

Giving a shape to the dream is always a challenge. It was so in this book project too. But the encouragement and positive response from Springer deserves mention here, as it was the real ignition that pushed us forward with this book project. The consideration we enjoyed with them could not be easily described. The overall frame of the book has reached a shape with their consent on which we could work swiftly.

Identification of right person would assure half success in any venture. In this project, the authors have been identified after a thorough analysis of the subject matter relevance and their research standing. This has greatly helped us to twine the chapters like energetic aspects of spermatozoa, factors affecting the quality of gametes, quality evaluation procedure and process, cryopreservation of spermatozoa of marine fish, crustaceans, bivalves, oocytes, embryos, and stem cells. In short, the researchers identified for the chapters have done their excellence, which has doubtlessly elevated the referral quality of the book.

As the Editors of this book, we have assumed greater responsibility of placing the chapters appropriately and twining the subject matter so as to maintain the harmony of science in the book. We left no stone unturned in bringing the wholesomeness for the book so as to make the readers feel the gourmet of scientific lead. We have also authored chapters like reproductive physiology of fishes, endocrine regulation of reproduction in fishes, and history of cryopreservation, which are basic and needed to assist in understanding the complex and high end research aspects in the coming chapters.

The difficulties and hardships that were met during our early days of research in the cryopreservation of fish spermatozoa motivated us to compile and present an array of chapters to impart knowledge and an insight into the reproductive aspects of different cultivable fishes together with the advanced knowledge on the cryopreservation and artificial fertilization that have been done in different species in the world. This book besides serving as guide to the entry level researchers and farmers also will help them understand the gametes and their production in fishes.

The sincere efforts that have been put in by the authors of different chapters in this book are gratefully acknowledged. The support that has been rendered by the Springer's, Inc. is awesome and enjoyable. We are indebted.

In conclusion, we welcome the comments and feedback from the readers on the book and the chapters. The authors with the corresponding status can be contacted as well. Constructive criticism will help us make the next edition with more contributions.

Tuticorin, Tamil Nadu, India — Judith Betsy
Thanjavur, Tamil Nadu, India — Stephen Kumar

Contents

About the Editors

Judith Betsy is an Assistant Professor at the Department of Aquaculture, Fisheries College and Research Institute of Tamil Nadu Dr. J. Jayalalithaa Fisheries University, Tuticorin, India. Holding a doctorate in Aquaculture, she is a recipient of the Maulana Azad National Fellowship from the University Grants Commission, New Delhi, India. Having worked in the area of fish gamete cryopreservation and spermatology for the past 8 years, she has been honored with 10 awards including gold medals for her outstanding academic performance during her graduate studies and has published over 25 research papers in peer-reviewed journals around the world, more than 70 technical articles, and 4 books.

Stephen Kumar is a University Officer and Director of Sustainable Aquaculture at Tamil Nadu Dr. J. Jayalalithaa Fisheries University, Thanjavur, India. He has been involved in aquaculture research since 1988 and has over 300 publications, including 75 research papers in national and international journals, to his credit. He has been honored with various awards, citations, and commendations for his academic, research, and administrative performance. His research interests include fish and crustacean physiology, breeding, nutrition, bioenergetics, cryopreservation of fish gametes, and artificial fertilization. Dr. Kumar also serves as a consultant to a number of firms involved in aquaculture around the globe. He has been involved in the development of fish and shrimp health care products in various countries, has more than 30 years of field experience in freshwater and brackish water aquaculture, and has coordinated more than 10 funded projects.

Reproductive Physiology in Fishes

1

Judith Betsy and Stephen Kumar

Abstract

Natural selection, which is the basic of theory of evolution as proposed by Charles Darwin and Alfred Wallace, is widely accepted as the central paradigm in biology. Any individual who performs better at surviving and reproduction alone will be selected by nature, which leads to the concept of survival of the fittest. Hence reproduction is an important criteria any individual should possess so that their genes will be disproportionably represented in the next generation. Thus, studying reproduction and the physiological mechanism behind reproduction becomes very essential. In this chapter, different types of reproduction and fertilization in fishes, spawning dynamics, mating systems and factors affecting reproduction are discussed.

Keywords

Reproduction · Mating · Physiology · Spawning · Fish

1.1 Introduction

Reproduction is a process by which organisms replicate themselves wherein genetic material is passed from one generation to the next. There are wide variety of organisms and animals on earth and each group of organism has a unique mode of reproduction. The reproduction mode ranges from simple division to sexual method.

J. Betsy
Department of Aquaculture, Fisheries College and Research Institute, Tamil Nadu Dr. J. Jayalalithaa Fisheries University, Tuticorin, India

S. Kumar (✉)
Directorate of Sustainable Aquaculture, Tamil Nadu Dr. J. Jayalalithaa Fisheries University, Thanjavur, India

J. Betsy, S. Kumar (eds.), *Cryopreservation of Fish Gametes*,
https://doi.org/10.1007/978-981-15-4025-7_1

Single-celled organisms such as bacteria and protozoans exhibit wide range of reproductive methods. In some unicellular algae, the youngones can be produced by multiple fission, while in case of yeasts, the organism itself turns into a gamete and fuses its nucleus with the opposite sex through a process called conjugation. Ciliate protozoans reproduce by binary fission.

Reproductive modes differ among multicellular organisms also. Most of the plants follow asexual or vegetative method of reproduction by giving off asexual spores or by budding off and giving a new individual or by spreading rhizoid or giving off new sprouts from rhizome or producing new plant from shoot. Similarly, lower animals like sponges produce gemmules, which later become sponges; coelenterates can reproduce by budding; worms can divide into two and each half can then regenerate. However, higher animals can reproduce only by sexual method, which produce gametes which are haploid in nature and after fertilization becomes a diploid, uninucleate zygote. The gametes may be equal in size as in isogamy, or one may be slightly larger than the other as in anisogamy, but the majority of forms exhibit oogamy wherein there is a large egg and a minute sperm.

1.2 Reproduction in Fishes

As in higher animals, reproduction in fishes takes place by sexual method. Sex differentiation in fishes is determined by interactions between genes and the environment. According to Patzner (2008), about 88% of fishes are gonochoristic, the case where fishes may be either male or female. However, about 2% of teleost species representing approximately 30 families exhibits functional hermaphroditism. Avise and Mank (2009) have stated that hermaphroditism in fishes can be synchronus or asynchronus. In synchronus hermaphroditism, male and female gonads mature at the same time as it can be seen in *Serranus scriba*. Asynchronus hermaphroditism includes protandrous, in which fishes function as males first and then becomes females as in *Lates calcarifer* and *Sparus auratus*; and protogynous in which fishes function as females first and then become males as in *Epinephelus* sp. Fishes belonging to the four families of the order Perciformes such as Serranidae, Sparidae, Centracanthidae and Labridae; four families of the order Myctophiformes; few species of the order Cyprinodontiformes, stomiatioids of Salmoniformes and few species of order Synbranchiformes and Scorpaeniformes exhibit hermaphroditism (Atz 1964). The reproduction processes and their behaviours in fishes are given in Table 1.1.

1.3 Spawning Dynamics

Fishes exhibit two major temporal patterns of reproductive activity such as semelparity and iteroparity (Roff 1992, 2002; Stearns 1992; Wootton 1998). In semelparity, fishes breed once and then die because in these organisms, the physiological changes associated with reproduction result in consequences that inevitably

Table 1.1 Reproduction in fishes

A. Non-guarders	
(a) Open substrate spawners	
Pelagic spawners (Pelagophils)	*Lates niloticus*
Rock and gravel spawners with pelagic larvae (Lithopelagophils)	*Prochilodus sp.*
Rock and gravel spawners with benthic larvae (Lithophils)	*Labeo sp.*
Nonobligatory plant spawners (Phytolithophils)	*Rutilus rutilus*
Obligatory plant spawners (Phytophils)	*Puntius gonionotus*
Sand spawners (Psammophils)	*Gobio gobio*
Terrestrial spawners (Aerophils)	*Brycon petrosus*
(b) Brood hiders	
Beach spawners (Aeropsammophils)	*Leuresthes tenuis*, puffer fish, stickleback
Annual fishes (Xerophils)	*Nothobranchias sp.*
Rock and gravel spawners (Lithophils)	Salmonid sp.
Cave spawners (Speleophils)	*Julidochromis* sp.
Spawning in live invertebrates (Ostracophils)	*Rhodeus sericeus*
B. Guarders	
(a) Substrate choosers	
Pelagic spawners	*Channa sp.*
Above water spawners (Aerophils)	*Copeina arnoldi*
Plant spawners	*Polypterus sp.*
Rock spawners	*Loricaria parva*
(b) Nest spawners	
Bubble nest builders (Aphrophils)	Gouramis
Hole nester	*Cottus aleoticus*
Miscellaneous nest builders (Polyphils)	*Notopterus chitala*
Rock and gravel nesters	*Ambloplites rupestris*
Anemone nesters	*Amphiprion sp.*
Plant material nesters—glue making	*Gasterosteus aculeatus*
Plant material nesters—non glue making	*Micropterus salmoides*
C. Bearers	
(a) External bearers	
Transfer brooders	*Oryzias latipes*
Skin brooders	*Bunocephalus sp.*
Forehead brooders	*Kurtius gullivers*
Gill chamber brooders	*Typhlichthys subterraneus*
Pouch brooders	*Syngnathus sp.*
Mouth brooders	*Oreochromis sp.*
Intestinal brooders	*Tachysurus barbus*
(b) Internal bearers	
Facultative internal bearers	*Rivulus marmoratus*
Obligate lecithotrophic live bearers	*Poecilia reticulata*
Viviparous trophoderms	*Anableps dowii*

end in death. Iteroparous fishes survive after reproduction to breed again where reproduction takes place at yearly intervals. Similarly, within a breeding season, two temporal patterns of spawning occur in female fishes namely total and batch spawners (Wootton 1998; Patzner 2008). In total spawners, the female releases all her eggs in single spawning, whereas in batch spawners, the female releases her eggs in batches at intervals during the breeding season.

1.4 Mating Systems

Shuster and Wade (2003) defined mating system as the species-specific pattern of male–female associations. Mating systems imply the number of mates individuals acquire with a description of how those mates are acquired, the characteristics of pair bonds and patterns of parental care by each sex and resolves the sexual conflicts since both the sexes have different roles in reproduction, the optimal behaviors and traits for each are likely to be different (Berglund 1997).

In promiscuity, both the sexes have multiple partners during the breeding system with little or no mating choice (*Clupea harengus*). A form of promiscuity, where a male will spawn with several females and a female with several males with mate choice is polygynandry (*Rhodeus amarus*). In polygamy, an individual of one sex has multiple partners during the breeding system, but individuals of the opposite sex have only one partner. Polygamous mating system is further divided into polygyny and polyandry. In polygynous, male has multiple female partners during the breeding season and it is the common form and can be seen in many cichlids like *Oreochromis niloticus.* Polyandry mating system is one in which a female has multiple male partners during the breeding season. However, it is an uncommon form of polygamy and has only been reported in anemone fish (*Amphiprio ninae*). In contradictory to all these, in monogamy, a single male and female form a mating pair and show some degree of bonding with the mate (*Cichlasoma*).

1.5 Fertilization in Fishes

Fertilization is union of a spermatozoa nucleus, of paternal origin, with an egg nucleus, of maternal origin, to form the primary nucleus of an embryo which results in fusion of the hereditary material of two different gametes, each of which carries half the number of chromosomes typical of the species. Gilkey (1981) defined fertilization as a process which encompasses all the events between release of gametes and fusion of male and female pronuclei.

In fishes, generally, females have separated ovaries. However, in some fishes, the right and left ovaries are fused and resemble as single organ. The ovary connects to the body wall by a short oviduct and opens posteriorly to the anus. Ovulated eggs are spawned with the ovarian fluid out of the genital pore. Males with internal fertilization have gonopodium which is a modified anal fin to introduce spermatozoa into the female genital tract. However, such a copulation organ is not found in fishes with external fertilization (Iwamatsu 2000).

At the time of breeding, fishes exhibit secondary sexual characteristics with which a male and female can be identified. Permanent dimorphism can be noticed in *Betta splendens,* whereas temporary dimorphism can be seen in many fish species (e.g., *Gasterosteus aculeatus*) only during breeding season.

1.5.1 External Fertilization

Most of the fishes are external spawners. In this mode of fertilization, female releases the egg either in the water or on any substrate over which males release their sperm. Water serves as activating medium and induces motility in spermatozoa. Thus, spermatozoa find their way to the eggs and fertilize them by entering through their micropyle. In this type of fertilization, female fertility can be limited by sperm availability.

1.5.2 Internal Fertilization

In this mode of fertilization, gametes fuse inside the female reproductive tract. In order to transfer sperm from males to females, males have a specialized copulatory structure like gonopodium (*Poecilia reticulata*). It has been reported that in teleosts, male intromittent organs are found in a number of families (Wootton and Smith 2015).

In species with internal fertilization, spermatozoa are often organized into bundles termed as spermatophores but in others, they are arranged into unencapsulated sperm bundles termed as spermatozeugmata, e.g., in viviparous kelp blennies (Clinidae) (Fishelson et al. 2006) and *P. reticulata* (Magurran 2005). As in the case of external fertilization, mucins also occur in the ejaculates of internal fertilizers, whose function is to increase fertilization efficiency by extending the period over which motile spermatozoa are released (Ginzburg 1968).

As in *P. reticulata,* some internal fertilizers can store sperm (Magurran 2005), which helps females to produce broods in the absence of males for up to 8 months. However, the capacity to store sperm appears to vary among populations. Stored spermatozoa are used to fertilize eggs even if the female mates with additional males, but the fertilization success of stored spermatozoa is typically lower than that from recently deposited sperm (Wootton and Smith 2015).

1.5.3 Quasi-Internal Fertilization

Fertilization which is basically external, but takes place in a site so restricted that it is analogous to internal fertilization is called quasi-internal fertilization. In mouth-brooding cichlids, buccal fertilization takes place where females deposit eggs on a substrate and collect them immediately into her mouth. In many species, males possess coloured spots on their anal or pelvic fins, termed egg spots, which resemble the female's egg in size and colour (Fryer and Iles 1972). Female cichlids respond to egg spots by nipping at the male's fins, and the male releases sperm into her buccal cavity where the eggs are fertilized.

Bronze corydoras catfish (*Corydoras aeneus*, Callichthyidae) have a specialized mating system, sperm drinking. In this type of fertilization, female collects sperm in her mouth from the male's genital opening. The live spermatozoa pass rapidly through the length of her digestive tract and are discharged to her eggs, which she releases into a pouch created by her paired pelvic fins (Kohda et al. 1995). The rapid passage of spermatozoa through the female's digestive tract is possible because this fish exhibits intestinal breathing, whereby air is gulped at the surface and passes through the stomach to the intestine where it is absorbed. This adaptation appears to facilitate the rapid passage of water and spermatozoa through the intestine and to the eggs (Kohda et al. 1995).

In the bitterling fishes (Cyprinidae), fertilization occurs in the gill chambers of living freshwater mussels. Female bitterling deposits their eggs into the gill chamber of a mussel with an unusually long ovipositor and males fertilise the eggs by releasing sperm into the inhalant siphon of the mussel, so that water filtered by the mussel carries the sperm to the eggs (Smith et al. 2004).

1.6 Factors Affecting Reproduction

Reproduction in fishes is affected by internal and external factors. According to Lowerre-Barbieri et al. (2011), environmental factors that show seasonal changes have two possible roles in determining the temporal pattern of reproduction viz., ultimate factors and proximate factors.

1.6.1 Ultimate Factors

The timing of reproduction maximises the individual fitness of the reproducing fishes either by the effect of timing on the survival of the progeny as in most teleosts (Wingfield 2008; Lowerre-Barbieri et al. 2011) or by the effect of timing on the future reproductive success of the reproducing adult as in the case of iteroparous species. Lowerre-Barbieri et al. (2011) stated that ultimate factors act as selective agents operating on heritable changes between generations. The important ultimate factors for the progeny are usually assumed to be availability of suitable food, no risk of predation and avoidance of adverse physical conditions.

1.6.2 Proximate Factors

Factors that affect the physiological processes in order to regulate reproduction are called as proximate factors which include photoperiod, temperature, the lunar and tidal cycles, the chemical composition of water and the flow rate of water. These factors act during the ontogeny of individuals and must be detectable by the sensory systems of the individual fish so that the information can be passed to the central nervous system (CNS). Proximate factors can affect the timing of reproduction by their effects on the initiation of gametogenesis, the stimulation of vitellogenesis and the induction of the final maturation, ovulation and oviposition (Wang et al. 2010). In addition to predictive, proximate cues that initiate gametogenesis, and any local or synchronising cues that are required for ovulation and spawning, there may also be terminating cues, which bring spawning to a close. In some species, the end of spawning may occur even in constant environmental conditions because of an endogenous process (Wootton and Smith 2015).

1.6.2.1 Lunar Related Reproductive Cycles

Lunar cycle influences the seasonal variation in spawning activity. The spawning of estuarine and shallow water marine species is closely related to the lunar cycle (Taylor 1984; Takemura et al. 2010) because of the close association between the lunar cycle and the tidal cycle. Lunar cycles influence spawning in few freshwater species also.

1.6.2.2 Photoperiod

Light appears to be an important factor in controlling the reproduction of fishes. The photoperiod is the most reliable of geophysical cues, and temperate and higher latitudes frequently act as a predictive cue for teleost species, sometimes in association with water temperature. Harrington (1956, 1957) suggested that long photoperiods initiate the gonadal activity and the fishes can spawn in advance of the natural time. However, the influence of light in activating the productive cycle varies from species to species.

Hazard and Eddy (1951) reported that functional maturity can be advanced in *Salvelinus fontinalis* first by exposing them to an increasing photoperiod and later to a decreasing photoperiod. An accelerated light regime can enhance the time of functional maturity in trout (Henderson 1963).

Early maturation under short photoperiod and delayed maturation under long photoperiod were observed in salmonids by Shiraishi and Fukuda (1966). When exposed to longer photoperiods, *Cirrhina reba* attained early maturity (Verghese 1967). According to Sanwal and Khanna (1972), both very long and very short photoperiods are unfavourable for maturation of ovaries in early stages in *Channa gachus*, and there is a delay in the appearance of yolk. But during vitellogenesis, a short photoperiod accelerates the formation of mature oocytes.

1.6.2.3 Temperature

Temperature has direct effects on metabolic rates and can modulate the rate at which all physiological process can occur, including reproduction (Wang et al. 2010). Temperature is known to influence the maturation of gonads in fishes. There is an optimum temperature for breeding of fishes, above and below which they may not reproduce (Hoar and Robertson 1959; Ahsan 1966).

Ahsan (1966) showed that warm temperature stimulates maturation of gonads and spermiation in lake chub. During induced breeding of Indian carps, spawning occurs at a temperature ranging from 24 to 34 °C with an optimum water temperature of about 27 °C (Chaudhuri 1968). Similarly, spawning in *Cirrhinus reba* was observed by increasing water temperature from 25 to 29 °C at the time of injection, and then allowing the water to cool (Bhowmick 1969). Thus, temperature appears to be a triggering factor for spawning in carps.

Hontela and Stacey (1990) reported that decreasing photoperiod and high water temperatures cause gonadal regression in rose bitterling (*Rhodeus ocellatus*, Cyprinidae). Such gonadal regression is frequently followed by a refractory period (Baggerman 1990) during which period, even the usual initiating proximate cues have no effect on gonadal recrudescence. Only at the end to the refractory period proximate cues initiate gonadal recrudescence. This refractory period provides a basic pattern of timing on the reproductive cycle, and probably prevents recrudescence at times unfavourable for reproduction (Wootton and Smith 2015).

1.6.2.4 Habitat and Repressive Factors

The nature of the habitat, like the presence of stones, plants, suitable substratum to lay eggs, also serves as an important stimulus for the fish to breed. Swingle (1956) reported that the excretory matter of fish released into the water constitutes the 'repressive factor', a hormone-like substance which inhibits reproduction in fishes. Similar observation was made by Tang et al. (1963) who observed that *Cyprinus carpio*, *Carassius auratus* and *Tilipia mossambica* do not spawn when overcrowded but spawn when transferred to fresh water and spawning takes place when the repressive factor is sufficiently diluted by the flood water in bunds or ponds. They also stated that ammonia itself may not be the repressive factor, but many other excretory substances may bring about inhibition of spawning in carps.

1.6.2.5 Petrichor

Factor which stimulates the fish to spawn when rain water comes into contact with dry soil is known as "Petrichor" (Lake 1967). This can be obtained by steam distillation of silicate minerals and rocks available in the breeding habitat of the fish. Fishes possess a remarkable olfactory sense, and the odour of petrichor works as a stimulant for spawning. Nikolsky (1962) reported that ripe male secrete a steroid hormone called "copulin" which induces females to spawn followed by mating. Petrichor and copulin stimulate gonad-stimulating hormones after which neurosecretory material (GnRH) is transported from the hypothalamus to the hypophysis through the nerve fibres of the hypothalamico-hypophyseal axis and hypothalamus exercises primary control over the secretion of hormones from the pituitary gland.

1.6.2.6 Food

Availability of required quantity of food at appropriate times during the reproductive cycle is a major factor affecting reproduction. It was reported that high food levels support the growth of fishes, leading to larger body size (Wootton 1998). High level of food intake leads to high fecundity in female fishes. Some fishes stop feeding during spawning season and for reproduction it utilises reserves built during active feeding. Houston et al. (2006) and Stephens et al. (2009) described about capital breeders and income breeders. In capital breeders, prior to breeding fishes consume more feed than its regular feeding level in order to compensate starvation during reproduction. In case of income breeders, energy for reproduction is met from regular feeding.

References

Ahsan SN (1966) Some effects of temperature and light on the cyclical changes in the spermatogenetic activity of the lake chub, in (Agassiz). Can J Zool 44:161–171

Atz JW (1964) Intersexuality in fishes. In: Intersexuality in vertebrates including man. Academic, New York, pp 145–232

Avise JC, Mank JE (2009) Evolutionary perspectives on hermaphroditism in fishes. Sex Dev 3:152–163

Baggerman B (1990) Sticklebacks. In: Reproductive seasonality in teleost: environmental influences. CRC Press, Boca Raton, pp 79–107

Berglund A (1997) Mating systems and sex allocation. In: Behavioural ecology of teleost fishes. Oxford University Press, New York, pp 237–265

Bhowmick RM (1969) Economics of induced breeding of Indian major carps and Chinese carps-FAO/U.N.D.P. Regional seminar on induced breeding of cultivated fishes, India. FR1/1BCF/20, p 13

Chaudhuri H (1968) Breeding and selection of cultivated warm water fishes in Asia and Far East. FAO Fish Rep 4(44):30–66

Fishelson L, Gon O, Holdengraber V, Delarea Y (2006) Comparative morphology cytology of the male sperm-transmission organs in viviparous species of Clinid fishes (Clinidae: Teleostei, Perciforms). J Morphol 267:1406–1414

Fryer G, Iles TD (1972) The cichlid fishes of the great lakes of Africa. Oliver and Boyd, Edinburgh

Gilkey JC (1981) Mechanisms of fertilization in fishes. Am Zool 21:359–375

Ginzburg AS (1968) Fertilization of fishes and the problem of polyspermy. Academy of Science of the USSR, Moscow

Harrington RW (1956) An experiment on the effects of contrasting daily photoperiods on gametogenesis and reproduction in the centrarchid fish, *Enneacanthus obesus* (Girard). J Exp Zool 131:203–223

Harrington RW (1957) Sexual photoperiodicity of the cyprinid fish, *Notropis bifrenatus* in relation to the phase of its annual reproductive cycle. J Exp Zool 135:529–553

Hazard TP, Eddy RE (1951) Modification of the sexual cycle brook trout (*Salvelinus fontinalis*) by control of light. Trans Am Fish Soc 80:158–162

Henderson NE (1963) Influence of light and temperature on the reproductive cycle of the eastern brook trout, *Salvelinus fontinalis* (Michill). J Fish Res Bd Canada 20:859–897

Hoar WS, Robertson GB (1959) Temperature resistance of goldfish maintained under controlled photoperiods. Can J Zool 37:419–428

Hontela A, Stacey NE (1990) Cyprinidae. In: Reproductive seasonality in teleost: environmental influences. CRC Press, Boca Raton, pp 53–77

Houston AI, Stephens PA, Boyd IL et al (2006) Capital or income breeding? A theoretical model of female reproductive strategies? Behav Ecol 18:241–250

Iwamatsu T (2000) Fertilization in fishes. In: Fertilization in protozoa and metazoan animals. Springer, Berlin, pp 89–145

Kohda M, Tanimura M, Kikue-Nakamura M, Yamagishi S (1995) Sperm drinking by female catfishes: a novel mode of insemination. Environ Biol Fish 42:1–6

Lake JS (1967) Rearing experiments with five species of Australian freshwater fishes. I. Inducement of spawning. Aust J Mar Freshwater Res 18:137–153

Lowerre-Barbieri SK, Ganias K, Saborido-Rey F, Murua H, Hunter JR (2011) Reproductive timing in marine fishes: variability, temporal scales, and methods. Mar Coast Fish Dyn Manag Ecosyst Sci 3:71–91

Magurran AE (2005) Evolutionary ecology: the Trinidadian guppy. Oxford University Press, Oxford

Nikolsky GV (1962) On participation of geneticists in working out biological fishery problems. Vestn Mosk Univ (Biol) 6:3–17

Patzner RA (2008) Reproductive strategies in fish. In: Fish reproduction. Science Publishers, Enfield, pp 311–350

Roff DA (1992) The evolution of life histories. Chapman and Hall, London

Roff DA (2002) Life history evolution. Sinauer, Sunderland

Sanwal R, Khanna SS (1972) Seasonal changes in the testes of a freshwater fish, *Channa gachua*. Acta Anat 83:139–148

Shiraishi Y, Fukuda Y (1966) The relation between the daylength and the maturation in four species of salmonid fish. Bull Freshwater Fish Res Lab 16:103–111

Shuster SM, Wade MJ (2003) Mating systems and strategies. Princeton University Press, Princeton

Smith C, Reichard M, Jurajda P, Przbylski M (2004) The reproductive ecology of the European bitterling (*Rhodeus sericeus*). J Zool 262:107–124

Stearns SC (1992) The evolution of life histories. Oxford University Press, Oxford

Stephens PA, Boyd IL, McNamara JM, Houston AI (2009) Capital breeding and income breeding: their meaning, measurement and worth. Ecology 90:2057–2067

Swingle HS (1956) A repressive factor controlling reproduction in fishes. Proc Pacific Sci Congr 8:865–871

Takemura A, Rahman MS, Park Y (2010) External and internal controls of lunar-related reproductive rhythms in fishes. J Biol 76:7–26

Tang YP, Hwang YW, Liu CK (1963) Preliminary report on injection of pituitary hormone to induce spawning of Chinese carps. FAO, Indo-Pacific Fisheries Council, Occasional Paper 63/14

Taylor MH (1984) Lunar synchronization of fish reproduction. Trans Am Fish Soc 113:484–493

Verghese PU (1967) Prolongation of spawning season in the carp *Cirrhina reha* (Ham) by artificial light treatment. Curr Sci 36:465–467

Wang N, Teletchea F, Kestemont P, Milla S, Fontaine P (2010) Photothermal control of the reproductive cycle in temperate fishes. Rev Aquacult 2:209–222

Wingfield JC (2008) Organization of vertebrate annual cycles: implications for control mechanisms. Philos Trans R Soc Lond B Biol Sci 363:425–441

Wootton RJ (1998) Ecology of teleost fishes, 2nd edn. Elsevier, Dordrecht

Wootton RJ, Smith C (2015) Reproductive biology of teleost fishes. Wiley Blackwell, Hoboken

2 Endocrine Regulation of Reproduction with Special Emphasis on Gametogenesis in Fishes

Stephen Kumar and Judith Betsy

Abstract

The aim of this chapter is to present the information about the brain–pituitary–gonad reproductive axis and their control on endocrinology of fishes through production of gonadotropins like FSH and LH. It also discusses different stages involved in the process of oogenesis and spermatogenesis and the endocrinological mechanism.

Keywords

Endocrinology · Reproduction · Spermatogenesis · Oogenesis · Fish

2.1 Introduction

Regulation of the complex processes, such as oogenesis and spermatogenesis is achieved by the neuroendocrine and endocrine systems in the fish (Kah and Dufour 2011; Yaron and Levavi-Sivan 2011). Neuroendocrine refers to hormones that are released by neurons of the nervous system, while endocrine refers to tissues and their hormonal products that do not form part of the nervous system (Adkins-Regan 2005).

Hormones are chemical substances that are transported in the bloodstream from the endocrine tissues producing the hormones to the target tissues of those hormones. Their production and their effects are modulated by local chemical

S. Kumar (✉)
Directorate of Sustainable Aquaculture, Tamil Nadu Dr. J. Jayalalithaa Fisheries University, Thanjavur, India

J. Betsy
Department of Aquaculture, Fisheries College and Research Institute, Tamil Nadu Dr. J. Jayalalithaa Fisheries University, Tuticorin, India

J. Betsy, S. Kumar (eds.), *Cryopreservation of Fish Gametes*,
https://doi.org/10.1007/978-981-15-4025-7_2

messengers. These local factors may have paracrine effects, influencing neighbouring cells, or autocrine effects, in which the producing cell is also the target of its product.

In teleosts, as in other vertebrates, once the gonads have differentiated, hormonal stimulation is required for the differentiated gonads to produce functional eggs and sperm. Hormones may also act directly on non-gonadal somatic tissues, influencing characteristics such as morphology and colour. The hormones that regulate gametogenesis also regulate the formation of secondary sexual characters.

2.2 Brain–Pituitary–Gonad Reproductive Axis in Vertebrates

In vertebrates, neurohormones produced in the hypothalamus influence the pituitary to control reproduction. These neurohormones induce pituitary to produce hormones, which are secreted into the bloodstream to be transported to the gonads which are the target organs. This arrangement is called the brain–pituitary–gonad reproductive axis (BPG axis), or the hypothalamus–pituitary–gonad (HPG) axis. In case of ovary, the thecal and granulosa cells of the ovarian follicle complex are the target cells, whereas in testes, the Leydig cells are the target cells. These target cells are known to be steroidogenic producing steroids.

Brain produces a neurohormone gonadotrophin-releasing hormone (GnRH) which stimulates the pituitary to release reproductive hormones such as follicle-stimulating hormone (FSH; GTH-I) and the luteinizing hormone (LH; GTH-II) which are together called as gonadotrophins (GTHs) (Sherwood and Adams 2005). GTH release is stimulated by GnRH and is repressed by dopamine (Azuadi et al. 2011). In the hypothalamus, these GTHs act directly on the GnRH receptors to reduce or modify the GnRH release. In addition to GnRHs, some peptides like pituitary adenylate cyclase activating (PACAP), neuropeptide-Y (NPY), galanin, endothelin, oxytocin, orexin, vasoactive intestinal polypeptide (VIP) and substance-P are also involved in the regulation of gonadotropins (Khazali and Behzadfar 2009; Levavi-Sivan et al. 2010).

FSH and LH are transported to the target cells in the gonads via bloodstream where they stimulate cells to produce steroid hormones, including androgens, oestrogens and progestogens, which regulate gametogenesis (Planas and Swanson 2008). These steroid hormones also provide information to the brain on the present state of the gonads. The steroids have both positive and negative feedback effects on the BPG reproductive axis.

Three classes of steroids can be recognized in fishes (Pankhurst 2008). The first class contains those steroids that have clear biological effects such as oestradiol-17β (E_2) and testosterone (T). A second class contains steroids that are produced as intermediate products in the synthesis of functional steroids. A third class is formed by steroids that are modified for the purposes of inactivation and excretion. These classes are not mutually exclusive. For example, T has both biological effects and can act as an intermediate product in the synthesis of E_2.

FSH stimulates the production of T by theca cells (Fard et al. 2013). Testosterone undergoes aromatization and changes to E_2 in the granulosa cells. Upon receiving the E_2 stimulus, liver produces vitellogenin (VTG) and secretes into the blood where they form a complex with Ca^{2+} (Nagahama 1994). VTGs are sequestered by the follicles to be incorporated into the oocyte as yolk.

2.3 Sex Steroid Synthesis

In steroidogenic cells (Young et al. 2005; Pankhurst 2008), the first step in the synthesis of functional sex steroids requires the transfer of cholesterol from the cytoplasm to the inner membrane of a mitochondrion. This transfer requires presence of the steroidogenic acute regulatory protein. After the transfer, the cholesterol is converted to pregnenolone by the enzyme $P450_{scc}$ (side-chain cleavage enzyme). Pregnenolone acts as the starting point for a number of steroidogenic pathways. A key enzyme in these pathways is $P450_{c17}$ which exists in two forms, one with C_{17-20} lyase activity ($P450_{c17\text{-}I}$) and the other with 17α-hydroxylase activity ($P450_{c17\text{-}II}$). When the activity of $P450_{c17\text{-}I}$ dominates, the pathway leads to androgens, whereas when $P450_{c17\text{-}II}$ dominates, the pathway leads to progestogens. Because of the ability of aromatase to convert T to E_2, the androgen pathway can lead to oestrogens.

2.4 Oogenesis

The developmental process by which immature oogonia are transformed into mature oocytes is called oogenesis (Grier and Aranzabel 2009; Lubzens et al. 2010). According to Tavosanis and Gonzalez (2003), oogenesis starts with the differentiation of germ line stem cell to generate a cyst of 16 cells in which one of them will become oocyte and the remaining 15 will supply material to the oocyte. Generally stem cells proliferate and undergo changes that turn them into oogonia (Fard et al. 2013). After an egg is fertilized by a spermatozoon, the zygote's genes will not be activated for a certain period of time during its development (Lyman-Gingerich and Pelegri 2007). The start of the zygotic genes programme typically occurs during the blastula stage of embryonic development at the mid-blastula transition (MBT) before which, the developmental programme of the embryo depends on products deposited in the developing oocyte by the female. Most of the products are synthesized endogenously in the oocyte, including mRNAs, growth factors and other biomolecules. However certain inevitable products are synthesized exogenously in other organs of the female apart from ovaries. One such product is VTG, which is synthesized in the liver and transported to the ovary in the blood supply and is the main source of the proteins and lipids needed by the developing embryo. Another product is the zona pellucid proteins (ZPs), which are synthesized in the liver of the female. It forms the protective envelope, the zona pellucid, around the oocyte. The quality of the maternal products partly determines the quality of the egg in terms of its capacity to be fertilized and develop as a zygote until it becomes a successful, actively feeding young fish (Brooks et al. 1997).

The timing of events during oogenesis is largely controlled by GTHs produced by the pituitary. FSH is important in the earlier stages of oogenesis, while LH is important in the later stages that lead to the maturation and ovulation of the oocyte. During egg development, primordial germ cells (PGCs) are formed and is transformed into oogonia and then into primary oocytes, with the onset of meiosis (Patiño and Sullivan 2002). Then during vitellogenesis, the oocyte grows massively and stores nutrition required for embryo development. Apart from nutritional reserves, the oocyte also stores maternal RNA to complete its cellular and non-cellular envelope differentiation and remains in meiotic arrest. The process of maturation can be characterized by reduced endocytosis, meiosis resumption, germinal vesicle breakdown (GVBD), formation of a monolayer of cortical alveoli under the oolemma, dissolution of yolk and hydration of pelagophil oocytes. During first meiotic division, two cells differing in size are formed. Among the two cells, degeneration of small cell with first polar body occurs, whereas formation of the large secondary oocyte happens followed by ovulation. This completes the maturation process. Later the secondary oocyte is extruded from its surrounding follicular cell layers and moves into the ovarian lumen. At this stage, the female gamete is known as an ovum which is haploid in nature because of second meiotic division and the formation and degeneration of the second polar body. During fertilization, the nucleus of haploid ovum fuses with the nucleus of haploid spermatozoon and forms the diploid egg.

2.4.1 Primordial Germ Cells and Formation of Oogonia

Once the egg is fertilized, some non-dividing PGCs are produced during the early stage of embryogenesis. PGCs contain components known as germplasm which has polar granules or electron dense structural organelles associated with mitochondria, RNA and proteins called as "nuage" or "ciments" (Braat et al. 1999; Lyman-Gingerich and Pelegri 2007; Le Menn et al. 2007). Nuage is found in PGC, oogonia, oocytes, spermatogonia, spermatocytes and spermatids. Due to the structural changes that happen within PGCs, oogonia are produced. Oogonial nests are formed in association with pre-granulosa cells through the multiplication of oogonium by mitotic division. A layer of somatic granulosa cells that secrete a basement lamina surrounds each oogonium thus separating it from the ovarian stroma cells. Outside the basement lamina, a layer is formed by somatic cells constituting the theca associated with blood vessels. The oocyte with its surrounding granulosa cells, basement lamina and theca somatic layer constitutes the ovarian follicle and forms the primary oocyte. The transition from oogonium to a primary oocyte is characterized by the initiation of the first meiotic division, before leaving the oogonial nest (Selman et al. 1993).

2.4.2 Oogonial Proliferation and Oogonial Nest Formation

The oogonia proliferate by mitotic division which is stimulated by E_2 (Miura et al. 2007). Oogonia are connected by cytoplasmic links in oogonial nests where daughter cells aggregate. The oogonial nest is surrounded by a single layer of somatic pre-granulosa cells. Then oogonium enters the prophase of the first meiotic division after isolating itself from other oogonia in a nest. When oogonium develops into oocyte, each chromosome divides to produce two daughter chromatids linked by a centromere. Then the cell enters the first prophase of meiosis, becoming a primary oocyte and individual oocytes become enveloped by a single layer of somatic cells, which form the granulosa layer. These somatic cells secrete a basement lamina that separate the oocyte and its granulosa from the stroma (Le Menn et al. 2007; Grier and Aranzabel 2009; Lubzens et al. 2010).

2.4.3 Chromatin Nucleolus Stage

Oocyte is completely surrounded by granulosa layer and the formation of an ovarian follicle consisting of oocyte, granulosa and basement lamina is completed. Each follicle may remain attached to germinal epithelium by a section of shared basement lamina. Outside a follicle, a single layer of somatic cells forms the theca, closely associated with blood vessels. The follicle and its associated theca form the ovarian follicular complex (Le Menn et al. 2007; Grier and Aranzabel 2009). The key step is that the oogonium becomes surrounded by the granulosa and thecal layers forming the ovarian follicle complex oocyte. When the germ cells progress through early oogenesis, E_2 is involved in the process. E_2 acts on oogonial proliferation and 17, 20β-dihydroxy-4-pregnen-3-one (17,20βP) which initiates the first meiotic division. This leads to the development of chromatin nucleolar oocytes (Miura et al. 2007).

2.4.4 Primary Growth

The formation of the follicle complex is completed early in the primary growth phase of the oocyte. In primary growth stage, synthesis of components for further development of the oocyte takes place. mRNA is required for protein synthesis and the receptors are required for the uptake of VTG, the precursor of yolk (Babin et al. 2007). These mRNA accumulates in the ooplasm and gives it a basophilic nature. The germinal vesicle increases in size and multiple nucleoli become arranged around its periphery. Then the cell organelles and associated RNA will form a complex called Balbiani body (Le Menn et al. 2007; Grier and Aranzabel 2009).

During primary growth of oocytes, cortical alveoli appears which are vesicles bounded by a membrane. Cortical alveoli are membrane-limited vesicles of variable size that stain with dyes for protein and carbohydrates. They appear in proximity to Golgi complexes which are involved in the synthesis of their contents. They increase in number and size as the oocyte grows and fills the oocyte cytoplasm. Cortical alveoli are eventually displaced to the oocyte periphery during the late stages of oocyte development, due to the centripetal accumulation of yolk proteins. During fertilization, the content of the cortical alveoli will be released to the egg surface because of the "cortical reaction" which leads to formation of chorion by restructuring of egg envelop proteins (Selman et al. 1993). The expression of GTH-I receptor genes and the genes responsible for other hormones in granulosa cells such as anti-mullerian hormone drastically increases cortical alveoli stage and decreases during vitellogenesis (Fard et al. 2013). Cortical alveoli stage can also be characterized by increased ability to produce steroids which leads to the accumulation and increase of oocyte diameter (Fard et al. 2013).

Similarly oil droplet also appears during this phase which contribute to the lipid content of the oocyte (Grier and Aranzabel 2009). The production of cortical alveoli and oil droplets is stimulated by FSH and E_2. The oocyte and the granulosa cells begin to separate, with the intervening space becoming filled with extracellular matrix. The oocyte extends microvilli into the space and around these microvilli, a proteinaceous structure, the zona pellucida appears (Le Menn et al. 2007; Modig et al. 2007). With the completion of the zona pellucida, the oocyte is protected by this acellular envelope. The growth of the oocytes is mediated by IGF-I.

2.4.5 Secondary Growth: Vitellogenesis

Secondary growth phase begins when the oocyte reaches a critical size (Tyler and Sumpter 1996). It is the major growth period of the oocyte, when growth is measured in terms of the dry mass of the oocyte. This growth primarily reflects the accumulation of yolk by the oocyte. In this phase, the zona pellucida continues to thicken and oil droplets will continue to accumulate. The major process in this phase is vitellogenesis which is the synthesis, transport, uptake and processing of VTG by the oocyte during which process VTG is transformed into the lipoprotein and protein stored in yolk globules. VTG is the main precursor of yolk (Babin et al. 2007; Mommsen and Korsgard 2008) and is synthesized in the liver by the stimulation of E_2.

In this stage of oocyte development, ovaries will have a major endocrine influence on the liver (Ding 2005; Pankhurst 2008; Lubzens et al. 2010). The thecal cells of the follicle complex synthesize T under the influence of pituitary FSH. Synthesized T passes to the granulose cells of the follicle where it is converted to E_2 with the help of aromatase enzyme. The E_2 is transported to the liver in the bloodstream where it exhibits the following effects (Wootton and Smith 2015):

- At low levels E_2 stimulates the synthesis of oestrogen receptors (ERs) which increases the presence of ERs in the liver cell allowing formation of more E_2-ER complexes.
- E_2–ER complexes stimulate the production of vtg genes which is responsible for synthesis of vitellogenin.

- E_2 stimulates the proteins responsible for formation of zona pellucida, which separates the granulose layer from the oocyte.
- Circulating E_2 has feedback effects on E_2 sensitive cells in the brain and pituitary, and plays a role in regulating the BPG reproductive axis.

There is an increasing evidence that the effects of FSH and E_2 are modulated by non-steroid growth factors, including IGFs, IGF-I which can influence steroid synthesis in the thecal and follicular layers (Lubzens et al. 2010).

2.4.6 Polarity of the Oocyte

Polarity in the oocyte can be noticed during secondary growth phase (vitellogenesis) (Lyman-Gingerich and Pelegri 2007). The oocyte polarizes along an animal–vegetal axis. The animal pole is indicated by the developing micropyle, whereas the vegetal pole contains the yolk mass as development of the embryo proceeds.

2.4.7 Oocyte Maturation

When the oocyte reaches its final dry mass, it terminates the uptake of VTG and the transfer of material resources from the female. Then the oocyte enters the phase of maturation, which ends with the ovulation of the oocyte into the ovarian lumen. According to Bobe et al. (2008) oocyte maturation occurs by the following two major events,

1. The arrest of the oocyte in diplotene I of meiosis I ends and meiosis proceeds until a second arrest in metaphase II.
2. Hydration of the oocyte by the uptake of water which results in an increase in the volume of the oocyte, but no increase in its dry mass. Hydration of oocyte depends on molecular water channels or aquaporins.

Oocyte maturation occurs when the oocyte reaches a state called oocyte maturation competence (Bobe et al. 2008). Only when this state is achieved, the follicular complex responds to GTHs. LH stimulates oocyte maturation (Suwa and Yamashita 2007; Bobe et al. 2008) and in response to this, the follicular cells synthesize and release a progestin steroid, the maturation-inducing hormone (MIH). In turn, this progestin stimulates the oocyte to produce a maturation-promoting factor (MPF).

E_2 binds to a membrane ER called G protein-coupled ER which maintains the arrest of meiosis I through series of events. As the levels of E_2 decline and the levels of MIH increase, the inhibitory effect of E_2 declines. The binding of MIH with the membrane progestogen receptors leads to resumption of meiosis I. During oocyte maturation, meiosis I is resumed which depends on the physiological condition of the female and suitable environment for spawning. Early effect of the LH surge may include upregulation of membrane progestogen receptors in the follicle cells which contributes to the development of oocyte maturation competence (Wootton and Smith 2015).

Paracrine growth factors, including epidermal growth factor, activin and follistatin, are also implicated in the onset of oocyte maturation competence and the fine control of maturation by cross-talk between the oocyte and the follicular layers (Yaron and Levavi-Sivan 2011). Thus according to Fard et al. (2013) oocyte maturation can be characterized by the following,

- Acquisition of oocyte maturation competence
- Production of MIH
- Production of MPF and resumption of meiosis
- Cytoplasmic maturation involving changes in the yolk proteins and lipids

2.4.8 Ovulation

During ovulation, the mature oocyte is released into the ovarian lumen for which the basement lamina must break down, opening the follicle to the ovarian lumen. The breakdown of the basement lamina involves proteases and other factors. It is to be noted that even during ovulation, some limited oocyte hydration continues, but the increase in oocyte volume is small (Wootton and Smith 2015).

2.4.9 Fertilization

After ovulation into the ovarian lumen, or in some species the coelomic cavity, the oocytes are carried along the gonoducts to the genital papilla for release into the water. The exception is viviparous species, with internal fertilization (Wootton and Smith 2015).

2.5 Spermatogenesis

Spermatogenesis is also under the control of FSH and LH and the process involves two "female" sex steroids such as E_2 and $17\alpha20\beta$-P both of which are synthesized in the Leydig cells under the influence of the GTHs. At the beginning of spermatogenesis, E_2 level in the plasma increase for a short period of time which promotes the slow proliferation of undifferentiated spermatogonia. The progestogen, $17\alpha20\beta$-P stimulates the onset of meiosis, which marks the transition between spermatogonia and spermatocytes (Roosen-Runge 1977; Billard 1986; Blum 1986; Schulz et al. 2010) and its level increases during spermiation and the maturation of spermatozoa.

Table 2.1 Details of spermatogonial proliferation

Proliferation type	Stimulating hormone	Proliferation action
SP-I	E_2	Proliferation is common in iteroparous species and it provides a reservoir of undifferentiated spermatogonia that can provide high number of recruits within a short period thereby generating spermatozoa for successive breeding seasons
SP-II	11-KT	Proliferation occurs within the spermatogenic cysts. A single spermatogonium is surrounded by Sertoli cells and goes through a special–typical number of mitotic divisions

2.5.1 Spermatogonial Proliferation

Spermatogonial proliferation (SP) by mitotic division takes two forms such as SP-I and SP-II (Wootton and Smith 2015), the details of which are given in Table 2.1.

2.5.2 Spermatocysts

Spermatocysts are formed by the association of spermatogonia with Sertoli cells. Spermatocysts consists of a group of isogenic germ cells surrounded by Sertoli cells (Uribe et al. 2014). Sertoli cells are derived from somatic cell precursors in the germinal epithelium and are essential for the successful development of a spermatogenic spermatocyst. The germ cells are derived from the PGCs. Sertoli cells exhibit the following functions during spermatogenesis,

- Essential for the survival, development and functioning of germ cells.
- Secrete the fluid that fills out the lumen of the tubules.
- Act as phagocytes, efficiently engulfing any spermatozoa left over after spawning.
- Engulf residual bodies that are ejected by the spermatids as they transform into spermatozoa.

2.5.3 Stages of Spermatogenesis

During the development of germ cells within a spermatogenic spermatocyst, the cells pass through successive stages such as, differentiated spermatogonia, primary spermatocyst, secondary spermtocysts, spermatids and spermatozoa.

2.5.3.1 Differentiated Spermatogonia

Testis of teleosts contains a reserve of spermatogonia in the germinal epithelium throughout the year which is called as spermatogonia A. They are diploid stem cells that proliferate by mitotic divisions, increasing the number of germ cells within the germinal compartment. They measure 12–16 μm in diameter. Spermatogonia B are enclosed within a spermatocyst and can be differentiated when they are surrounded by Sertoli cell processes. They measure 9–12 μm in diameter. They replicate their chromosomes and enter in meiosis, becoming primary spermatocytes (Uribe et al. 2014).

When a type A spermatogonium becomes associated with a Sertoli cell, it is one of the largest cells in the testis. In *Oreochromis* spp., it has a volume of about 2300 μm^3. When a type A spermatogonium converts to the type B, a series of mitotic division is initiated. A type B spermatogonium in *Oreochromis* spp., has a volume of about 160 μm^3, but after the series of mitotic divisions, there are about 120 per spermatocyst, so the total volume is around 19,200 μm^3. This is the reason for large increase in GSI as the testes mature (Wootton and Smith 2015). In female, the increase in GSI is largely the result of the accumulation of yolk in the oocytes. Type B spermatogonia are still diploid cells.

The transition from mitotic to meiotic divisions of the germ cells in a spermatogenic cyst is triggered by a brief surge in LH, which leads to the production of 17α20β-P which is essential in the initiation of meiosis and formation of primary spermatocytes (Scott et al. 2010). Once meiosis has started, further development of the spermatocytes is dependent on the presence of androgens, usually 11-KT, again controlled by FSH acting on the Leydig cells. In contrast to oogenesis, there is no arrest of meiosis I in the primary spermatocytes. The transition from meiosis I to meiosis II marks the transition from primary to the short-lived phase of secondary spermatocytes, which is completed with the transformation into spermatids. Spermiogenesis, the process of the transformation of spermatids into the spermatozoa, continues to be androgen-dependent.

2.5.3.2 Primary Spermatocysts

At the end of mitotic divisions, each spermatogonium in the clone enters meiosis I and becomes a primary spermatocyte. Primary spermatocytes are spherical cells, with diameters of 8–12 μm (Uribe et al. 2014). Meiosis I doubles each chromosome to form chromatids and crossing over between homologous chromosomes, which are linked by synaptonemal complexes. At the end of meiosis I, two secondary spermatocytes will be produced, each with a haploid number of chromosomes, but with two chromatids per chromosome.

2.5.3.3 Secondary Spermatocysts

Secondary spermatocysts are also spherical cells with diameters of 4–7 μm (Uribe et al. 2014). These secondary spermatocysts rapidly pass through meiosis II, which generates four haploid spermatids. Thus for each spermatogonium entering meiosis I, four haploid spermatids emerge. They divide rapidly after a short interphase between the two divisions of meiosis. There is probably some cell death as the cells of a clone of a spermatocyst increase in abundance and these are phagocytosed.

2.5.3.4 Spermatids

Haploid spermatids measure 2–4 μm diameter with a spherical nucleus. They do not divide further and only undergo morphological transformation, becoming

Table 2.2 Different noticeable stages of spermiogenesis in fishes

Stages in spermiogenesis	Features
Stage I	• Early spermatids are connected with cytoplasmic bridges • Each spermatid has a spherical nucleus • Nuclear membrane has a shallow groove on one side of the nucleus otherwise called as fossa • Fossa contains centrioles
Stage II	• Chromatin in the nucleus of the spermatid is more compact • Spermatid is smaller • Two centrioles develop and are located in the fossa • A flagellum starts to form, with the distal centriole forming the basal body of the flagellum
Stage III	• Chromatin in the nucleus is even more compact • The flagellum is linked to the distal centriole and is surrounded by mitochondria
Stage IV	• Nucleus forms a regular sphere, with condensed chromatin • The flagellum is attached at about 110° to the spermatozoon head • Spermatid ejects the residual body which is the excess cytoplasm and is phagocytosed by Sertoli cells

spermatozoa through spermiogenesis. This process is characterized by the formation of the sperm head and its condensed nucleus, with midpiece and flagellum. According to Rupik et al. (2011), following four arbitrary stages in spermiogenesis can be observed in many of the teleosts (Table 2.2).

During the third and fourth stages of spermiogenesis, the spermatids lose their intercellular links and their association with the Sertoli cells. At the completion of spermiogenesis, the links between Sertoli cells weaken and the spermatocyst ruptures, releasing spermatozoa into the tubule lumen. In teleosts, three types of spermiogenesis (Table 2.3) have been defined based on the position of the flagellum in relation to the nucleus (Quagio-Grassiotto and Oliveira 2008; Schulz et al. 2010).

2.5.3.5 Spermiation and Capacitation

Once a spermatogenic cyst is full of spermatozoa, spermiation occurs with the breakdown of the wall of the spermatogenic cyst. Plasma levels of LH increase and result in the synthesis of 17α20β-P, while 11-KT levels continue to be high. Under the influence of these steroids, the cyst walls rupture, releasing the spermatozoa into the testis tubules (Schulz and Miura 2002; Schulz et al. 2010). Spermiation is a process in which connections between Sertoli cells breakdown and spermatozoa are released into the tubule lumen and further into the efferent sperm ducts (Schulz et al. 2010).

When spermiation occurs, milt can be pressed out by gently squeezing the abdomen of the male. The gonoducts and testicular ducts are filled with spermatozoa. At this stage, the spermatozoa lack motility and this state is maintained by the ionic composition of the seminal fluid. The final stage in the development of a spermatozoon capable of fertilizing an egg is the process of capacitation, including

Table 2.3 Types of spermiogenesis in fishes

Type	Features	Example
I	• Early spermatid has a central nucleus, with two centrioles and one or more mitochondria • Centrioles migrate towards the nucleus by drawing the flagellum and the cell membrane along with them creating a cytoplasmic canal • The nucleus develops a fossa and rotates through 90° with respect to the axoneme of the flagellum, which is perpendicular to the nucleus • The centrioles move, either fully or partially, into the fossa • The centriole closest to the nucleus is the proximal centriole and the other is the distal centriole	*Danio rerio*
II	• No nuclear rotation • Centrioles migrate in the direction of the nucleus	Perciform fishes
III	• The flagellum is central, but there is no nuclear rotation	Siluriform catfishes and Characiform fishes

the acquisition of motility (Longo 1987). This final maturation of the spermatozoa can take place in the testicular ducts, the gonoduct and the external environment. The final acquisition of motility can be a response to a change in the osmotic environment, the pH or the concentration of inorganic ions such as K^+ or Ca^{2+} (Kinsey et al. 2007). Chemical cues from the egg may also play a role in promoting the motility of spermatozoa.

References

Adkins-Regan E (2005) Hormones and animals social behavior. Princeton University Press, Princeton

Azuadi NM, Siraj SS, Daud SK et al (2011) Enhancing ovulation of Malaysian mahseer (*Tor tambroides*) in captivity by removal of dopaminergic inhibition. J Fish Aquat Sci 6:740–750

Babin PJ, Carnevale O, Lubzens E et al (2007) Molecular aspects of oocyte vitellogenesis in fish. In: The fish oocyte: from basic studies in biotechnological applications. Springer, Berlin, pp 39–76

Billard R (1986) Spermatogenesis and spermatology of some teleost species. Reprod Nutr Dev 26:877–920

Blum V (1986) Vertebrate reproduction. Springer, Berlin

Bobe J, Jalabert B, Fostier A (2008) Oogenesis: post vitellogenic events leading to a fertilizable oocyte. In: Fish reproduction. Science Publishers, Enfield, pp 1–36

Braat AK, Speksnijder JE, Zivkovic D (1999) Germ line development in fishes. Int J Dev Biol 43:745–760

Brooks S, Tyler CR, Sumpter JP (1997) Egg quality in fish. What makes a good egg? Rev Fish Biol Fish 7:387–416

Ding JL (2005) Vitellogensis and vitellogenin uptake in oocytes. In: Hormones and their receptors in fish. World Scientific, Hackensack, pp 254–276

Fard ER, Kamarudin MS, Harmin SA (2013) Endocrine control of oogenesis in teleosts. Asian J Anim Vet Adv 8(2):205–215. https://doi.org/10.3923/ajava.212

Grier HJ, Aranzabel MCU (2009) The testis and spermatogenesis in teleosts. In: Reproductive biology and phylogeny of fishes (Agnathans and bony fishes). Science Publishers, Enfield, pp 121–142

Kah O, Dufour S (2011) Conserved and divergent features of reproductive endocrinology of teleost fishes. In: Hormones and reproduction of vertebrates, Fishes, vol I. Academic, London, pp 15–42

Khazali H, Behzadfar M (2009) Effect of orexin infusion into third ventricle on the GnRH and LH secretions in the prepubertal rat. J Appl Sci 10:1936–1942

Kinsey WH, Sharma D, Kinsey SC (2007) Fertilization and egg activation in fishes. In: The fish oocyte: from basic studies to biotechnological application. Springer, Berlin, pp 397–409

Le Menn F, Cerda J, Babin PJ (2007) Ultrastructural aspects of ontogeny and differentiation of ray-finned fish ovarian follicles. In: The fish oocyte: from basic studies to biotechnological applications. Springer, Berlin, pp 1–37

Levavi-Sivan B, Bogerd J, Mananos EL et al (2010) Perspectives on fish gonadotropins and their receptors. Gen Comp Endocrinol 165:412–437

Longo FJ (1987) Fertilization. Chapman and Hall, London

Lubzens E, Young G, Bobe J (2010) Oogenesis in teleosts: how eggs are formed. Gen Comp Endocrinol 165:367–389

Lyman-Gingerich J, Pelegri F (2007) Maternal factors in fish oogenesis and embryonic development. In: The fish oocyte: from basic studies to biotechnological applications. Springer, Berlin, pp 141–174

Miura C, Higashino T, Miura T (2007) A progestin and an estrogen regulate early stages of oogenesis in fish. Biol Reprod 27:822–828

Modig C, Westerlund L, Olsson PE (2007) Oocyte zona pellucida proteins. In: The fish oocyte: from basic studies to biotechnological applications. Springer, Dordrecht, pp 113–139

Mommsen TP, Korsgard B (2008) Vitellogenesis. In: Fish reproduction. Science Publishers, Enfield, pp 115–169

Nagahama Y (1994) Endocrine regulation of gametogenesis in fish. Int J Dev Biol 38:217–229

Pankhurst NW (2008) Gonadal steroids: functions and patterns of change. In: Fish reproduction. Science Publishers, Enfield, pp 67–111

Patiño R, Sullivan CV (2002) Ovarian follicle growth, maturation, and ovulation in teleost fish. Fish Physiol Biochem 26:57–70

Planas JV, Swanson P (2008) Physiological function of gonadotropins in fish. In: Fish reproduction. Science Publishers, Enfield, pp 37–68

Quagio-Grassiotto I, Oliveira C (2008) Sperm ultrastructure and a new type of spermiogenesis in two species of Pimelodidae, with a comparative review of sperm ultrastructure in siluriformes (Teleostei: Ostariophysi). J Comp Zool 247:55–56

Roosen-Runge EC (1977) The process of spermatogenesis in animals. Cambridge University Press, Cambridge

Rupik W, Huszno J, Klag J (2011) Cellular organization of the mature testes and stages of spermiogenesis in *Danio rerio* (Cyprinidae: Teleosei)—structural and ultrastructural studies. Micron 42:833–839

Schulz RW, Miura T (2002) Spermatogenesis and its endocrine regulation. Fish Physiol Biochem 26:43–56

Schulz RW, de Franca LR, Lareyre JJ et al (2010) Spermatogenesis in fish. Gen Comp Endocrinol 166:390–411

Scott AP, Sumpter JP, Stacey N (2010) The role of the maturation-inducing steroid, 17, 20-β-dihydroxypregn-4-en-3-one, in male fishes: a review. J Fish Biol 76:183–224

Selman S, Wallace RA, Sarka A et al (1993) Stages of oocyte development in the zebrafish, *Brachydanio rerio*. J Morphol 218:203–224

Sherwood NM, Adams BA (2005) Gonadotropin-releasing hormone in fish: evolution, expression and regulation of the GnRH gene. In: Hormones and their receptors in fish reproduction. World Scientific, Hackensack, pp 1–39

Suwa K, Yamashita M (2007) Regulatory mechanisms of oocyte maturation and ovulation. In: The fish oocyte: from basic studies to biotechnological applications. Springer, Berlin, pp 323–347
Tavosanis G, Gonzalez C (2003) γ-Tubulin function during female germ-cell development and oogenesis in Drosophila. Proc Natl Acad Sci U S A 100:10263–10268
Tyler CR, Sumpter JP (1996) Oocyte growth a development in teleosts. Rev Fish Biol Fish 6:287–318
Uribe MC, Grier HJ, Roa VM (2014) Comparative testicular structure and spermatogenesis in bony fishes. Spermatogenesis 4(3):e983400
Wootton RJ, Smith C (2015) Reproductive biology of teleost fishes. Wiley Blackwell, Hoboken
Yaron Z, Levavi-Sivan B (2011) Endocrine regulation of fish reproduction. In: Encyclopedia of fish physiology: from genome to environment. Academic, San Diego, pp 1500–1508
Young G, Kusakabe M, Nakamura I et al (2005) Gonadal steroidiogenesis in teleost fish. In: Hormones and their receptors in fish reproduction. World Scientific, Singapore, pp 155–223

Factors Influencing Milt Quality in Fishes and Its Usefulness to Cryopreservation

3

Andrzej Ciereszko, Sylwia Judycka, Joanna Nynca, Mariola Słowińska, and Mariola A. Dietrich

Abstract

Sperm quality is primarily defined as its capability of fertilising an egg and consequently allowing for the development of a normal embryo. Sperm quality can be highly variable; therefore, gamete quality assessment is a prerequisite for fertilisation trials. Fish semen analysis includes several methods, such as measurements of sperm motility parameters, sperm concentration and viability, mitochondrial potential and chromatin structure integrity. Recently, 'omics' methodologies, including genomics, transcriptomics, proteomics and metabolomics, have been utilised to search for potential molecular indicators that predict sperm quality, which could be used in practice. Several analytical parameters appear to be useful for the evaluation of changes in sperm quality during cryopreservation. Major factors that influence sperm quality are related to genetics, biology and environmental effects, both natural and anthropogenic. Among the genetic factors, inbreeding is the most common factor that usually leads to a decrease in sperm quality, and biological factors include, for example, social status and the age of the fish. Environmental factors markedly shape fish reproduction and encompass season, water temperature and salinity, nutrition, living conditions in captivity, fish handling and the effects of various environmental contaminants. Factors securing the high quality of fresh sperm usually determine its usefulness for predicting cryopreservation success.

Keywords

Semen · Sperm quality · Cryopreservation · Fish · Reproduction

A. Ciereszko (✉) · S. Judycka · J. Nynca · M. Słowińska · M. A. Dietrich
Department of Gametes and Embryo Biology, Institute of Animal Reproduction and Food Research, Polish Academy of Sciences, Olsztyn, Poland
e-mail: a.ciereszko@pan.olsztyn.pl

J. Betsy, S. Kumar (eds.), *Cryopreservation of Fish Gametes*,
https://doi.org/10.1007/978-981-15-4025-7_3

3.1 Introduction

Good quality semen is a prerequisite for the reproductive success of fish through the control of reproduction under practical hatchery conditions. The development of efficient reproductive biotechnics, such as the hormonal stimulation of males, short- and long-term semen storage and fertilisation protocols, rely on reliable sperm quality tests. Sperm quality is primarily defined as its capability of fertilising an egg and consequently allowing for the development of a normal embryo. Therefore, the ultimate sperm quality test is the measurement of sperm fertilising ability via fertilisation trials. Unfortunately, this method is labourious, time-consuming and takes weeks or months (for cold-water species) to obtain results. For this reason, routine fertilisation trials are rarely used; however, they are often performed to validate the usefulness of a wide range of alternative sperm quality tests aiming to distinguish the different sperm structures and functions that are important for semen quality.

The quality of sperm can be highly variable; therefore, assessing sperm quality is a prerequisite for fertilisation trials. Fish semen analysis includes several methods, including measurements of sperm motility parameters, sperm concentration and viability, mitochondrial potential and chromatin structure integrity; the latter is often measured using flow cytometry, which appears to be especially useful as it allows for the measurement of parameters at the level of individual sperm cells and it can distinguish specific subpopulations of spermatozoa. Recently, 'omics' methodologies, including genomics, transcriptomics, proteomics and metabolomics, have been utilised to search for potential molecular indicators that predict sperm quality and could be used in practice.

There are a number of excellent recent reviews concerning sperm quality and cryopreservation. These articles focus on, among other aspects, the evaluation of fish gamete quality and broodstock management (Bobe and Labbé 2010; Fauvel et al. 2010; Ciereszko et al. 2011; Migaud et al. 2013; Valdebenito et al. 2015), factors enhancing fish sperm quality and emerging tools for sperm analysis (Cabrita et al. 2014; Figueroa et al. 2016a), the molecular basis of spermatogenesis and sperm quality (Robles et al. 2017), challenges and perspectives of fish gamete cryopreservation (Cabrita et al. 2010; Li et al. 2010a; Asturiano et al. 2017), endocrine-disrupting chemicals in the aquatic environment and potential risks to fish gamete quality (Hatef et al. 2013; Carnevali et al. 2018). Moreover, Figueroa et al. (2016a) described technologies used for the study of sperm function in cryopreserved fish spermatozoa. Hormonal manipulations for the enhancement of sperm production in cultured fish were recently reviewed by Mylonas et al. (2017). Moreover, Gallego and Asturiano (2018, 2019) provided comprehensive and up-to-date reviews of sperm motility assessment for basic physiological studies and as a tool for aquaculture research.

The main aim of this chapter was to present the recent insights into factors that influence milt quality in fish and their usefulness for monitoring changes in spermatozoa during cryopreservation and for predicting cryopreservation success. Hence this chapter is made of three parts, focusing on (1) brief descriptions of the major methods for the evaluation of semen quality; (2) factors influencing sperm quality and its usefulness for the prediction of cryopreservation success; and (3) evaluation of changes in sperm quality during cryopreservation.

3.2 Major Methods for the Evaluation of Semen Quality

Semen analysis is a prerequisite to evaluate the mechanisms underlying semen functions and semen quality, and to identify factors influencing semen quality and its usefulness for short- and long-term storage (cryopreservation). Several analyses aimed at assessing different semen structures, functions and biochemistry have been developed and are described in detail in recent excellent and comprehensive reviews (Cabrita et al. 2014; Valdebenito et al. 2015; Figueroa et al. 2016a). Hence this chapter will only briefly summarise the existing and emerging methods, with some emphasis on recent developments in sperm concentration measurements, flow cytometry methods and emerging methodologies of 'omics' technologies and epigenetic changes in spermatozoa.

3.2.1 Collection and Macroscopic Evaluation of Good Quality, Contaminant-Free Semen

The correct collection of semen, free of contaminants, is a prerequisite for high-quality semen. Any contamination with mucus, blood, water, urine and faeces will seriously affect semen physiology and composition, leading to a deterioration of quality and the discarding of such semen samples. Some contamination can be inspected visually (blood, mucus, faeces), whereas other types (urine, water) are difficult to quickly assess. Urine contamination is unavoidable while collecting milt by stripping. The use of a catheter or cannula is the preferred method, allowing for a great reduction in urine contamination (Glogowski et al. 2000; Sarosiek et al. 2016; Król et al. 2018); however, this method is difficult to implement due to the fragility of the spermatic duct and species-specific differences in the anatomy of spermatic and ureic ducts. Measurement of the osmolality of seminal plasma is the preferred method for the evaluation of the urine/water contamination of semen.

3.2.2 Sperm Concentration

The measurement of sperm concentration is the most basic measurement of sperm quality, allowing for an estimation of the number of sperm cells in semen samples. The main method for measurement seems to be straightforward and is based on counting cells under a microscope using different counting chambers, such as double Neubauer or Bürke chambers. Unexpectedly, in andrological studies, high variability in sperm concentration has been observed, which may be related to differences in the particular microscopic counting methods (Tomlinson 2016). For

Table 3.1 Advantages and disadvantages of different methods for the determination of sperm concentration

Instrument	Advantages	Disadvantages
Haemocytometer	Reliable, low cost	Time-consuming, prone to observer bias, not appropriate for the analysis of numerous samples, differences between different types of haemocytometers
Spectrophotometer	Reliable, fast, low cost	Initial preparation of standard curve requires time, not accurate for samples that contain excessive debris or particulate matter or are contaminated with cells other than sperm
NucleoCounter	Reliable, fast, no standardisation required, repeatable, simultaneous determination of viability, can be applied to cryopreserved semen	Cost of instrument and disposable cassettes, potential for excessive debris from other nucleated cells to create non-specific fluorescence
Flow cytometer	Reliable, repeatable, can be applied to cryopreserved semen, simultaneous determination of other sperm parameters, e.g. viability, mitochondrial potential, oxidative stress	Cost of instrument and its maintenance, requires highly skilled personnel

this reason, the search for additional alternative and accurate methods for the measurement of sperm concentration is highly justified. For fish sperm, the spectrophotometric method has been used for some time (Suquet et al. 1992; Ciereszko and Dabrowski 1993; Cuevas-Uribe and Tiersch 2011), and new approaches based on fluorescent microscopy (Nynca and Ciereszko 2009) or flow cytometry are currently being tested as well (Nynca et al. 2016). The methods for the measurement of sperm concentration that are used for fish, and their advantages and disadvantages, are summarised in Table 3.1. Our recent studies clearly indicated that the final sperm concentration in straw significantly affects sperm cryopreservation success (Nynca et al. 2017; Judycka et al. 2018, 2019a, b). For this reason, accurate sperm concentration measurements are critical for the development and implementation of successful cryopreservation procedures for fish semen.

3.2.3 Sperm Motility

The percentage of motile sperm is the most useful parameter for assessing sperm quality in fish because it allows for the prediction of fertilisation success, since high values of sperm motility parameters often strongly correlate with fertilisation ability (Rurangwa et al. 2004). In the past, the most commonly used technique was the subjective microscopic evaluation of sperm movement, which only enables the determination of the percentage of sperm motility and is strongly dependent on the observer, often leading to the over- or underestimation of true values. For this reason, employing qualified staff is a key factor to reach accurate motility estimations using subjective methods (Gallego and Asturiano 2018; Gallego et al. 2018). The computer-assisted sperm analysis (CASA) system is an advanced technique that enables the fast, objective and accurate measurement of sperm motility (Kime et al. 2001). Moreover, it also allows for the determination of sperm velocity and trajectory parameters (curvilinear velocity (VCL), straight line velocity (VSL), average path velocity (VAP), amplitude of lateral head displacement (ALH) and linearity (LIN $= 100 \times$ VSL/VCL)] as well as improves reproducibility and facilitates the documentation of results. The main challenge of sperm motility measurements in fish is the very short duration of sperm motility, which requires high skill in quickly and accurately pipetting samples and microscope operation. It is also important to remember that CASA outcomes are highly influenced by the different systems, counting chambers and settings (Boryshpolets et al. 2013; Bompart et al. 2018).

3.2.4 Fluorescence Microscopy and Flow Cytometry Analysis

For measurements of different aspects of sperm physiology and structures, several fluorescent probes have been developed. For example, Hoechst, PI/SYBR, acridine orange (AO), DAPI and YO-PROI have been developed for assessing sperm viability (Cabrita et al. 2014). For DNA fragmentation analysis, comet and TUNEL assays, with the use of fluorescent DNA dyes, were developed (Cabrita et al. 2014). These measurements are performed using fluorescent microscopic analysis; however, this method enables the measurement of a small number of spermatozoa within a population, is time-consuming, can be subjective and generally measures sperm attributes individually (Hossain et al. 2011). On the other hand, flow cytometry is a technique that is gaining popularity for the evaluation of sperm quality in aquatic organisms, including fish (Cabrita et al. 2001, 2005; He and Woods III 2004; Paniagua-Chávez et al. 2006). A flow cytometer, an instrument that measures the intensity of fluorescent stains associated with individual cells, has been used to evaluate the growing numbers of sperm characteristics that are related to sperm quality and/or function. This analysis is objective and accurate, and a large number of spermatozoa can be analysed in a small volume of samples in a short period of time. To date, flow cytometry has been used to assess several fish sperm quality parameters, such as plasma membrane integrity, fluidity, mitochondrial function, ROS levels, apoptosis, DNA damage, as well as lipid peroxidation, both in fresh and cryopreserved sperm (Cabrita et al. 2014). Recently introduced portable flow cytometers make it possible to perform this analysis in field conditions (Nynca et al. 2016; Mostek et al. 2018).

3.2.5 Emerging Methods

3.2.5.1 Genome, Transcriptome and Epigenetic Analyses

There are a number of emerging methods that aim to extend our basic knowledge of fish sperm physiology, with the potential to develop new parameters to assess quality. Epigenetics concerns phenotype changes, which are heritable, but do not involve changes to the DNA sequence. As such, these changes reflect a large array of environmental constraints, such as photothermal changes, nutrition, exposure to xenobiotics, and, recently, cryopreservation (Herráez et al. 2017; Labbé et al. 2017). Major epigenetic changes are based on molecular mechanisms, including DNA methylation, histone modifications and the action of non-coding RNAs. Currently, DNA methylation is a widely studied epigenetic feature and is measured through the evaluation of cytosine DNA methylation (Woods III et al. 2018). The most common analytical approach is the use of fluorescence or luminometric measurements of DNA treated with specific restriction enzymes (de Mello et al. 2017; Herranz-Jusdado et al. 2019a, b). Non-coding RNAs include both small non-coding RNAs (micro RNAs, small interfering RNAs, piwi interacting RNAs) and long non-coding RNAs (Robles et al. 2017). Although it is generally accepted that there is no protein biosynthesis activity in spermatozoa, several transcripts (mRNA) are present in these cells and are linked to sperm quality as well (Riesco et al. 2017a).

3.2.5.2 Proteomic Analysis

The comprehensive analysis of molecules is now being introduced to fish semen analysis. Two-dimensional difference gel electrophoresis (2D-DIGE) appears to be especially useful in quantitative approaches, allowing for the co-separation of the proteins of control samples and the proteins of treatment/disease samples on the same gel, eliminating gel-to-gel variability (Ciereszko et al. 2017; Nynca et al. 2018). The principle of 2D-DIGE is to label proteins prior to isoelectric focusing using three spectrally resolvable fluorescent dyes, allowing for the independent labelling of control and experimental samples. Proteins are digested and identified using MALDI-TOF or LC-MS/MS mass spectrometry. Gel-free proteomics is an alternative and complementary proteomics methodology that allows the researcher to obtain information that is not accessible by 2-DE. The basis of the methodological approach of gel-free mass spectrometry-based high throughput proteomics is the coupling of micro-scale chromatographic separation with automated MS/MS (Nynca et al. 2018). Peptides, after digestion with trypsin, are fractionated on a strong cation exchange column and then further separated by reverse phase chromatography coupled in line with electrospray mass spectrometers. The analysis of post-translational modifications is now being developed and introduced into studies of semen quality, such as protein phosphorylation (Dumorné et al. 2018) and protein oxidation (Mostek et al. 2018).

3.3 Major Factors Influencing Sperm Quality

Major factors influencing sperm quality are related to genetics, biology and environmental effects, both natural and anthropogenic (caused by human intervention). Examples are presented in Table 3.2 and briefly described below.

3.3.1 Biological Factors

3.3.1.1 Genetics

Among genetic factors, inbreeding is the most common factor that influences sperm quality, usually leading to a decrease in sperm quality (Mehlis et al. 2012); this effect may be more pronounced with ageing (Langen et al. 2017). Hatchery selection was found to intensify genetic selection, which was demonstrated in terms of the swimming speed of spermatozoa (Kekäläinen et al. 2013).

3.3.1.2 Social Status

The social status of fish is an important factor that determines semen quality. This status is due to female mate choice as a driving force influencing male characteristics via precopulatory sexual selection, leading to the evolution of divergent male phenotypes or alternative reproductive tactics (ARTs; Fitzpatrick et al. 2016). For example, for Chinook salmon, two ARTs were found, one involving large males that participate in dominance-based hierarchies for access to spawning females, known as hooknoses, and the second including small males that attempt to sneak fertilisations during spawning events from peripheral positions, known as jacks (Flannery et al. 2013). The consequences of ARTs are profound differences in sperm characteristics; for example, the spermatozoa of Chinook salmon jacks appear to have significantly faster sperm than hooknoses (Flannery et al. 2013).

3.3.1.3 Fish Age

The age of fish is an important factor that determines semen quality; generally, lower quality parameters are recorded for older males due to the ageing process, accompanied by a high level of oxidative stress, as was recently reported for rainbow trout (Risopatrón et al. 2018). However, there is high variability in the effects of age on sperm quality related to species and strains, which may be additionally modulated by genetic and environmental factors; for example, fish in captivity usually live longer than in the wild (Chalde et al. 2016). Under hatchery conditions, the selection of younger fish as brooders is often justified by the lower costs of fish maintenance. Moreover, interactions between the optimal age of females and males have been identified; for example, for common carp, 2 and 3 years of age were found to be the optimal reproductive ages for males and females, respectively (Aliniya et al. 2013). The condition of fish during increased sperm competition can positively influence the swimming speed of lake whitefish spermatozoa (Blukacz et al. 2010).

Table 3.2 Major factors that influence the sperm quality of fresh semen

Factor	Species	Main study findings	Reference
Biology			
Inbreeding	Slender krib (*Pelvicachromis taeniatus*)	Age-dependent inbreeding effects were identified	Langen et al. (2017)
	Three-spined stickleback (*Gasterosteus aculeatus*)	The number of sperm from inbred males was negatively correlated with sperm quality	Mehlis et al. (2012)
Social status/sperm competition	Grass goby (*Zosterisessor ophiocephalus*)	Cross interaction between seminal plasma and sperm from males with different tactics	Locatello et al. (2013)
	Chinook salmon (*Oncorhynchus tshawytscha*)	Differences in sperm motility, which can be an important part of the mechanism of sperm competition	Flannery et al. (2013)
	Plainfin midshipman (*Porichthys notatus*)	Female choice amplified the selection gradient acting on males, both between and within reproductive tactics	Fitzpatrick et al. (2016)
	Three-spined stickleback (*G. aculeatus*)	Sperm quality traits were under strong sexual selection, presumably driven by the high risk of sperm competition under natural conditions	Mehlis et al. (2015)
		Expression of stickleback male breeding colouration was positively associated with the linearity of sperm movement, whereas sperm morphology was negatively related to the trajectory of sperm movement	Mehlis et al. (2013)
Condition	Lake whitefish (*Coregonus clupeaformis*)	Larger males produced faster swimming spermatozoa	Blukacz et al. (2010)
Age	Rainbow trout (*Oncorhynchus mykiss*)	The sperm quality of 2- and 3-year-old fish was superior to that of 4-year-old fish	Risopatrón et al. (2018)
	Rainbow trout (*O. mykiss*)	2- to 3-year-old sex-reversed females produced the best quality semen	İnanan and Yılmaz (2018)
	Heterobranchus longifilis	Age was the main component affecting sperm quality	Otoh and Udoh (2018)

(continued)

Table 3.2 (continued)

Factor	Species	Main study findings	Reference
	Bighead carp (*Hypophthalmichthys nobilis*)	4-year-old males were more suitable for reproduction than 3-year-old males	Dadras et al. (2017b)
	Common carp (*Cyprinus carpio*)	12-month-old brooders produced good quality milt	Betsy et al. (2016)
		12-month-old brooders produced good quality milt, at a 1:40 dilution ratio	Betsy et al. (2017)
	Pejerrey (*Odontesthes bonariensis*)	Broodstock age can be reduced to mid-age fish	Chalde et al. (2016)
	Silver catfish (*Rhamdia quelen*)	1- or 2-year old males were preferred	Hilbig et al. (2017)
Environmental factors			
Season	Rainbow trout Common carp (*C. carpio*)	High values of sperm quality markers were found in the beginning of the reproductive season	Shaliutina-Kolešová et al. (2016)
	Sea bass (*Dicentrarchus labrax*)	A high level of sperm lipid peroxidation was noted at the beginning of the reproductive season	Martínez-Páramo et al. (2012c)
	Senegalese sole (*Solea senegalensis*)	The months for the best quality sperm were identified; individual inter-male variability was observed as well	Beirão et al. (2011)
	Rainbow trout (*O. mykiss*)	Several season-related changes were found	İnanan and Yılmaz (2018)
		The semen of sex-reversed females was characterised by higher total antioxidant capacity and lower lipid peroxidation	İnanan et al. (2016)
		Combined long and short artificial photoperiod regimes can advance ovulation and spermiation, and can also affect the gamete quality and hatchery performance of rainbow trout during out-of-season spawning	Atasever and Bozkurt (2015)
	Eurasian perch (*Perca fluviatilis*)	Semen characteristics changed from the beginning (November) to the end (April) of the spermiation period	Alavi et al. (2010)

(continued)

Table 3.2 (continued)

Factor	Species	Main study findings	Reference
	Caspian roach (*Rutilus rutilus caspicus*)	Sperm motility was the parameter that was most affected by season	Golpour et al. (2013)
	Northern pike (*Esox lucius*)	The optimal time for semen collection was March (middle of the spawning period)	Bondarenko et al. (2018)
	Atlantic cod (*Gadus morhua*)	Several changes in relation to the month and origin (wild-caught vs. cultivated) of spawners were observed	Butts et al. (2010)
	Goldfish (*Carassius aureus*)	Evidence of seasonal variation in sperm characteristics	Zadmajid et al. (2013)
Out-of-season	Rainbow trout (*O. mykiss*)	Low fertilisation rates in out-of-season groups	Momin and Memiş (2018)
Water temperature Temperature regimes	European grayling (*Thymalus thymallus*)	Regression functions were established between temperatures and spawning	Lahnsteiner and Kletzl (2012)
	Brown trout (*Salmo trutta*)	Differences in sperm quality in relation to temperature regimes	Lahnsteiner and Leitner (2013)
Length of cold water treatments	Pike perch (*Sander lucioperca*)	The length of cold water treatments had an influence on sperm production	Blecha et al. (2015)
Heated water	Pumpkinseed (*Lepomis gibbosus*)	Fish maintained in a heated thermal reservoir developed precocial maturity and early season reproduction	Valente et al. (2016)
Acclimation to water temperatures	Brown trout (*S. trutta*)	Novel insights into the complex impacts of climate change on fish sperm, with implications for the reproduction and management of hatchery and wild trout populations in future climate scenarios	Fenkes et al. (2017)
Changes in ambient temperature	Three-spined stickleback (*G. aculeatus*)	Sperm velocity and linearity increased, whereas the percentage of motile sperm decreased at higher temperatures	Mehlis and Bakker (2014)
Salinity	Tilapia (*Sarotherodon melanotheron heudelotii*)	Adaptive responses to salinity were completed within 2 months or less	Legendre et al. (2016)
Ultraviolent light	Three-spined stickleback (*G. aculeatus*)	Enhanced ambient UVA levels had detrimental effects on both male breeding colouration and sperm velocity	Rick et al. (2014)

(continued)

Table 3.2 (continued)

Factor	Species	Main study findings	Reference
Environmental pollution			
Cadmium and mercury ions	Rainbow trout (*O. mykiss*)	Inhibition of sperm motility and viability after short-term exposure	Dietrich et al. (2010)
Mercury ions	Eurasian perch (*P. fluviatilis*)	Mercury ions acted on sperm through disrupting the function of the plasma membrane, axoneme and ATP content	Hatef et al. (2011)
	Silver catfish (*R. quelen*)	$HgCl_2$ increased primary pathologies and reduced sperm motility, vigour and motility time	Rocha et al. (2018)
	Tuvira (*Gymnotus carapo*)	$HgCl_2$ caused progressive damage to testicular tissue, reduced sperm count and altered sperm morphology	Vergílio et al. (2014)
Cadmium ions	Common carp (*C. carpio*)	Impairment of sperm motility by cadmium and a protective effect of transferrin	Dietrich et al. (2011)
	Zebrafish (*Danio rerio*)	Negative effects on path speed, straight speed, curvilinear velocity, motility time, progressive and total motility, and plasma and DNA integrity	Acosta et al. (2016)
	Tuvira (*G. carapo*)	Alterations in sperm number and morphology after exposure to cadmium ions	Vergilio et al. (2015)
Crude oil	Capelin (*Mallotus villosus*)	Dispersant, but not the dispersant oil, decreased sperm fertilising ability	Beirão et al. (2018)
Duroquinone	Sterlet (*Acipenser ruthenus*)	Spermatozoa motility, the content of carbonyl proteins and superoxide dismutase activity were shown to be sensitive biomarkers, exhibiting strong responses to low concentrations of the xenobiotic	Linhartova et al. (2013)
Carbamazepine	Common carp (*C. carpio*)	The use of fish spermatozoa in vitro assays may provide a novel and efficient means for monitoring residual pharmaceutical in the aquatic environment	Li et al. (2010b)

(continued)

Table 3.2 (continued)

Factor	Species	Main study findings	Reference
Decabrominated diphenyl ether (BDE-209)	Zebrafish (*D.rerio*)	Parental chronic low dose BDE-209 exposure affected F0 growth and reproduction	He et al. (2011)
Endocrine disruptors			
Bisphenol A (BPA)	Eurasian perch (*P. fluviatilis*)	Bisphenol A decreased both the motility and velocity of exposed sperm	Hatef et al. (2010)
	Goldfish (*Carassius auratus*)	Adverse effects of BPA on sperm motility and velocity via modifications of testicular steroidogenesis, which might correspond to alternation in sperm maturation	Hatef et al. (2012a)
	Zebrafish (*D.rerio*)	Low-level BPA exposures for two continuous generations affected the sex ratio and sperm quantity/quality in F1 and F2 adults, and reproductive success in offspring from F2 parents	Chen et al. (2015)
		BPA exposure during three different developmental periods impaired zebrafish reproductive development	Chen et al. (2017)
	Sterlet (*A. ruthenus*)	Concentrations of BPA that can be encountered in nature were capable of inducing oxidative stress, leading to impaired sperm quality, DNA fragmentation and intracellular ATP content	Hulak et al. (2013)
Tetrabrombisphenol A (TBBPA)	Sterlet (*A. ruthenus*)	Fish spermatozoa can be used in in vitro assays for monitoring residual pollution in aquatic environments	Linhartova et al. (2015)
Vinclozolin	Goldfish (*C. auratus*)	Different dose-dependent effects, which led to impairment in sperm quality via the disruption of steroidogenesis	Hatef et al. (2012b)
	Sterlet (*A. ruthenus*)	Impairment of hormone production and semen quality	Gazo et al. (2013)

(continued)

Table 3.2 (continued)

Factor	Species	Main study findings	Reference
2,2′-dithiobis-pyridine	Guppy (*Poecilia reticulata*)	2,2′-dithiobis-pyridine can induce reproductive toxicity, which would provide a basis for the future assessment of its ecological risk	Li et al. (2019)
Nonylphenol (NP)/ propranolol (PN)/ diethylstilbestrol (DES)	Sterlet (*A. ruthenus*)	NP, PN and DES can induce reactive oxygen species stress in fish spermatozoa, which could impair sperm quality and the antioxidant defence system	Shaliutina et al. (2016)
Di-(2-ethylhexyl)-phthalate (DEHP)	Goldfish (*C. auratus*)	DEHP interfered with testis and pituitary hormonal functions to reduce sperm quality in goldfish	Golshan et al. (2015)
Levonorgestrel	Fathead minnow (*Pimephales promelas*)	Males exposed to levonorgestrel exhibited decreases in multiple sperm motion characteristics	Frankel et al. (2018)
Herbicides			
Atrazine	Zebrafish (*D. rerio*)	Impairments in sperm motility and concentration	Bautista et al. (2018)
Roundup	*Jenynsia multidentata*	Slight effects on sperm motility; reduction of egg swelling	Sánchez et al. (2017)
	Common carp (*C. carpio*)	Losses in several sperm quality parameters, consequently decreasing the fertilisation potential of spermatozoa	Lugowska (2018)
	Silverside (*Odontesthes humensis*)	A decrease in the quality parameters of spermatozoa	Silveira et al. (2019)
Rearing in captivity	Greater amberjack (*Seriola dumerili*)	Disturbances to spermatogenesis and sperm quality	Zupa et al. (2017)
	Lebranche mullet (*Mugil liza*)	Significant changes in sperm quality characteristics in response to captivity rearing	Magnotti et al. (2018)
	Arctic charr (*Salvelinus alpinus*)	Hatchery selection may depress the number of motile sperm, but intensify swimming velocity	Kekäläinen et al. (2013)
	European eel (*Anguilla anquilla*)	Wild males produced spermatozoa with longer longevity of motion	Locatello et al. (2018)

(continued)

Table 3.2 (continued)

Factor	Species	Main study findings	Reference
	Burbot (*Lota lota*)	Fish grown at natural conditions may be more suitable as broodstock than RAS-cultured fish	Blecha et al. (2018)
	Atlantic cod (*G. morhua*)	Significant discriminations in the sperm quality characteristics between wild and cultivated fish were recorded	Butts et al. (2011a, b)
Nutrition			
Fatty acids	European eel (*A. anquilla*)	Semen characteristics were differentially affected by particular fatty acids	Butts et al. (2015)
	Rainbow trout (*O. mykiss*)	Dietary fatty acid levels affected semen quality, but not fertilisation ability in rainbow trout	Hajiahmadian et al. (2016)
n-3 long-chain polyunsaturated fatty acids (n-3 LC-PUFA)	Siberian sturgeon (*Acipenser baerii*)	n-3 LC-PUFA can improve the reproductive performance of male and female Siberian sturgeon broodstock	Luo et al. (2017)
Phospholipids	Zebrafish (*D. rerio*)	Improvement in sperm motility.	Diogo et al. (2015)
Nutritional status	Pike perch (*S. lucioperca*)	The use of males with initial fat stores was beneficial for sperm quality	Teletchea et al. (2009)
Diet supplements			
Vitamin K	Senegalese sole (*S. senegalensis*)	Lower sperm DNA fragmentation after vitamin K supplementation	Fernández et al. (2019)
Vitamin E + Se	Senegalese sole (*S. senegalensis*)	Improvement in sperm motility after vitamin E + Se supplementation	Beirão et al. (2015)
Probiotics	Goldfish (*C. auratus*)	Dietary supplementation of probiotics resulted in improvements in the percentage and duration of sperm motility, absolute fecundity and fertilisation success	Mehdinejad et al. (2018)
Resveratrol	Medaka (*Oryzias latipes*)	Dietary resveratrol improved immunity, but reduced the reproduction of broodstock medaka	Kowalska et al. (2017)

(continued)

Table 3.2 (continued)

Factor	Species	Main study findings	Reference
Hormonal manipulations			
	Barbel (*Barbus barbus*)	Ovopel was optimal for the hormonal stimulation of barbel	Cejko et al. (2014)
	Chub (*Leuciscus cephalus*)	Ovaprim was effective for obtaining higher amounts of semen	Cejko and Krejszeff (2016)
	Eurasian perch (*P. fluviatilis*)	Longer spermiation with the use of gonadoliberine	Żarski et al. (2017)
	European eel (*A. anquilla*)	Specific recombinant gonadotropin was effective in inducing spermatogenesis and spermiation	Peñaranda et al. (2018)
	European eel (*A. anquilla*)	Economic analysis favoured the use of recombinant hormone	Herranz-Jusdado et al. (2019b)
	Northern pike (*E. lucius*)	Ovaprim was optimal for the hormonal stimulation of pike	Cejko et al. (2018a)
	Levantine scraper (*Capoeta damascina*)	Ovaprim treatment successfully induced steroidogenesis and the maturation of spermatogenic germ cells, leading to spermiation and milt production	Zadmajid et al. (2018)
	Longspine scraper (*Capoeta trutta*)	Hormonal treatments stimulated steroidogenesis and spermiation in male *C. trutta*, with the highest efficiency with the Ovaprim™ treatment	Zadmajid (2016)
Semen collection and handling			
Semen collection	Rainbow trout (*O. mykiss*)	Consistent, good quality semen when collected with a catheter	Glogowski et al. (2000)
	Pike perch (*S. lucioperca*)	It is recommended that pike perch sperm be collected using a catheter	Sarosiek et al. (2016)
	Eurasian perch (*Perca fluviatilis*)	Detrimental effects of urine on sperm motility characteristics	Król et al. (2018)
Sequential stripping	Sterlet (*A. ruthenus*)	Sequential stripping and spermatozoa cryopreservation in combination could improve the efficiency of sturgeon aquaculture	Dzyuba et al. (2012)

(continued)

Table 3.2 (continued)

Factor	Species	Main study findings	Reference
		Sequential stripping had no negative effect on the percentage of motility and spermatozoa velocity	Shaliutina et al. (2012)
Anaesthesia	South American catfish (*R. quelen*)	Stress reduction can be reconciled with reproductive management, without compromising reproductive performance	Corso et al. (2019)
Short-term storage	Patagonian blenny (*Eleginops maclovinus*)	ROS generation and ATP depletion were important factors determining storage success	Ulloa-Rodríguez et al. (2018)
	Patagonian blenny (*E. maclovinus*)	Flow cytometry was useful for evaluating sperm quality during in vitro storage	Contreras et al. (2017)
	Meagre (*Argyrosomus regius*)	Optimal conditions for storage were defined	Santos et al. (2018)

3.3.2 Environmental Factors

3.3.2.1 Photoperiod and Temperature

Environmental factors have markedly shaped fish reproduction throughout 500 million years of evolution. Temperate fish spawning is programmed for an optimal seasonal period, aiming to increase the chance of progeny survival in favourable environmental conditions and the abundance of food (Migaud et al. 2013). The optimal period for spawning is characterised by the highest possible gamete quality and is characteristic for particular species; for example, March–May for Senegalese sole (Beirão et al. 2011), January–February for perch (Alavi et al. 2010) and March for the Northern pike (Bondarenko et al. 2018). Ensuring optimal reproductive behaviour and producing the highest possible quality gametes are amongst the most important challenges in aquaculture for both normal and sex-reversed females (Lahnsteiner and Leitner 2013; İnanan and Yılmaz 2018). Several strategies are employed to reach this goal, mostly involving photoperiod manipulations, which are also used to induce out-of-season spermiation (Nascimento et al. 2010; Atasever and Bozkurt 2015; Momin and Memiş 2018). Usually, sperm production precedes ovulation, so males are ready to spawn ahead of females. Spawning time is shaped and fine-tuned by several environmental factors, natural or manmade, such as heated thermal reservoirs (Valente et al. 2016), the controlled temperature of ponds or tanks (Blecha et al. 2015) or a wide range of water salinity levels for euryhaline fish species (Legendre et al. 2016). Moreover, ambient ultraviolet light may affect sperm quality as well (Rick et al. 2014). The effects of temperature on sperm quality have been recently discussed in the context of climate change and the reproduction of fish populations (Fenkes et al. 2017).

3.3.2.2 Environmental Pollution

Xenobiotics

Contamination of the aquatic environment with water pollutants due to anthropogenic activity is a serious threat for fish reproduction, including gamete quality (Devaux et al. 2015). Heavy metals (mercury and cadmium ions) clearly impair sperm motility, membrane and DNA integrity and fertilising ability (Dietrich et al. 2010; Hatef et al. 2011; Acosta et al. 2016; Rocha et al. 2018). Moreover, damage to the male reproductive system caused by heavy metals, including testis morphology and germ cells and different stages of differentiation, is well documented (Vergílio et al. 2014; Vergilio et al. 2015). Several organic compounds present in the environment, such as decabrominated diphenyl ether, a flame retardant, impair the growth and reproduction of parental fish, with further complications in F1 offspring (He et al. 2011). Chemically dispersed crude oil affects sperm membrane functionality, probably due to its surfactant action (Beirão et al. 2018).

Moreover, herbicides, including the popular Roundup, decrease sperm quality, including effects on sperm motility and concentration, leading to a decrease in fish reproductive potential (Sánchez et al. 2017; Silveira et al. 2019). For some herbicides, such as atrazine, impairments of the expression of genes related to spermatogenesis have been described (Bautista et al. 2018). There is growing concern regarding the rise in human pharmaceuticals as a source of environmental pollutants, including the aquatic environment, affecting fish reproduction. For example, carbamazepine (an antiepileptic drug) was found to impair the semen quality and antioxidant defence system of carp semen (Li et al. 2010b).

Endocrine-Disrupting Chemicals

Endocrine-disrupting chemicals (EDCs) present in anthropised hydrosystems are mainly organic agents of a natural or industrial origin, which interfere with endocrine functions. Bisphenol A (BPA) and its derivative tetrabrombisphenol A are common EDCs present in aquatic environments, with several potential threats to sperm quality, including motility characteristics, DNA fragmentation due to oxidative stress and the disruption of reproductive development (Hatef et al. 2010, 2012a; Hulak et al. 2013; Linhartova et al. 2015; Chen et al. 2017). The toxic effects of BPA may result in reproductive dysfunctions in the next generation of fish (Chen et al. 2015). Other relevant EDCs are nonylphenol, propranolol and diethylstilbestrol, which are capable of inducing oxidative stress in fish spermatozoa, leading to the impairment of the antioxidant defence system and a decrease in the number of intact sperm cells (Shaliutina et al. 2016). Di-(2-ethylhexyl)-phthalate disrupts pituitary and testicular hormonal functions, leading to reduced sperm quality (Golshan et al. 2015). Vinclozolin is an anti-androgenic compound that leads to increases in oxidative stress-related damage to fish spermatozoa and disturbances to steroidogenesis (Hatef et al. 2012b; Gazo et al. 2013). Levonorgestrel, a component of human contraceptives that frequently enters the aquatic environment via wastewater treatment plant effluent, impairs the sperm motility and reproductive behaviour of male fathead minnows (Frankel et al. 2018).

3.3.2.3 Rearing of Fish in Captivity

Activities related to the rearing of fish and controlling their reproduction have profound effects on semen quality. Major factors include domestication, nutrition, hormonal stimulation and semen collection and handling. Usually, wild-caught males are characterised by higher quality spermatozoa, compared to farmed fish (Locatello et al. 2018), which is reflected in differences in sperm composition, including lipids (Blecha et al. 2018).

Nutrition

The development of formulated diets with similar efficiencies as natural foods is an on-going challenge for aquaculture. Several nutritional and non-nutritional compounds have been tried and their effects on reproductive characteristics have been recorded. Among nutritional compounds, different unsaturated fatty acids were found to affect the milt composition and sperm performance of European eel (Butts et al. 2015), Senegalese sole (Beirão et al. 2015) and Siberian sturgeon (Luo et al. 2017). The positive effects of phospholipids on the reproductive performance of zebrafish have also been shown (Diogo et al. 2015). High initial fat storage, before the start of a reproductive cycle, was found to be a critical factor influencing the semen quality of pike perch (Teletchea et al. 2009). The supplementation of diets with various additives often produces an improvement in semen quality. The positive effects of probiotics and nucleotides on the reproductive performance of goldfish (Mehdinejad et al. 2018), resveratrol on medaka sperm motility (Kowalska et al. 2017) and vitamin E on sole sperm quality (Beirão et al. 2015) have been recorded.

Hormonal Manipulations

Fish reproduction under aquaculture conditions is often seriously impaired, leading to reproductive dysfunctions, such as a lack of breeding behaviour, disturbances to spermatogenesis and sperm maturation, and low and variable quality semen (Mylonas et al. 2017). The hormonal stimulation of spermiation is the most commonly used method for fish kept in captivity. The main spawning agents include carp pituitary extracts, gonadotropins of a mammalian origin and gonadoliberins, mostly synthetic GnRH analogues; the latter are often used in combination with dopamine antagonists. Carp (or other fish) pituitary extracts have become less popular in recent years due to difficulties with standardisation and concerns of their potential for disease transmission (Viveiros et al. 2015). Gonadoliberine analogues alone have been found to be effective spawning agents for captive fish, including burbot (Kucharczyk et al. 2018) and Eurasian perch (Żarski et al. 2017). Also, mammalian gonadotropins, purified or recombinant, are effective, especially for European eel (Peñaranda et al. 2018; Herranz-Jusdado et al. 2019b). Two commercial preparations, Ovaprim and Ovopel, containing different GnRH analogues and dopamine antagonists, have been successfully tested for several fish species, leading to improvements in both semen quantity and quality (Cejko et al. 2014, 2018a; Cejko and Krejszeff 2016; Zadmajid 2016; Zadmajid et al. 2018). For some species, dopamine antagonists alone can be as effective as Ovopel, which has been demonstrated for crucian carp (Cejko and Kucharczyk 2015).

Semen Collection and Handling

The necessity of collecting non-contaminated semen to maintain good semen quality has been outlined above (Glogowski et al. 2000; Sarosiek et al. 2016; Król et al. 2018). An important factor related to changes in semen quality in relation to collection procedures is multiple collections of semen because significant variability in common carp sperm quality has been found to be related to the collection period (Cejko et al. 2018b). Sequential semen collection was found to be useful for sturgeon fish breeding (Dzyuba et al. 2012; Shaliutina et al. 2012). Some conditions of anaesthesia may also affect semen parameters, as demonstrated in South American catfish (Corso et al. 2019). The short-term storage of semen after collection is associated with progressive changes in sperm quality over time (Ulloa-Rodríguez et al. 2018). Therefore, monitoring sperm quality over storage time is necessary (Contreras et al. 2017) and conditions for the improvement of storage conditions need to be defined (Santos et al. 2018).

3.4 Factors Influencing Fresh Sperm Quality and Their Usefulness for Predicting Cryopreservation Success

Excellent quality fresh semen is a prerequisite for successful cryopreservation. Consequently, the use of existing knowledge regarding factors influencing sperm quality (see Table 3.3) is extremely useful for the prediction of cryopreservation success. Any improvement in semen quality may be critical for spermatozoa to withstand the stress related to freezing–thawing. Table 3.3 presents examples of major factors that are important for improving sperm cryoresistance.

3.4.1 Season

For temperate fish, season is the major factor that determines sperm quality. Indeed, for sex-reversed female rainbow trout, seasonality clearly affects the success of the cryopreservation process; it was recommended that sperm should also be obtained in the winter season, during the natural breeding period (Robles et al. 2003). This recommendation has been challenged because we have demonstrated that the efficient cryopreservation procedure of sex-reversed female rainbow trout semen from autumn spawning is also very effective for spring spawning fish (Ciereszko et al. 2015). Perhaps the use of winter and spring spawning strains is prerequisite to counteract the seasonal effects on sperm cryoresistance. The great challenges in aquaculture are obtaining gametes out-of-season to perform spawning and

Table 3.3 Factors that influence fresh sperm quality, in relation to their usefulness for predicting cryopreservation success

Factor	Species	Main findings	References
Season	Rainbow trout sex-reversed females (*O. mykiss*)	Seasonality clearly affected the success of the cryopreservation process, which should always be carried out with sperm obtained in the winter season, the natural breeding period Spring semen from sex-reversed rainbow trout were successfully cryopreserved and used for the fertilisation of an elevated number of eggs	Robles et al. (2003) Ciereszko et al. (2015)
	P. lineatus, *B. orbignyanus*	The quality and freezing ability of sperm from both species were sustained over the spawning season, thus fish farmers can reproduce these species and freeze their sperm at any time throughout the spawning season	Di Chiacchio et al. (2017)
	Pirapitinga (*P. brachypomus*)	Out-of-season sperm cryopreserved in glucose and methylglycol was of high quality and can be used to facilitate artificial reproduction	Nascimento et al. (2010)
	Eurasian perch (*P. fluviatilis*)	The applicability of off-season collected perch sperm for cryopreservation and fertilisation was demonstrated	Bernáth et al. (2016a)
	Siberian sturgeon (*A. baerii*)	Methods of the short- and long-term storage (cryopreservation) of sperm obtained in December (out-of-season) can be implemented for the artificial reproduction of Siberian sturgeon	Judycka et al. (2015)
Rearing in captivity	Senegalese sole (*S. senegalensis*)	Living in captivity was related to decreased sperm viability in cryopreserved semen samples from the F1 generation	Valcarce and Robles (2016)
	Zebrafish (*D. rerio*)	Cryopreserved zebrafish sperm from a farm population and an AB research line showed significant differences in post-thaw fertility	Yang et al. (2016)
Sequential stripping	Sterlet (*A. ruthenus*)	There were differences in sperm cryoresistance (post-thaw motility and fertilisation ability) in relation to sperm collection time after hormonal stimulation	Dzyuba et al. (2012)

(continued)

Table 3.3 (continued)

Factor	Species	Main findings	References
Hormonal stimulation	*P. lineatus*, *B. orbignyanus*	Sperm obtained from Ovaprim™-treated males can be cryopreserved without any loss of quality, compared to sperm collected after carp pituitary therapy	Viveiros et al. (2015)
In vitro storage	*P. lineatus*	*P. lineatus* sperm can be transported/shipped to the laboratory without decreasing its suitability for cryopreservation. Sperm should be kept undiluted during storage and be frozen within 3 h	Viveiros et al. (2017)
Microbial contamination	Silver barb (*Barbodes gonionotus*)	Microbial contamination had a deleterious effect on cryopreserved silver barb sperm, based on a reduction in sperm motility and viability and an alteration of sperm morphometry, especially flagellum width	Boonthai et al. (2016)
Nutrition			
Fatty acids	Thailand tilapia (*Oreochromis* spp.)	Significant differences were observed between fish, corn and linseed oil treatments on the rate of motility relative to other different oil sources. Fish fed a diet supplemented with fish oil presented a significantly higher rate and duration of post-thaw sperm motility	Navarro et al. (2014)
Lipid composition	Common carp (*C. carpio*)	Increasing concentrations of saturated fatty acids and monounsaturated acids, and decreasing concentrations of polyunsaturated n-6 fatty acids were negatively correlated with post-thaw motility	Horokhovatskyi et al. (2016)
	Sea bass (*D. labrax*)	Motility and plasma membrane lipid composition (CHO/PL) were the most desirable variables determined in fresh samples to predict cryoresistance in European sea bass sperm	Martínez-Páramo et al. (2012a)
Semen characteristics	Atlantic salmon (*S. salar*)	There were significant positive relationships between frozen-thawed motility, velocity, track crossing frequency and subsequent fertilisation success. The spermatocrit and pH of fresh semen were negatively related, whereas osmolality and antioxidant capacity were positively related to frozen-thawed motility and velocity	Butts et al. (2011a, b)

(continued)

Table 3.3 (continued)

Factor	Species	Main findings	References
	Curimba (*P. lineatus*)	The rate and motility duration of fresh semen may be used to predict the quality of thawed *P. lineatus* sperm	Carvalho et al. (2017)
Protein composition of seminal plasma	Common carp (*C. carpio*)	Seminal plasma with good freezability was characterised by a higher abundance of three proteins involved in the maintenance of sperm membrane integrity and oxidative protection, whereas poor freezability was characterised by a higher amount of seven proteins involved in the innate immune response	Dietrich et al. (2017)
	Common carp (*C. carpio*)	Good semen freezability was related to a higher concentration of 20 sperm proteins involved in the maintenance of flagella structure, membrane fluidity, energy production and oxidative protection. The poor freezability of spermatozoa was related to 18 proteins involved in the control of proper protein folding, energy metabolism and cytoskeletal organisation	Dietrich and Ciereszko (2018)

cryopreserving spermatozoa; such an approach was successfully developed for some fish species, including the South American fish *Prochilodus lineatus*, *Brycon orbignyanus* and pirapitinga (*Piaractus brachypomus*; Nascimento et al. 2010; Di Chiacchio et al. 2017), Eurasian perch (Bernáth et al. 2016a) and Siberian sturgeon (Judycka et al. 2015).

3.4.2 Obtaining and Handling Semen

Obtaining and handling semen are of great importance to semen quality and, consequently, to cryopreservation success. Rearing in captivity itself poses a significant challenge to sperm quality and its usefulness for cryopreservation (Valcarce and Robles 2016; Yang et al. 2016). Since semen is often obtained after hormonal stimulation, the efficacy of this procedure in terms of the type of treatment agent and the time of semen collection has been proven to significantly determine the outcome of cryopreservation (Dzyuba et al. 2012; Viveiros et al. 2015). The in vitro storage time of semen is an important factor that determines its usefulness for cryopreservation. Under practical hatchery conditions, it is common to collect semen in remote sites and transport it to a laboratory for cryopreservation. Since the quality of semen declines with the time of storage, it is very important to determine the time for usefulness to freezing; for example, for *P. lineatus*, this time is 3 h (Viveiros et al. 2017). Recently, awareness has been raised about the threat of microbial contamination to semen quality and cryopreservation success. Boonthai et al. (2016) demonstrated that microbial contamination has a deleterious effect on cryopreserved silver barb sperm, leading to a reduction in sperm motility and viability and an alteration of sperm morphometry.

3.4.3 Nutrition

Fatty acids, especially polyunsaturated fatty acids, are very important for maintaining the high quality of the sperm membrane, which is critical for cryoresistance, as has been demonstrated for common carp (Horokhovatskyi et al. 2016) and sea bass (Martínez-Páramo et al. 2012a). Therefore, developing fish diets that support the desired lipid composition of spermatozoa is a great challenge; such an approach was successfully developed by Navarro et al. (2014), who indicated that Thailand tilapia fed a diet supplemented with fish oil presented a significantly higher rate and duration of post-thaw sperm motility, compared to other sources. The search for an efficient replacement for fish oil is an important challenge for the aquaculture industry due to the shortage of fish oil; the consequences of such replacements on reproductive performance need to be tested each time a new replacement is introduced.

3.4.4 Biochemical and Physiological Characteristics of Fresh Semen

The identification of biochemical or physiological characteristics of fresh semen as predictors for cryopreservation success would greatly accelerate the implementation of cryopreservation into hatchery practice because the predictability of the procedure would be enhanced. As indicated above, the concentration of particular lipids of sperm membranes and their ratio are promising candidates for evaluating cryoresistance. Butts et al. (2011a, b) found that among several potential characteristics, spermatocrit, pH, osmolality and antioxidant capacity were correlated with Atlantic salmon semen fertility and motility. For curimba, the rate and duration of the motility of fresh semen were found to be useful for the prediction of post-thaw sperm quality (Carvalho et al. 2017). Recently, using a proteomic approach, specific proteins in both the seminal plasma and spermatozoa of common carp were identified as characteristics of semen with good freezability (Dietrich et al. 2017; Dietrich and Ciereszko 2018). A comprehensive approach may be useful to predict the biochemical and physiological pathways that are critical for cryopreservation success.

3.5 Changes in Sperm Quality During Cryopreservation

During cryopreservation, spermatozoa have to withstand several serious stresses caused by freezing–thawing (Bailey et al. 2000). The main factors affecting the survival of spermatozoa are susceptibility to cold shock, cooling and freezing rates, diluent composition and osmotic stress (Watson 2000). Cryodamage to sperm cells is complex and several cryoinjuries have been identified. In addition to a dramatic reduction in sperm motility, damage to plasma membranes and the nucleus, mitochondrial function and ROS production, and numerous changes to phosphorylated proteins have been described for mammals (Yeste 2016). Many of these changes are on the sublethal level, which requires the application of subtle sperm quality tests (Pini et al. 2018) to assess semen quality after cryopreservation (Palomar Rios and Molina Botella 2017). This approach has been incorporated into the analysis of fish sperm quality, in which numerous specific tests for several specific structures and functions are used (Cabrita et al. 2014).

Examples of the changes in sperm quality parameters following freezing–thawing are shown in Table 3.4. Sperm motility and viability are the most common parameters used for the evaluation of both fresh and cryopreserved semen, with the percentage of motile spermatozoa as a fundamental measurement that dominates almost every cryopreservation study. For most experiments, the percentage of sperm motility decreases after cryopreservation, with a parallel decrease in sperm curvilinear velocity (VCL) (Martínez-Páramo et al. 2012a, b, c; Nynca et al. 2012; Bernáth et al. 2016b; Judycka et al. 2018; Xin et al. 2018; Herranz-Jusdado et al. 2019a). On the other hand, for the semen of sex-reversed females, the percentage of sperm motility is often not affected by cryopreservation; a rise in sperm motility is usually observed after equilibration, and then sperm motility decreases to the values recorded for fresh semen (Dietrich et al. 2014). CASA parameters also appear to be useful for evaluating the interaction between season and cryopreservation in Siberian sturgeon semen (Judycka et al. 2015). It should be underlined that there is still a need for the standardisation of CASA technology because systems may strongly influence sperm motility parameters (Boryshpolets et al. 2013); recently, such studies have been performed for Atlantic salmon, European eel and Siberian sturgeon (Caldeira et al. 2019).

The viability of spermatozoa appears to be subjected to similar changes, such as the percentage of sperm motility, with a clear decrease after cryopreservation (Pérez-Cerezales et al. 2010a; Figueroa et al. 2016b; Valcarce and Robles 2016). It should be underlined that viability values are often higher than motility values; for example, values for the viability of Senegalese sole cryopreserved sperm were higher than 70%, whereas the percentage of sperm motility amounted to 40–60% (Riesco et al. 2017a). Natural membrane fluidity is related to the coexistence of fluid and gel phase areas in the sperm membranes and reflects the interactions and distributions of specific membrane components (Müller et al. 2008). The observation of changes in membrane fluidity was found to be useful in selecting optimal cryoprotectant for the cryopreservation of Suravi spermatozoa (Pereira et al. 2019). Loss of viability as a result of cryopreservation is often accompanied by the death of sperm cells, which

Table 3.4 Examples of changes in fish sperm quality indices that occur during cryopreservation

Quality index	Assay method	Species	Main outcome	References
Sperm motility parameters	CASA	Sea bass (*D. labrax*)	The values of motility, progressive motility, VCL, VSL and LIN decreased after cryopreservation	Martínez-Páramo et al. (2012a)
		Whitefish (*Coregonus lavaretus*)	The percentage of sperm motility, VCL, BCF and ALH decreased after cryopreservation; however, VSL and LIN increased	Nynca et al. (2012)
		Common carp (*C. carpio*)	Progressive motility and VCL decreased after cryopreservation; however, straightness values increased	Bernáth et al. (2016b)
		Rainbow trout (*O. mykiss*)	Percentage of sperm motility decreased; however, VCL, VAP, VSL, ALH and LIN were not affected by cryopreservation	Judycka et al. (2018)
		Sterlet (*A. ruthenus*)	Sperm motility and velocity decreased after cryopreservation	Xin et al. (2018)
		European eel (*A. anguilla*)	The percentage of sperm motility and VCL decreased after cryopreservation	Herranz-Jusdado et al. (2019a)
		Sex-reversed females rainbow trout (*O. mykiss*)	The percentage of sperm motility was not affected by cryopreservation. VCL and ALH decreased after cryopreservation; however, VAP, VSL and LIN increased	Dietrich et al. (2014)
		Siberian sturgeon (*A. baerii*)	The percentage of sperm motility decreased after cryopreservation for semen obtained in December; the other sperm velocity and trajectory parameters (VCL, VSL, VAP, LIN, ALH) were not affected. However, for semen collected in April, a decrease after cryopreservation was recorded for MOT, VCL, VAP and ALH. The values of VSL and LIN were not affected by cryopreservation	Judycka et al. (2015)

(continued)

Table 3.4 (continued)

Quality index	Assay method	Species	Main outcome	References
Sperm membrane integrity/viability	Flow cytometry (with SYBR-14/PI)	Atlantic salmon (*Salmo salar*)	Viability decreased after cryopreservation, from 90% in fresh to 75% in frozen semen.	Figueroa et al. (2016b)
		Rainbow trout (*O. mykiss*)	Viability decreased after cryopreservation. There was a positive effect of LDL on sperm viability (99% in fresh vs. 13.7% in frozen semen in the presence of egg yolk, 38% in the presence of LDL)	Pérez-Cerezales et al. (2010a)
		Suruvi (*Steindachneridion scriptum*)	5, 7.5 and 10% DMSO were the most effective treatments for cryopreserved sperm integrity	Pereira et al. (2019)
	PI and flow cytometry	Senegalese sole (*Solea selegalensis*)	Captivity and cryopreservation decreased viability after cryopreservation	Valcarce and Robles (2016)
		Senegalese sole (*S. selegalensis*)	No effect of different extenders on post-thaw viability; Viability in the range of 75–88% was observed after cryopreservation	Riesco et al. (2017a)
	PI/SYBR-14 and fluorescence microscopy	Gilthead seabream (*Sparus aurata*)	Post-thaw sperm viability in the range of 34–58%	Cabrita et al. (2011)
		European Sea bass (*D. labrax*)	Cholesterol/phospholipids ratio was a good predictor of cryoresistance	Martínez-Páramo et al. (2012a)
Membrane fluidity	Flow cytometry (with M540 and YO PRO-1)	Suruvi (*S. scriptum*)	12.5% DMSO was the most effective treatment for the maintaining cryopreserved sperm membrane fluidity	Pereira et al. (2019)
Caspase detection	Caspase 3/7 assay	Dusky grouper (*Epinephelus marginatus*)	Detection of sperm necrotic damage due to cryopreservation	Riesco et al. (2017b)
	Flow cytometry (with muse Caspase-3/7)	Senegalese sole (*S. selegalensis*)	Detection of early and late apoptosis in cryopreserved semen	Riesco et al. (2017a)

Mitochondrial membrane potential (MMP)	Flow cytometry (with JC-1)	Atlantic salmon (*S. salar*)	MMP decreased after cryopreservation, from 91% in fresh to 52% in frozen semen	Figueroa et al. (2016b)
	Flow cytometry (with Rhodamine 123)	Suruvi (*S. scriptum*)	12.5% and 15% DMSO were the most effective treatments for maintaining cryopreserved sperm MMP	Pereira et al. (2019)
Oxidative stress				
ROS in sperm	Flow cytometry (with H_2DCFDA)	Rainbow trout (*O. mykiss*)	Increased ROS+ production after cryopreservation (2.3% in fresh vs. 3.2% in frozen semen)	Pérez-Cerezales et al. (2010a)
		Suruvi (*S. scriptum*)	5% methylglycol was the most effective treatment for the cryopreservation of the *S. scriptum* sperm and the lowest ROS production	Pereira et al. (2019)
	Flow cytometry (with DHE)	Carp (*C. carpio*)	An elevated amount of ROS was generated during cryopreservation	Mostek et al. (2018)
	Flow cytometry (with DCFH-DA)	Senegalese sole (*S. selegalensis*)	Captivity and cryopreservation enhanced ROS production after cryopreservation	Valcarce and Robles (2016)
Lipid peroxidation	MDA measurement	European sea bass (*D. labrax*)	Greater lipid peroxidation in cryopreserved sperm	Martínez-Páramo et al. (2012a)
		Senegalese sole (*S. selegalensis*)	Different extenders had variable effects on post-thaw lipid peroxidation	Riesco et al. (2017a)
		Rainbow trout/brook trout (*O. mykiss*/ *Salvelinus fontinalis*)	Supplementation of extenders with antioxidants was not recommended	Lahnsteiner et al. (2011)
		Beluga (*Huso huso*)	Demonstration of the cryoprotective effect of synthetic antioxidant	Osipova et al. (2014)
		Russian sturgeon (*Acipenser guldenstadti*)	Demonstration of the cryoprotective effect of heterocyclic derivatives of butylated hydroxytoluene	Osipova et al. (2016)
		Atlantic salmon (*S. salar*)	Supplementation of extender with α-tocopherol and ascorbic acid increased sperm function and fertility rate	Figueroa et al. (2018)

(continued)

Table 3.4 (continued)

Quality index	Assay method	Species	Main outcome	References
	Flow cytometry (with Bodipy C11)	Suruvi (*S. scriptum*)	7.5% methylglycol was the most effective treatment for the cryopreservation of *S. scriptum* sperm	Pereira et al. (2019)
Protein oxidation	DNPH colorimetric	Sea bass (*D. labrax*)	Lack of protein oxidation due to cryopreservation	Martínez-Páramo et al. (2012b)
		Common carp (*C. carpio*)	Reduction of carbonyl derivatives of proteins in response to antioxidant supplementation	Shaliutina-Kolešová et al. (2014)
	2DE-Oxybot, MALDI-TOF/TOF MS	Common carp (C. carpio)	Cryopreservation triggered the carbonylation of 19 spermatozoa proteins, mainly structural, as well as those involved in calcium ion binding, signal transduction, protein folding and intracellular transport	Mostek et al. (2018)
Protein phosphorylation	2DE-Westernblot	Sea bream (*S. aurata*)	Cryopreservation had a strong effect on the phosphorylation state of sperm proteins	Zilli et al. (2008)
DNA fragmentation	Comet assay	European seabass (*D. labrax*)	Reduction of DNA damage during the reproductive season; protective effect of LDL against cryodamage	Pérez-Cerezales et al. (2010a)
		European seabass (*D. labrax*)	Positive effect of taurine and hypotaurine supplementation of extender	Martínez-Páramo et al. (2013)
		Common carp (*C. carpio*)	Positive effects of the supplementation of extender with cysteine	Öğretmen et al. (2015)
		Sterlet (*A. ruthenus*)	Ethylene glycol was not effective for cryopreservation	Shaliutina-Kolešová et al. (2015)
	SCGE or commet assay	Rainbow trout (*O. mykiss*)	Determination of the relationship between DNA damage and embryo development success	Pérez-Cerezales et al. (2010b)

	TUNEL and comet	Gilthead seabream (*S. aurata*)	Identification of species-specific changes in DNA, fragmentation in response to the supplementation of extenders with antioxidants	Cabrita et al. (2011)
	TUNEL	Brown-marbled grouper (*Epinephelus fuscoguttatus*)	Evaluation of the effects of seminal plasma on sperm vitrification	Figueroa et al. (2015)
	DNA laddering analysis	Zebrafish (*D. rerio*)	Optimisation of cryopreservation methodology	Yusoff et al. (2018)
	Flow cytometry (by TUNEL)	Atlantic salmon (*S. salar*)	No effect of cryopreservation on DNA damage (2.2% in fresh vs. 4.8% in frozen semen)	Figueroa et al. (2016b)
	SCSA Flow cytometry (with acridine orange)	Suruvi (*S. scriptum*)	No effect of different cryoprotectants on DNA fragmentation after cryopreservation	Pereira et al. (2019)
Transcript abundance	Real-time PCR	Senegalese sole (*S. selegalensis*)	Transcript degradation was the best predictor of sperm status after cryopreservation	Riesco et al. (2017a)
Global DNA methylation	The use of MspI and HpaII restriction enzymes	Tambaqui (*Colossoma macropomum*)	Alterations in methylation levels in relation to different cryoprotectant agents	de Mello et al. (2017)
	Luminometric methylation assay	European eel (*A. anguilla*)	Induced hypermethylation in the presence of DMSO, but not methanol in the extender	Herranz-Jusdado et al. (2019a)
Lipid composition	Gas chromatography	Common carp (*C. carpio*)	Lipid composition of sperm was affected by temperature	Dadras et al. (2017a)
			Lipid composition of sperm was related to cryoresistance	Horokhovatskyi et al. (2016)
	Gas chromatography, HPLC for phospholipids	European Sea bass (*D. labrax*)	The cholesterol/phospholipids ratio was a good predictor of cryoresistance	Martínez-Páramo et al. (2012a)
		Gilthead seabream (*S. aurata*)	Antifreeze proteins stabilised the plasma membrane during cryopreservation.	Beirão et al. (2012)
Proteome changes	2D-DIGE, MALDI-TOF/TOF MS	Sea bass (*D. labrax*)	Cryopreservation caused the degradation of 21 sperm proteins, 2 of which were identified	Zilli et al. (2005)

(continued)

Table 3.4 (continued)

Quality index	Assay method	Species	Main outcome	References
		Rainbow trout (*O. mykiss*)	63 spermatozoal proteins involved in cellular movement, protein folding, antioxidant protection and metabolism enriched in extracellular fluid after cryopreservation	(Nynca et al. 2015)
		Common carp (*C. carpio*)	Cryopreservation affected 11 sperm protein spots involved in membrane trafficking and organisation, cell movement, metabolism and signal transduction	Li et al. (2010a)
			Identification of 183 proteins released from sperm to the extracellular medium, involved in metabolism and energy production, as well as in response to stress, apoptosis, small GTPase-mediated signal transduction, transcription, translation, protein folding and turnover, reproduction and DNA repair	Dietrich et al. (2015)
		Sterlet (*A. ruthenus*)	Cryopreservation induced changes in 20 protein spots of spermatozoa involved in metabolism and energy production, stress response and flagella assembly and organisation Six altered protein spots in seminal plasma and 13 altered spots in sperm were detected in fresh and thawed sperm. These proteins were involved in sperm metabolism, cytoskeleton and stress response	Horokhovatskyi et al. (2018) Xin et al. (2018)
	2DE, nanoHPLC ESI-Q-TOF MS/MS	Sea bream (*S. aurata*)	Identification of six proteins (four from the flagella and two present in both the flagella and head plasma membranes) that were involved in the sperm bioenergetic system	Zilli et al. (2014)

has been demonstrated as a necrotic mechanism for dusky grouper (Riesco et al. 2017b) or early and late apoptosis for Senegalese sole (Riesco et al. 2017a).

The bioenergy mechanisms responsible for the activation and maintenance of sperm motility are related to the functioning of sperm mitochondria. Measurements of mitochondrial membrane potential provide valuable information regarding the energetics of spermatozoa and have proven to be useful in the optimisation of the cryopreservation protocol and the evaluation of cryoinjuries (Figueroa et al. 2016a, b, 2017; Pereira et al. 2019).

Reactive oxygen species (ROS) are important for redox signalling pathways, including the activation of sperm motility in fish. However, excessive ROS concentrations are harmful and cause much damage to sperm cells, including those caused by cryopreservation (Pini et al. 2018). For this reason, the determination of ROS levels in fresh and cryopreserved spermatozoa is important for the evaluation of cryopreservation procedures (Pérez-Cerezales et al. 2010a; Mostek et al. 2018; Pereira et al. 2019). Enhanced ROS production after cryopreservation has also been described in relation to captivity (Valcarce and Robles 2016). Excessive ROS production leads to the peroxidation of polyunsaturated fatty acids, resulting in the production of aldehydes, such as 4-hydroxynonenal, acrolein and malondialdehyde (Pini et al. 2018). Lipid peroxidation has been demonstrated in the cryopreserved spermatozoa of several fish species, including European sea bass (Martínez-Páramo et al. 2012a) and Senegalese sole (Riesco et al. 2017a). Measurements of lipid peroxidation appeared to be especially useful for the evaluation of the usefulness of the supplementation of extenders with antioxidative compounds for the improvement of cryopreservation technology (Lahnsteiner et al. 2011; Osipova et al. 2014, 2016; Figueroa et al. 2018). The oxidation of proteins, which is usually measured as global protein carbonylation, is an important cryoinjury to fish spermatozoa (Martínez-Páramo et al. 2012b; Shaliutina-Kolešová et al. 2014). Recently, using a proteomic approach, it was possible to identify the individual proteins of common carp spermatozoa for which carbonylation was triggered by cryopreservation (Mostek et al. 2018).

The results of recent work clearly demonstrate that cryopreservation affects the sperm genome in several ways. DNA fragmentation (together with lipid peroxidation) is the commonly studied lesion to the genome, caused mostly by ROS, as the supplementation of antioxidants to extenders may decrease the levels of this cryoinjury (Pérez-Cerezales et al. 2010a; Cabrita et al. 2011; Martínez-Páramo et al. 2013; Öğretmen et al. 2015). Analysis of DNA fragmentation appeared to be useful for the development and improvement of cryopreservation protocols (Figueroa et al. 2015; Shaliutina-Kolešová et al. 2015; Yusoff et al. 2018; Pereira et al. 2019). Damage to sperm DNA can be, to some extent, repaired by the egg, therefore studies of the fertilisation capacity of DNA-damaged sperm and embryo developmental success are of utmost importance for the evaluation of fish sperm quality (Pérez-Cerezales et al. 2010b). Besides DNA fragmentation, other emerging technologies are aimed at assessing changes in genetic material in response to cryopreservation. Riesco et al. (2017a) indicated that the degradation of transcripts could be predictive of sperm status after cryopreservation. Global methylation is used as a potential marker in epigenetic studies and has been recently used to evaluate the impact of particular cryoprotectants on DNA methylation patterns (de Mello et al. 2017; Herranz-Jusdado et al. 2019a).

Cryopreservation influences the biochemical composition of spermatozoa. Changes in lipid composition have been described and attributed to cryoresistance (Horokhovatskyi et al. 2016), the prediction of cryopreservation success (Martínez-Páramo et al. 2012a) and the evaluation of the stabilisation of the plasma membrane (Beirão et al. 2012). Changes in proteins have also been characterised during cryopreservation, mainly for individual proteins, mostly enzymes (Ciereszko 2008; Ciereszko et al. 2011), and recently for several proteins with the use of a proteomic approach; for the latter, both an increase and decrease of particular proteins have been reported for several fish species, including see bass (Zilli et al. 2005, 2014), rainbow trout (Nynca et al. 2015), common carp (Li et al. 2010a; Dietrich et al. 2015) and sterlet (Horokhovatskyi et al. 2018; Xin et al. 2018). An increase in the abundance of particular sperm proteins is likely related to the post-translational modification of proteins, such as proteolysis, phosphorylation or oxidation (see above). The comprehensive analysis of cryodamage to spermatozoa, with the use of emerging 'omic' technologies, offers a new opportunity for better understanding the mechanisms of biochemical changes, especially damage, in spermatozoa during cryopreservation.

3.6 Concluding Remarks

The last decade brought significant progress with the development of several sensitive and specific tests for the evaluation of semen quality, especially the introduction of portable flow cytometers, which may have an important impact on the evaluation of gametes obtained in field conditions. Also, the rapid development of 'omic' technologies offers new opportunities in the comprehensive analysis of sperm structure and function. Linking sperm quality indices obtained with advanced analytical methods with sperm fertilising ability is still a challenge. Most importantly, such tests would need to be scaled to simple and high-throughput procedures, allowing for the fast and simple evaluation of semen. This would be a prerequisite for the better control of fish reproduction, especially for the perfection of procedures for the short-term and long-term (cryopreservation) storage of semen and protocols for in vitro fertilisation.

Acknowledgements This work was supported by Projects 2015/17/B/NZ9/01542, 2016/21/B/NZ9/03620 and 2018/29/N/NZ9/00761 from the National Science Centre, as well as funds appropriated to the Institute of Animal Reproduction and Food Research, Polish Academy of Sciences.

References

Acosta IB, Junior ASV, de Silva EF, Cardoso TF, Caldas JS, Jardim RD, Corcini CD (2016) Effects of exposure to cadmium in sperm cells of zebrafish, *Danio rerio*. Toxicol Rep 3:696–700

Alavi SMH, Rodina M, Hatef A, Stejskal V, Policar T, Hamáčková J, Linhart O (2010) Sperm motility and monthly variations of semen characteristics in *Perca fluviatilis* (Teleostei: Percidae). Czeh J Anim Sci 55(4):174–182

Aliniya M, Khara H, Noveiri SB, Dadras H (2013) Influence of age of common carp (*Cyprinus carpio*) broodstock on reproductive traits and fertilization. Turk J Fish Aquat Sci 13(1):19–25

Asturiano JF, Cabrita E, Horváth Á (2017) Progress, challenges and perspectives on fish gamete cryopreservation: a mini-review. Gen Comp Endocrinol 245:69–76

Atasever M, Bozkurt Y (2015) Effect of different photoperiod regimes on sperm quality, fecundity and fertilization in rainbow trout (*Oncorhynchus mykiss*). Turk J Fish Aquat Sci 15(2):517–523

Bailey JL, Bilodeau J, Cormier N (2000) Minireview: semen cryopreservation in domestic animals: a damaging and capacitating phenomenon. J Androl 21(1):1–7

Bautista FEA, Varela Junior AS, Corcini CD, Acosta IB, Caldas SS, Primel EG, Zanette J (2018) The herbicide atrazine affects sperm quality and the expression of antioxidant and spermatogenesis genes in zebrafish testes. Comp Biochem Physiol Part C Toxicol Pharmacol 206-207:17–22

Beirão J, Soares F, Herráez MP, Dinis MT, Cabrita E (2011) Changes in *Solea senegalensis* sperm quality throughout the year. Anim Reprod Sci 126(1–2):122–129

Beirão J, Zilli L, Vilella S, Cabrita E, Schiavone R, Herráez MP (2012) Improving sperm cryopreservation with antifreeze proteins: effect on gilthead seabream (*Sparus aurata*) plasma membrane lipids. Biol Reprod 86(2):1–9

Beirão J, Soares F, Pousão-Ferreira P, Diogo P, Dias J, Dinis MT, Herráez MP, Cabrita E (2015) The effect of enriched diets on *Solea senegalensis* sperm quality. Aquaculture 435:187–194

Beirão J, Litt MA, Purchase CF (2018) Chemically-dispersed crude oil and dispersant affects sperm fertilizing ability, but not sperm swimming behaviour in capelin (*Mallotus villosus*). Environ Pollut 241:521–528

Bernáth G, Bokor Z, Zarski D, Várkonyi L, Hegyi A, Staszny A, Urbányi B, Radóczi Ifj J, Horváth A (2016a) Commercial-scale out-of-season cryopreservation of Eurasian perch (*Perca fluviatilis*) sperm and its application for fertilization. Anim Reprod Sci 170:170–177

Bernáth G, Żarski D, Kása E, Staszny Á, Várkonyi L, Kollár T, Hegyi Á, Bokor Z, Urbányi B, Horváth Á (2016b) Improvement of common carp (*Cyprinus carpio*) sperm cryopreservation using a programable freezer. Gen Comp Endocrinol 237:78–88

Betsy CJ, Stephen Sampath Kumar J, Thilak Pon Jawahar K, Athithan S, Ahilan B, Karal Marx K (2016) Captive broodstock development of common carp *Cyprinus carpio* Linnaeus, 1758: influence of age on spermatogenesis and fertilisation parameters. Indian J Fish 63(4):75–81

Betsy J, Kumar S, Jawahar T (2017) Effect of age on the spermatological and fertilisation parameters of common carp *Cyprinus carpio* (Linnaeus 1758) brooders cryopreserved at three dilution ratios. Asian Fish Sci 30(3):195–205

Blecha M, Kristan J, Samarin AM, Rodina M, Policar T (2015) Quality and quantity of pikeperch (*Sander lucioperca*) spermatozoa after varying cold water treatments. J Appl Ichthyol 31:75–78

Blecha M, Dzyuba B, Boryshpolets S, Horokhovatskyi Y, Dadras H, Malinovskyi O, Sampels S, Policar T (2018) Spermatozoa quality and sperm lipid composition in intensively cultured and wild burbot (*Lota lota*). Anim Reprod Sci 198:129–136

Blukacz EA, Koops MA, Sutton TM, Arts MT, Fitzsimons JD, Muir AM, Claramunt RM, Johnson TB, Kinnunen RE, Ebener MP, Suski C, Burness G (2010) Linking lake whitefish (*Coregonus clupeaformis*) condition with male gamete quality and quantity. J Great Lakes Res 36(Suppl. 1):78–83

Bobe J, Labbé C (2010) Egg and sperm quality in fish. Gen Comp Endocrinol 165(3):535–548

Bompart D, García-Molina A, Valverde A, Caldeira C, Yániz J, Núñez De Murga M, Soler C (2018) CASA-Mot technology: how results are affected by the frame rate and counting chamber. Reprod Fertil Dev 30:810–819

Bondarenko V, Blecha M, Policar T (2018) Changes of sperm morphology, volume, density, and motility parameters in northern pike during the spawning period. Fish Physiol Biochem 44 (6):1591–1597

Boonthai T, Khaopong W, Sangsong J, Vuthiphandchai V, Nimrat S (2016) Influence of *Aeromonas hydrophila* and *Pseudomonas fluorescens* on motility, viability and morphometry of cryostored silver barb (*Barbodes gonionotus*) sperm. Cryobiology 73(2):140–146

Boryshpolets S, Kowalski RK, Dietrich GJ, Dzyuba B, Ciereszko A (2013) Different computer-assisted sperm analysis (CASA) systems highly influence sperm motility parameters. Theriogenology 80:758–765

Butts IAE, Litvak MK, Trippel EA (2010) Seasonal variations in seminal plasma and sperm characteristics of wild-caught and cultivated Atlantic cod, *Gadus morhua*. Theriogenology 73 (7):873–885

Butts IAE, Trippel EA, Ciereszko A, Soler C, Słowińska M, Alavi SMH, Litvak MK, Babiak I (2011a) Seminal plasma biochemistry and spermatozoa characteristics of Atlantic cod (*Gadus morhua* L.) of wild and cultivated origin. Comp Biochem Physiol A Mol Integr Physiol 159 (1):16–24

Butts IAE, Babiak I, Ciereszko A, Litvak MK, Słowińska M, Soler C, Trippel EA (2011b) Semen characteristics and their ability to predict sperm cryopreservation potential of Atlantic cod, Gadus morhua L. Theriogenology 75(7):1290–1300

Butts IAE, Baeza R, Støttrup JG, Krüger-Johnsen M, Jacobsen C, Pérez L, Asturiano JF, Tomkiewicz J (2015) Impact of dietary fatty acids on muscle composition, liver lipids, milt composition and sperm performance in European eel. Comp Biochem Physiol A Mol Integr Physiol 183:87–96

Cabrita E, Martínez F, Real M, Alvarez R, Herráez MP (2001) The use of flow cytometry to assess membrane stability in fresh and cryopreserved trout spermatozoa. Cryo Letters 22:263–272

Cabrita E, Robles V, Cuñado S, Wallace JC, Sarasquete C, Herráez MP (2005) Evaluation of gilthead sea bream, *Sparus aurata*, sperm quality after cryopreservation in 5ml macrotubes. Cryobiology 50:273–284

Cabrita E, Sarasquete C, Martínez-Páramo S, Robles V, Beirão J, Pérez-Cerezales S, Herráez MP (2010) Cryopreservation of fish sperm: applications and perspectives. J Appl Ichthyol 26 (5):623–635

Cabrita E, Ma S, Diogo P, Martínez-Páramo S, Sarasquete C, Dinis MT (2011) The influence of certain aminoacids and vitamins on post-thaw fish sperm motility, viability and DNA fragmentation. Anim Reprod Sci 125(1–4):189–195

Cabrita E, Martínez-Páramo S, Gavaia PJ, Riesco MF, Valcarce DG, Sarasquete C, Herráez MP, Robles V (2014) Factors enhancing fish sperm quality and emerging tools for sperm analysis. Aquaculture 432:389–401

Caldeira C, Hernández-Ibáñez S, Valverde A, Martin P, Herranz-Jusdado JG, Gallego V, Asturiano JF, Dzyuba B, Pšenička M, Soler C (2019) Standardization of sperm motility analysis by using CASA-Mot for Atlantic salmon (*Salmo salar*), European eel (*Anguilla anguilla*) and Siberian sturgeon (*Acipenser baerii*). Aquaculture 502:223–231

Carnevali O, Santangeli S, Forner-Piquer I, Basili D, Maradonna F (2018) Endocrine-disrupting chemicals in aquatic environment: what are the risks for fish gametes? Fish Physiol Biochem 44 (6):1561–1576

Carvalho AFS, Murgas LDS, Ferreira-Machado MR, Andrade ES, Felizardo VO, Allaman IB, De Paula FG (2017) Canonical correlation analysis to identify the semen characteristics used to forecast the freeze survival of curimba (*Prochilodus lineatus*) spermatozoa. Cryo Letters 38 (4):263–268

Cejko BI, Krejszeff S (2016) Sperm characteristics of chub *Leuciscus cephalus* (L.) collected in artificial condition after Ovopel and Ovaprim treatment. Aquac Res 47(3):847–856

Cejko BI, Kucharczyk D (2015) Application of dopaminergic antagonist: metoclopramide, in reproduction of crucian carp *Carassius carassius* (L.) under controlled conditions. Anim Reprod Sci 160:74–81

Cejko BI, Zarski D, Judycka S, Kucharczyk D, Sarosiek B, Kowalski RK (2014) Effect of two commercial preparations containing different GnRH analogues with dopamine antagonists on barbel *Barbus barbus* (L.) sperm quantity and quality. Aquac Int 22(1):97–109

Cejko BI, Krejszeff S, Żarski D, Judycka S, Targońska K, Kucharczyk D (2018a) Effect of carp pituitary homogenate (CPH) and sGnRHa (Ovaprim) on northern pike (*Esox lucius*) spermiation stimulation and its effect on quantity and quality of sperm. Anim Reprod Sci 193:217–225

Cejko BI, Sarosiek B, Krejszeff S, Kowalski RK (2018b) Multiple collections of common carp *Cyprinus carpio* L. semen during the reproductive period and its effects on sperm quality. Anim Reprod Sci 188:178–188

Chalde T, Gárriz A, Sanches EA, Miranda LA (2016) Influence of pejerrey *Odontesthes bonariensis* (Valenciennes, 1835) broodstock age on gamete quality, reproductive performance and plasma sex steroid levels during the spawning season. Aquac Res 47(3):969–982

Chen J, Xiao Y, Gai Z, Li R, Zhu Z, Bai C, Tanguay RL, Xu X, Huang C, Dong Q (2015) Reproductive toxicity of low level bisphenol A exposures in a two-generation zebrafish assay: evidence of male-specific effects. Aquat Toxicol 169:204–214

Chen J, Saili KS, Liu Y, Li L, Zhao Y, Jia Y, Bai C, Tanguay RL, Dong Q, Huang C (2017) Developmental bisphenol A exposure impairs sperm function and reproduction in zebrafish. Chemosphere 169:262–270

Ciereszko A (2008) Chemical composition of seminal plasma and its physiological relationship with sperm motility, fertilizing capacity, and cryopreservation success in fish. In: Alavi SMH, Cosson J, Coward K, Rafiee G (eds) Fish spermatology. Alpha Science International Ltd, Oxford, pp 215–240

Ciereszko A, Dabrowski K (1993) Estimation of sperm concentration of rainbow trout, whitefish and yellow perch using a spectrophotometric technique. Aquaculture 109(3–4):367–373

Ciereszko A, Glogowski J, Dabrowski K (2011) Biochemical characteristics of seminal plasma and spermatozoa of freshwater fish and the relation to semen biology, quality and cryopreservation. In: Tiersch TR, Green CC (eds) Cryopreservation in aquatic species, 2nd edn. World Aquaculture Society, Baton Rouge, pp 46–79

Ciereszko A, Dietrich GJ, Nynca J, Krom J, Dobosz S (2015) Semen from sex-reversed rainbow trout of spring strain can be successfully cryopreserved and used for fertilization of elevated number of eggs. Aquaculture 448:564–568

Ciereszko A, Dietrich MA, Nynca J (2017) Fish semen proteomics—New opportunities in fish reproductive research. Aquaculture 472:81–92

Contreras P, Ulloa P, Merino O, Valdebenito I, Figueroa E, Farías J, Risopatrón J (2017) Effect of short-term storage on sperm function in Patagonian blenny (*Eleginops maclovinus*) sperm. Aquaculture 481:58–63

Corso MN, Marques LS, Gracia LFG, Rodrigues RB, Barcellos LJG, Streit DP Jr (2019) Effects of different doses of eugenol on plasma cortisol levels and the quality of fresh and frozen-thawed sperm in South American catfish (*Rhamdia quelen*). Theriogenology 125:135–139

Cuevas-Uribe R, Tiersch TR (2011) Estimation of fish sperm concentration by use of spectrophotometry. In: Tiersch TR, Green CC (eds) Cryopreservation in aquatic species, 2nd edn. World Aquaculture Society, Baton Rouge, pp 162–200

Dadras H, Sampels S, Golpour A, Dzyuba V, Cosson J, Dzyuba B (2017a) Analysis of common carp *Cyprinus carpio* sperm motility and lipid composition using different in vitro temperatures. Anim Reprod Sci 180:37–43

Dadras H, Golpour A, Zahmatkesh M, Khara H, Noveiri SB, Siddique MAM (2017b) Effects of age on the reproductive performance of different males and females in bighead carp *Hypophthalmichthys nobilis* (Richardson, 1845). Comp Clin Pathol 26(5):1165–1171 de Mello F, Garcia JS, Godoy LC, Depincé A, Labbé C, Streit DP Jr (2017) The effect of cryoprotectant agents on DNA methylation patterns and progeny development in the spermatozoa of *Colossoma macropomum*. Gen Comp Endocrinol 245:94–101

Devaux A, Bony S, Plenet S, Sagnes P, Segura S, Suaire R, Novak M, Gilles A, Olivier J (2015) Field evidence of reproduction impairment through sperm DNA damage in the fish nase (*Chondrostoma nasus*) in anthropized hydrosystems. Aquat Toxicol 169:113–122

Di Chiacchio IM, Almeida ILG, Leal MC, Viveiros ATM (2017) Sperm quality and its freezing ability throughout the spawning season in *Prochilodus lineatus* and *Brycon orbignyanus*. Theriogenology 90:284–288

Dietrich MA, Ciereszko A (2018) Proteomic characterization of fresh spermatozoa and supernatant after cryopreservation in relation to freezability of carp (*Cyprinus carpio* L) semen. PLoS One 13:e0192972. https://doi.org/10.1371/journal.pone.0192972

Dietrich GJ, Dietrich M, Kowalski RK, Dobosz S, Karol H, Demianowicz W, Glogowski J (2010) Exposure of rainbow trout milt to mercury and cadmium alters sperm motility parameters and reproductive success. Aquat Toxicol 97(4):277–284

Dietrich MA, Dietrich GJ, Hliwa P, Ciereszko A (2011) Carp transferrin can protect spermatozoa against toxic effects of cadmium ions. Comp Biochem Physiol Part C Toxicol Pharmacol 153 (4):422–429

Dietrich GJ, Nynca J, Dobosz S, Zalewski T, Ciereszko A (2014) Application of glucose-methanol extender to cryopreservation of semen of sex-reversed females rainbow trout results in high post-thaw sperm motility and fertilizing ability. Aquaculture 434:27–32

Dietrich MA, Arnold GJ, Fröhlich T, Otte KA, Dietrich GJ, Ciereszko A (2015) Proteomic analysis of extracellular medium of cryopreserved carp (*Cyprinus carpio* L.) semen. Comp Biochem Physiol Part D Genomics Proteomics 15:49–57

Dietrich MA, Irnazarow I, Ciereszko A (2017) Proteomic identification of seminal plasma proteins related to the freezability of carp semen. J Proteome 162:52–61

Diogo P, Martins G, Gavaia P, Pinto W, Dias J, Cancela L, Martínez-Páramo S (2015) Assessment of nutritional supplementation in phospholipids on the reproductive performance of zebrafish, *Danio rerio* (Hamilton, 1822). J Appl Ichthyol 31(S1):3–9

Dumorné K, Figueroa E, Cosson J, Lee-Estevez M, Ulloa-Rodríguez P, Valdebenito I, Farías JG (2018) Protein phosphorylation and ions effects on salmonid sperm motility activation. Rev Aquac 10(3):727–737

Dzyuba B, Boryshpolets S, Shaliutina A, Rodina M, Yamaner G, Gela D, Dzyuba V, Linhart O (2012) Spermatozoa motility, cryoresistance, and fertilizing ability in sterlet *Acipenser ruthenus* during sequential stripping. Aquaculture 356-357:272–278

Fauvel C, Suquet M, Cosson J (2010) Evaluation of fish sperm quality. J Appl Ichthyol 26 (5):636–643

Fenkes M, Fitzpatrick JL, Ozolina K, Shiels HA, Nudds RL (2017) Sperm in hot water: direct and indirect thermal challenges interact to impact on brown trout sperm quality. J Exp Biol 220 (14):2513–2520

Fernández I, Fernandes JMO, Roberto VP, Kopp M, Oliveira C, Riesco MF, Dias J, Cox CJ, Leonor Cancela M, Cabrita E, Gavaia P (2019) Circulating small non-coding RNAs provide new insights into vitamin K nutrition and reproductive physiology in teleost fish. Biochim Biophys Acta Gen Subj 1863(1):39–51

Figueroa E, Merino O, Risopatrón J, Isachenko V, Sánchez R, Effer B, Isachenko E, Farias JG, Valdebenito I (2015) Effect of seminal plasma on Atlantic salmon (*Salmo salar*) sperm vitrification. Theriogenology 83(2):238–245.e2

Figueroa E, Valdebenito I, Farias JG (2016a) Technologies used in the study of sperm function in cryopreserved fish spermatozoa. Aquac Res 47(6):1691–1705

Figueroa E, Valdebenito I, Merino O, Ubilla A, Risopatrón J, Farias JG (2016b) Cryopreservation of Atlantic salmon *Salmo salar* sperm: effects on sperm physiology. J Fish Biol 89 (3):1537–1550

Figueroa E, Valdebenito I, Zepeda AB, Figueroa CA, Dumorné K, Castillo RL, Farias JG (2017) Effects of cryopreservation on mitochondria of fish spermatozoa. Rev Aquac 9(1):76–87

Figueroa E, Farias JG, Lee-Estevez M, Valdebenito I, Risopatrón J, Magnotti C, Romero J, Watanabe I, Oliveira RPS (2018) Sperm cryopreservation with supplementation of α-tocopherol and ascorbic acid in freezing media increase sperm function and fertility rate in Atlantic salmon (*Salmo salar*). Aquaculture 493:1–8

Fitzpatrick JL, Earn DJD, Bucking C, Craig PM, Nadella S, Wood CM, Balshine S (2016) Postcopulatory consequences of female mate choice in a fish with alternative reproductive tactics. Behav Ecol 27(1):312–320

Flannery EW, Butts IAE, Słowińska M, Ciereszko A, Pitcher TE (2013) Reproductive investment patterns, sperm characteristics, and seminal plasma physiology in alternative reproductive tactics of chinook salmon (*Oncorhynchus tshawytscha*). Biol J Linn Soc 108(1):99–108

Frankel T, Yonkos L, Ampy F, Frankel J (2018) Exposure to levonorgestrel increases nest acquisition success and decreases sperm motility in the male fathead minnow (*Pimephales promelas*). Environ Toxicol Chem 37(4):1131–1137

Gallego V, Asturiano JF (2018) Sperm motility in fish: technical applications and perspectives through CASA-Mot systems. Reprod Fertil Dev 30:820–832

Gallego V, Asturiano JF (2019) Fish sperm motility assessment as a tool for aquaculture research: a historical approach. Rev Aquacult. https://doi.org/10.1111/raq.12253

Gallego V, Herranz-Jusdado JG, Rozenfeld C, Pérez L, Asturiano JF (2018) Subjective and objective assessment of fish sperm motility: when the technique and technicians matter. Fish Physiol Biochem 44:1457–1467

Gazo I, Linhartova P, Shaliutina A, Hulak M (2013) Influence of environmentally relevant concentrations of vinclozolin on quality, DNA integrity, and antioxidant responses of sterlet *Acipenser ruthenus* spermatozoa. Chem Biol Interact 203(2):377–385

Glogowski J, Kwasnik M, Piros B, Dabrowski K, Goryczko K, Dobosz S, Kuzminski H, Ciereszko A (2000) Characterization of rainbow trout milt collected with a catheter: semen parameters and cryopreservation success. Aquacult Res 31:289–296

Golpour A, Akhoundian M, Khara H, Rahbar M, Dadras H (2013) Changes of sperm quality parameters in Caspian roach (Rutilus rutilus caspicus) during spawning migration. Czeh J Anim Sci 58(3):117–124

Golshan M, Hatef A, Socha M, Milla S, Butts IAE, Carnevali O, Rodina M, Sokołowska-Mikołajczyk M, Fontaine P, Linhart O, Alavi SMH (2015) Di-(2-ethylhexyl)-phthalate disrupts pituitary and testicular hormonal functions to reduce sperm quality in mature goldfish. Aquat Toxicol 163:16–26

Hajiahmadian M, Sarvi Moghanlou K, Agh N, Farrokhi Ardabili F (2016) Semen characteristics of rainbow trout (*Oncorhynchus mykiss*) following diets containing different vegetable fatty acid levels. Reprod Domest Anim 51(6):979–984

Hatef A, Alavi SMH, Linhartova Z, Rodina M, Policar T, Linhart O (2010) In vitro effects of bisphenol A on sperm motility characteristics in *Perca fluviatilis* L. (Percidae; teleostei). J Appl Ichthyol 26(5):696–701

Hatef A, Alavi SMH, Butts IAE, Policar T, Linhart O (2011) Mechanism of action of mercury on sperm morphology, adenosine triphosphate content, and motility in *Perca fluviatilis* (Percidae; Teleostei). Environ Toxicol Chem 30(4):905–914

Hatef A, Alavi SMH, Abdulfatah A, Fontaine P, Rodina M, Linhart O (2012a) Adverse effects of bisphenol A on reproductive physiology in male goldfish at environmentally relevant concentrations. Ecotoxicol Environ Saf 76(1):56–62

Hatef A, Alavi SMH, Milla S, Křišťan J, Golshan M, Fontaine P, Linhart O (2012b) Anti-androgen vinclozolin impairs sperm quality and steroidogenesis in goldfish. Aquat Toxicol 122-123:181–187

Hatef A, Alavi SMH, Golshan M, Linhart O (2013) Toxicity of environmental contaminants to fish spermatozoa function in vitro—A review. Aquat Toxicol 140-141:134–144

He S, Woods LC III (2004) Effects of dimethyl sulfoxide and glycine on cryopreservation induced damage of plasma membranes and mitochondria to striped bass (*Morone saxatilis*) sperm. Cryobiology 48:254–262

He J, Yang D, Wang C, Liu W, Liao J, Xu T, Bai C, Chen J, Lin K, Huang C, Dong Q (2011) Chronic zebrafish low dose decabrominated diphenyl ether (BDE-209) exposure affected parental gonad development and locomotion in F1 offspring. Ecotoxicology 20(8):1813–1822

Herráez MP, Ausió J, Devaux A, González-Rojo S, Fernández-Díez C, Bony S, Saperas N, Robles V (2017) Paternal contribution to development: sperm genetic damage and repair in fish. Aquaculture 472:45–59

Herranz-Jusdado JG, Gallego V, Morini M, Rozenfeld C, Pérez L, Kása E, Kollár T, Depincé A, Labbé C, Horváth Á, Asturiano JF (2019a) Comparison of European eel sperm cryopreservation protocols with standardization as a target. Aquaculture 498:539–544

Herranz-Jusdado JG, Rozenfeld C, Morini M, Pérez L, Asturiano JF, Gallego V (2019b) Recombinant vs purified mammal gonadotropins as maturation hormonal treatments of European eel males. Aquaculture 501:527–536

Hilbig CC, do Nascimento NF, Granville M, de Toledo CR, Neumann G, Bombardelli RA, Nakaghi LSO (2017) Effect of broodstock age on reproductive parameters in silver catfish *Rhamdia quelen*. Aquac Res 48(9):5150–5154

Horokhovatskyi Y, Sampels S, Cosson J, Linhart O, Rodina M, Fedorov P, Blecha M, Dzyuba B (2016) Lipid composition in common carp (*Cyprinus carpio*) sperm possessing different cryoresistance. Cryobiology 73(2):282–285

Horokhovatskyi Y, Dietrich MA, Lebeda I, Fedorov P, Rodina M, Dzyuba B (2018) Cryopreservation effects on a viable sperm sterlet (*Acipenser ruthenus*) subpopulation obtained by a Percoll density gradient method. PLoS One 13:e0202514

Hossain MS, Johannisson A, Wallgren M, Nagy S, Siqueira AP, Rodriguez-Martinez H (2011) Flow cytometry for the assessment of animal sperm integrity and functionality: state of the art. Asian J Androl 13(3):406–419

Hulak M, Gazo I, Shaliutina A, Linhartova P (2013) In vitro effects of bisphenol A on the quality parameters, oxidative stress, DNA integrity and adenosine triphosphate content in sterlet (*Acipenser ruthenus*) spermatozoa. Comp Biochem Physiol Part C Toxicol Pharmacol 158 (2):64–71

İnanan BE, Yılmaz F (2018) Seasonal and age-related changes in semen characteristics and composition of seminal plasma in sex-reverse female rainbow trout (*Oncorhynchus mykiss*) in comparison with normal males. Anim Reprod Sci 196:160–167

İnanan BE, Öğretmen F, İnanan T, Yılmaz F (2016) Total antioxidant capacity, catalase activity, and lipid peroxidation changes in seminal plasma of sex-reversed female and male rainbow trout (*Oncorhynchus mykiss*) during spawning season. Theriogenology 86(8):1975–1982

Judycka S, Szczepkowski M, Ciereszko A, Slowińska M, Bodek G, Dietrich GJ (2015) Characterization of Siberian sturgeon (*Acipenser baerii*, Brandt 1869) sperm obtained out of season. J Appl Ichthyol 31(S1):34–40

Judycka S, Nynca J, Liszewska E, Dobosz S, Grudniewska J, Ciereszko A (2018) Optimal sperm concentration in straws and final glucose concentration in extender are crucial for improving the cryopreservation protocol of salmonid spermatozoa. Aquaculture 486:90–97

Judycka S, Nynca J, Liszewska E, Mostek A, Ciereszko A (2019a) Comparative analysis of sperm freezability of sex-reversed female brook trout and sex-reversed female rainbow trout semen. Aquaculture 498:201–207

Judycka S, Żarski D, Dietrich MA, Palińska-Żarska K, Karol H, Ciereszko A (2019b) Standardized cryopreservation protocol of European perch (*Perca fluviatilis*) semen allows to obtain high fertilization rates with the use of frozen/thawed semen. Aquaculture 498:208–216

Kekäläinen J, Figenschou L, Janhunen M, Kortet R, Peuhkuri N, Rudolfsen G (2013) Hatchery selection may depress the number of motile sperm but intensify selection for their swimming velocity in the Arctic charr. Aquac Int 21(2):405–411

Kime DE, Van Look KJW, McAllister BG, Huyskens G, Rurangwa E, Ollevier F (2001) Computer-assisted sperm analysis (CASA) as a tool for monitoring sperm quality in fish. Comp Biochem Physiol Part C Toxicol Pharmacol 130:425–433

Kowalska A, Siwicki AK, Kowalski RK (2017) Dietary resveratrol improves immunity but reduces reproduction of broodstock medaka *Oryzias latipes* (Temminck & Schlegel). Fish Physiol Biochem 43(1):27–37

Król J, Żarski D, Bernáth G, Palińska-Żarska K, Krejszeff S, Długoński A, Horváth Á (2018) Effect of urine contamination on semen quality variables in Eurasian perch *Perca fluviatilis* L. Anim Reprod Sci 197:240–246

Kucharczyk D, Nowosad J, Kujawa R, Dietrich G, Biegaj M, Sikora M, Luczynski MJ (2018) Comparison of spontaneous and hormone-induced reproduction of burbot (*Lota lota* L.) under hatchery conditions. Aquaculture 485:25–29

Labbé C, Robles V, Herraez MP (2017) Epigenetics in fish gametes and early embryo. Aquaculture 472:93–106

Lahnsteiner F, Kletzl M (2012) The effect of water temperature on gamete maturation and gamete quality in the European grayling (*Thymalus thymallus*) based on experimental data and on data from wild populations. Fish Physiol Biochem 38(2):455–467

Lahnsteiner F, Leitner S (2013) Effect of temperature on gametogenesis and gamete quality in brown trout, *Salmo trutta*. J Exp Zool A Ecol Genet Physiol 319(3):138–148

Lahnsteiner F, Mansour N, Kunz FA (2011) The effect of antioxidants on the quality of cryopreserved semen in two salmonid fish, the brook trout (*Salvelinus fontinalis*) and the rainbow trout (*Oncorhynchus mykiss*). Theriogenology 76(5):882–890

Langen K, Bakker TCM, Baldauf SA, Shrestha J, Thünken T (2017) Effects of ageing and inbreeding on the reproductive traits in a cichlid fish I: The male perspective. Biol J Linn Soc 120(4):752–761

Legendre M, Alavi SMH, Dzyuba B, Linhart O, Prokopchuk G, Cochet C, Dugué R, Cosson J (2016) Adaptations of semen characteristics and sperm motility to harsh salinity: extreme situations encountered by the euryhaline tilapia *Sarotherodon melanotheron heudelotii* (Dumeril, 1859). Theriogenology 86(5):1251–1267

Li P, Hulak M, Koubek P, Sulc M, Dzyuba B, Boryshpolets S, Rodina M, Gela D, Manaskova-Postlerova P, Peknicova J, Linhart O (2010a) Ice-age endurance: The effects of cryopreservation on proteins of sperm of common carp, *Cyprinus carpio* L. Theriogenology 74(3):413–423

Li Z, Li P, Rodina M, Randak T (2010b) Effect of human pharmaceutical carbamazepine on the quality parameters and oxidative stress in common carp (*Cyprinus carpio* L.) spermatozoa. Chemosphere 80(5):530–534

Li X, Wang J, Yu M, Zhang X, Wang W, Tian H, Ru S (2019) 2,2′-dithiobis-pyridine induced reproductive toxicity in male guppy (*Poecilia reticulata*). Ecotoxicol Environ Saf 169:778–785

Linhartova P, Gazo I, Shaliutina A, Hulak M (2013) The in vitro effect of duroquinone on functional competence, genomic integrity, and oxidative stress indices of sterlet (*Acipenser ruthenus*) spermatozoa. Toxicol In Vitro 27(6):1612–1619

Linhartova P, Gazo I, Shaliutina-Kolesova A, Hulak M, Kaspar V (2015) Effects of tetrabrombisphenol A on DNA integrity, oxidative stress, and sterlet (*Acipenser ruthenus*) spermatozoa quality variables. Environ Toxicol 30(7):735–745

Locatello L, Poli F, Rasotto MB (2013) Tactic-specific differences in seminal fluid influence sperm performance. Proc R Soc B Biol Sci 280(1755):20122891

Locatello L, Bertotto D, Cerri R, Parmeggiani A, Govoni N, Trocino A, Xiccato G, Mordenti O (2018) Sperm quality in wild-caught and farmed males of the European eel (*Anguilla anguilla*). Anim Reprod Sci 198:167–176

Lugowska K (2018) The effects of roundup on gametes and early development of common carp (*Cyprinus carpio* L). Fish Physiol Biochem 44(4):1109–1117

Luo L, Ai L, Liang X, Hu H, Xue M, Wu X (2017) n-3 long-chain polyunsaturated fatty acids improve the sperm, egg, and offspring quality of Siberian sturgeon (*Acipenser baerii*). Aquaculture 473:266–271

Magnotti C, Figueroa E, Farias JG, Merino O, Valdebenito I, Oliveira RPS, Cerqueira V (2018) Sperm characteristics of wild and captive lebranche mullet *Mugil liza* (Valenciennes, 1836), subjected to sperm activation in different pH and salinity conditions. Anim Reprod Sci 192:164–170

Martínez-Páramo S, Diogo P, Dinis MT, Herráez MP, Sarasquete C, Cabrita E (2012a) Sea bass sperm freezability is influenced by motility variables and membrane lipid composition but not by membrane integrity and lipid peroxidation. Anim Reprod Sci 131(3–4):211–218

Martínez-Páramo S, Diogo P, Dinis MT, Herráez MP, Sarasquete C, Cabrita E (2012b) Incorporation of ascorbic acid and α-tocopherol to the extender media to enhance antioxidant system of cryopreserved sea bass sperm. Theriogenology 77(6):1129–1136

Martínez-Páramo S, Diogo P, Beirão J, Dinis MT, Cabrita E (2012c) Sperm lipid peroxidation is correlated with differences in sperm quality during the reproductive season in precocious european sea bass (*Dicentrarchus labrax*) males. Aquaculture 358-359:246–252

Martínez-Páramo S, Diogo P, Dinis MT, Soares F, Sarasquete C, Cabrita E (2013) Effect of two sulfur-containing amino acids, taurine and hypotaurine in European sea bass (*Dicentrarchus labrax*) sperm cryopreservation. Cryobiology 66(3):333–338

Mehdinejad N, Imanpour MR, Jafari V (2018) Combined or individual effects of dietary probiotic, Pediococcus acidilactici and nucleotide on reproductive performance in goldfish (*Carassius auratus*). Probiotics Antimicrob Proteins 2018:1–6

Mehlis M, Bakker TCM (2014) The influence of ambient water temperature on sperm performance and fertilization success in three-spined sticklebacks (*Gasterosteus aculeatus*). Evol Ecol 28 (4):655–667

Mehlis M, Frommen JG, Rahn AK, Bakker TCM (2012) Inbreeding in three-spined sticklebacks (*Gasterosteus aculeatus* L.): Effects on testis and sperm traits. Biol J Linn Soc 107(3):510–520

Mehlis M, Hilke LK, Bakker TCM (2013) Attractive males have faster sperm in three-spined sticklebacks *Gasterosteus aculeatus*. Curr Zool 59(6):761–768

Mehlis M, Rahn AK, Bakker TCM (2015) Sperm quality but not relatedness predicts sperm competition success in threespine sticklebacks (*Gasterosteus aculeatus*). BMC Evol Biol 15 (1):1–9

Migaud H, Bell G, Cabrita E, Mcandrew B, Davie A, Bobe J, Herráez MP, Carrillo M (2013) Gamete quality and broodstock management in temperate fish. Rev Aquac 5(Suppl. 1):S194–S223

Momin M, Memiş D (2018) Sperm quality analysis of normal season (NG) and out-season by photoperiod manipulation (PG) of male rainbow trout broodstock (*Oncorhynchus mykiss*). Fish Physiol Biochem 44(6):1551–1560

Mostek A, Slowinska M, Judycka S, Karol H, Ciereszko A, Dietrich MA (2018) Identification of oxidatively modified proteins due to cryopreservation of carp semen. J Anim Sci 96:1453–1465

Müller K, Müller P, Pincemy G, Kurz A, Labbe C (2008) Characterization of sperm plasma membrane properties after cholesterol modification: consequences for cryopreservation of rainbow trout spermatozoa. Biol Reprod 78(3):390–399

Mylonas CC, Duncan NJ, Asturiano JF (2017) Hormonal manipulations for the enhancement of sperm production in cultured fish and evaluation of sperm quality. Aquaculture 472:21–44

Nascimento AF, Maria AN, Pessoa NO, Carvalho MAM, Viveiros ATM (2010) Out-of-season sperm cryopreserved in different media of the Amazonian freshwater fish pirapitinga (*Piaractus brachypomus*). Anim Reprod Sci 118(2–4):324–329

Navarro RD, Navarro FKSP, Felizardo VO, Murgas LDS, Pereira MM (2014) Cryopreservation of semen of Thailand tilapia (*Oreochromis* spp.) fed diet with different oil sources. Acta Sci Technol 36(3):399–404

Nynca J, Ciereszko A (2009) Measurement of concentration and viability of brook trout (*Salvelinus fontinalis*) spermatozoa using computer-aided fluorescent microscopy. Aquaculture 292 (3–4):256–258

Nynca J, Dietrich GJ, Fopp-Bayat D, Dietrich MA, Slowinska M, Liszewska E, Karol H, Martyniak A, Ciereszko A (2012) Quality parameters of fresh and cryopreserved whitefish (*Coregonus lavaretus* L.) semen. J Appl Ichthyol 28:934–940

Nynca J, Arnold GJ, Fröhlich T, Ciereszko A (2015) Cryopreservation-induced alterations in protein composition of rainbow trout semen. Proteomics 15:2643–2654

Nynca J, Dietrich GJ, Liszewska E, Judycka S, Karol H, Dobosz S, Krom J, Ciereszko A (2016) Usefulness of a portable flow cytometer for sperm concentration and viability measurements of rainbow trout spermatozoa. Aquaculture 451:353–356

Nynca J, Judycka S, Liszewska E, Dobosz S, Ciereszko A (2017) Standardization of spermatozoa concentration for cryopreservation of rainbow trout semen using a glucose-methanol extender. Aquaculture 477:23–27

Nynca J, Dietrich MA, Ciereszko A (2018) Dige analysis of fish tissues. Methods Mol Biol 1664:203–219

Öğretmen F, Inanan BE, Kutluyer F, Kayim M (2015) Effect of semen extender supplementation with cysteine on postthaw sperm quality, DNA damage, and fertilizing ability in the common carp (*Cyprinus carpio*). Theriogenology 83(9):1548–1552

Osipova VP, Kolyada MN, Berberova NT, Milaeva ER, Ponomareva EN, Belaya MM (2014) Cryoprotective effect of phosphorous-containing phenolic anti-oxidant for the cryopreservation of beluga sperm. Cryobiology 69(3):467–472

Osipova VP, Berberova NT, Gazzaeva RA, Kudryavtsev KV (2016) Application of new phenolic antioxidants for cryopreservation of sturgeon sperm. Cryobiology 72(2):112–118

Otoh AJ, Udoh JP (2018) Age-related sperm quality of male *Heterobranchus longifilis* broodstock fed two different isonitrogenous feeds. Trop Freshw Biol 27:21–42

Palomar Rios A, Molina Botella I (2017) Sperm parameters that play a major role in the assessment of semen quality after cryopreservation. J Assist Reprod Genet 34(10):1271–1276

Paniagua-Chávez CG, Jenkins J, Segovia M, Tiersch TR (2006) Assessment of gamete quality for the eastern oyster (*Crassostrea virginica*) by use of fluorescent dyes. Cryobiology 53:128–138

Peñaranda DS, Gallego V, Rozenfeld C, Herranz-Jusdado JG, Pérez L, Gómez A, Giménez I, Asturiano JF (2018) Using specific recombinant gonadotropins to induce spermatogenesis and spermiation in the European eel (*Anguilla anguilla*). Theriogenology 107:6–20

Pereira JR, Pereira FA, Perry CT, Pires DM, Muelbert JRE, Garcia JRE, Corcini CD, Varela Junior AS (2019) Dimethylsulfoxide, methanol and methylglycol in the seminal cryopreservation of Suruvi, *Steindachneridion scriptum*. Anim Reprod Sci 200:7–13

Pérez-Cerezales S, Martínez-Páramo S, Beirão J, Herráez MP (2010a) Evaluation of DNA damage as a quality marker for rainbow trout sperm cryopreservation and use of LDL as cryoprotectant. Theriogenology 74(2):282–289

Pérez-Cerezales S, Martínez-Páramo S, Beirão J, Herráez MP (2010b) Fertilization capacity with rainbow trout DNA-damaged sperm and embryo developmental success. Reproduction 139 (6):989–997

Pini T, Leahy T, de Graaf SP (2018) Sublethal sperm freezing damage: manifestations and solutions. Theriogenology 118:172–181

Rick IP, Mehlis M, Eßer E, Bakker TCM (2014) The influence of ambient ultraviolet light on sperm quality and sexual ornamentation in three-spined sticklebacks (*Gasterosteus aculeatus*). Oecologia 174(2):393–402

Riesco MF, Oliveira C, Soares F, Gavaia PJ, Dinis MT, Cabrita E (2017a) *Solea senegalensis* sperm cryopreservation: new insights on sperm quality. PLoS One 12(10):e0186542

Riesco MF, Raposo C, Engrola S, Martínez-Páramo S, Mira S, Cunha ME, Cabrita E (2017b) Improvement of the cryopreservation protocols for the dusky grouper, *Epinephelus marginatus*. Aquaculture 470:207–213

Risopatrón J, Merino O, Cheuquemán C, Figueroa E, Sánchez R, Farías JG, Valdebenito I (2018) Effect of the age of broodstock males on sperm function during cold storage in the trout (*Oncorhynchus mykiss*). Andrologia 50(2):e12857

Robles V, Cabrita E, Cuñado S, Herráez MP (2003) Sperm cryopreservation of sex-reversed rainbow trout (*Oncorhynchus mykiss*): parameters that affect its ability for freezing. Aquaculture 224(1–4):203–212

Robles V, Herráez P, Labbé C, Cabrita E, Pšenička M, Valcarce DG, Riesco MF (2017) Molecular basis of spermatogenesis and sperm quality. Gen Comp Endocrinol 245:5–9

Rocha S, Streit DP Jr, Marques LS, Varela AS Jr, Corcini CD, Hoshiba MA (2018) Toxic effects of mercury chloride on silver catfish (*Rhamdia quelen*) spermatozoa. Aquac Res 49(2):963–968

Rurangwa E, Kime DE, Ollevier F, Nash JP (2004) The measurement of sperm motility and factors affecting sperm quality in cultured fish. Aquaculture 234:1–28

Sánchez JAA, Varela AS, Corcini CD Jr, da Silva JC, Primel EG, Caldas S, Klein RD, Martins CDMG (2017) Effects of roundup formulations on biochemical biomarkers and male sperm quality of the livebearing *Jenynsia multidentata*. Chemosphere 177:200–210

Santos M, Soares F, Moreira M, Beirão J (2018) Evaluation of different extenders for the cold storage of meagre (*Argyrosomus regius*) semen. Aquac Res 49(8):2723–2731

Sarosiek B, Dryl K, Krejszeff S, Zarski D (2016) Characterization of pikeperch (*Sander lucioperca*) milt collected with a syringe and a catheter. Aquaculture 450:14–16

Shaliutina A, Dzyuba B, Hulak M, Boryshpolets S, Li P, Linhart O (2012) Evaluation of spermiation indices with multiple sperm collections in endangered sterlet (*Acipenser ruthenus*). Reprod Domest Anim 47(3):479–484

Shaliutina O, Shaliutina-Kolešová A, Lebeda I, Rodina M, Gazo I (2016) The in vitro effect of nonylphenol, propranolol, and diethylstilbestrol on quality parameters and oxidative stress in sterlet (*Acipenser ruthenus*) spermatozoa. Toxicol In Vitro 43:9–15

Shaliutina-Kolešová A, Gazo I, Cosson J, Linhart O (2014) Protection of common carp (*Cyprinus carpio* L.) spermatozoa motility under oxidative stress by antioxidants and seminal plasma. Fish Physiol Biochem 40:1771–1781

Shaliutina-Kolešová A, Cosson J, Lebeda I, Gazo I, Shaliutina O, Dzyuba B, Linhart O (2015) The influence of cryoprotectants on sturgeon (*Acipenser ruthenus*) sperm quality, DNA integrity, antioxidant responses, and resistance to oxidative stress. Anim Reprod Sci 159:66–76

Shaliutina-Kolešová A, Kotas P, Štěrba J, Rodina M, Dzyuba B, Cosson J, Linhart O (2016) Protein profile of seminal plasma and functionality of spermatozoa during the reproductive season in the common carp (*Cyprinus carpio*) and rainbow trout (*Oncorhynchus mykiss*). Mol Reprod Dev 83 (11):968–982

Silveira T, Varela Junior AS, Corcini CD, Domingues WB, Remião M, Santos L, Barreto B, Lessa I, Martins D, Boyle RT, Costa PG, Bianchini A, Robaldo RB, Campos VF (2019) Roundup® herbicide decreases quality parameters of spermatozoa of silversides *Odontesthes humensis*. Bull Environ Contam Toxicol 102(1):1–6

Suquet M, Omnes MH, Normant Y, Fauvel C (1992) Assessment of sperm concentration and motility in turbot (*Scophthalmus maximus*). Aquaculture 101:177–185

Teletchea F, Gardeur J, Psenicka M, Kaspar V, Le Doré Y, Linhart O, Fontaine P (2009) Effects of four factors on the quality of male reproductive cycle in pikeperch *sander lucioperca*. Aquaculture 291(3–4):217–223

Tomlinson MJ (2016) Uncertainty of measurement and clinical value of semen analysis: has standardisation through professional guidelines helped or hindered progress? Andrology 4 (5):763–770

Ulloa-Rodríguez P, Contreras P, Dumorné K, Lee-Estevez M, Díaz R, Figueroa E, Valdebenito I, Risopatrón J, Farías JG (2018) Patagonian blenny (*Eleginops maclovinus*) spermatozoa quality after storage at 4 °C in Cortland medium. Anim Reprod Sci 197:117–125

Valcarce DG, Robles V (2016) Effect of captivity and cryopreservation on ROS production in *Solea senegalensis* spermatozoa. Reproduction 152(5):439–446

Valdebenito II, Gallegos PC, Effer BR (2015) Gamete quality in fish: evaluation parameters and determining factors. Zygote 23(2):177–197

Valente E, Masson G, Maul A, Fox MG, Meyer A, Pihan JC (2016) Seasonal gonadal development and age-related maturity patterns of introduced pumpkinseed (*Lepomis gibbosus* Linnaeus, 1758) in a heated thermal reservoir and an adjacent river reach. J Therm Biol 58:60–71

Vergílio CS, Moreira RV, Carvalho CEV, Melo EJT (2014) Effects of in vitro exposure to mercury on male gonads and sperm structure of the tropical fish tuvira *Gymnotus carapo* (L.). J Fish Dis 37(6):543–551

Vergilio CS, Moreira RV, Carvalho CEV, Melo EJT (2015) Evolution of cadmium effects in the testis and sperm of the tropical fish Gymnotus carapo. Tissue Cell 47(2):132–139

Viveiros ATM, Gonçalves ACS, Nascimento AF, Leal MC (2015) Fresh, equilibrated and post-thaw sperm quality of *Brycon orbignyanus* (Valenciennes, 1850) and *Prochilodus lineatus* (Valenciennes, 1837) treated with either salmon GnRHa and domperidone or pituitary extract. Neotrop Ichthyol 13(1):157–164

Viveiros ATM, Di Chiacchio IM, Almeida ILG, Taffarel TR, Leal MC (2017) Storage and transportation of *Prochilodus lineatus* (Characiformes) sperm prior to cryopreservation. Gen Comp Endocrinol 245:84–88

Watson PF (2000) The causes of reduced fertility with cryopreserved semen. Anim Reprod Sci 60-61:481–492

Woods LC III, Li Y, Ding Y, Liu J, Reading BJ, Fuller SA, Song J (2018) DNA methylation profiles correlated to striped bass sperm fertility. BMC Genomics 19:244

Xin M, Shaliutina-Kolesova A, Sterba J, Konik P, Boryshpolets S, Rodina M, Li P, Nian R, Linhart O (2018) Impact of cryopreservation on sterlet, *Acipenser ruthenus* sperm motility and proteome. Anim Reprod Sci 192:280–289

Yang H, Daly J, Carmichael C, Matthews J, Varga ZM, Tiersch T (2016) A procedure-spanning analysis of plasma membrane integrity for assessment of cell viability in sperm cryopreservation of zebrafish *Danio rerio*. Zebrafish 13(2):144–151

Yeste M (2016) Sperm cryopreservation update: cryodamage, markers, and factors affecting the sperm freezability in pigs. Theriogenology 85(1):47–64

Yusoff M, Hassan BN, Ikhwanuddin M, Sheriff SM, Hashim F, Mustafa S, Koh ICC (2018) Successful sperm cryopreservation of the brown-marbled grouper, *Epinephelus fuscoguttatus* using propylene glycol as cryoprotectant. Cryobiology 81:168–173

Zadmajid V (2016) Comparative effects of human chorionic gonadotropin (hCG) and ovaprim™ (sGnRHa + domperidone) on the reproductive characteristics of wild-caught male Longspine scraper, *Capoeta trutta* (Heckel, 1843). Aquaculture 463:7–15

Zadmajid V, Imanpoor MR, Shabani A, Baharlouei A (2013) Evaluation of sperm characteristics and plasma testosterone in the goldfish (*Carassius auratus*) during four consecutive seasons. Comp Clin Pathol 22(4):703–711

Zadmajid V, Bashiri S, Sharafi N, Butts IAE (2018) Effect of hCG and Ovaprim™ on reproductive characteristics of male Levantine scraper, *Capoeta damascina* (Valenciennes, 1842). Theriogenology 115:45–56

Żarski D, Bernáth G, Król J, Cejko BI, Bokor Z, Palińska-Żarska K, Milla S, Fontaine P, Krejszeff S (2017) Effects of hCG and salmon gonadoliberine analogue on spermiation in the Eurasian perch (*Perca fluviatilis*). Theriogenology 104:179–185

Zilli L, Schiavone R, Zonno V, Rossano R, Storelli C, Vilella S (2005) Effect of cryopreservation on sea bass sperm proteins. Biol Reprod 72:1262–1267

Zilli L, Schiavone R, Storelli C, Vilella S (2008) Effect of cryopreservation on phosphorylation state of proteins involved in sperm motility initiation in sea bream. Cryobiology 57:150–155

Zilli L, Beirão J, Schiavone R, Herraez MP, Gnoni A, Vilella S (2014) Comparative proteome analysis of cryopreserved flagella and head plasma membrane proteins from sea bream spermatozoa: effect of antifreeze proteins. PLoS One 6:e99992

Zupa R, Fauvel C, Mylonas CC, Pousis C, Santamaria N, Papadaki M, Fakriadis I, Cicirelli V, Mangano S, Passantino L, Lacalandra GM, Corriero A (2017) Rearing in captivity affects spermatogenesis and sperm quality in greater amberjack, *Seriola dumerili* (Risso, 1810). J Anim Sci 95(9):4085–4100

Energetics of Fish Spermatozoa

4

Vitaliy Kholodnyy, Sergii Boryshpolets, Borys Dzyuba, and Jacky Cosson

Abstract

Fish spermatozoa need to perceive a signal from the external milieu so as to activate their motility. The osmolarity and/or ionic composition of the marine or freshwater surrounding fish control the activation of fish sperm flagella at spawning. Flagella motility is energy dependent and the amount of energy stored in the spermatozoon prior to its activation is a main factor that will sustain their vigorous motility but for only a short period (one to several minutes) in fish until partial exhaustion of this energy stops motility. Storage of energy mostly results from mitochondrial respiration that generates ATP. Energy metabolism also involves other compounds such as creatine–phosphate that contribute to the maintenance of the intracellular energy level in connection with ATP. In few species, especially internally fertilizing ones, part of the energy comes from glycolysis. For efficient motility of swimming fish spermatozoa, the ATP usage (*chemical* hydrolysis) is mainly balanced by the *physical* power developed by the flagellum against viscosity forces: this balance sheet is also discussed in this chapter. The rest of ATP consumption is needed by fish spermatozoa for "house-keeping" tasks such as maintaining the ionic balance across the cell membrane as example.

The cryopreservation process induces deleterious perturbations at several levels of the energy metabolic network such as damage to mitochondria, leaking of cell membrane, and damage to the flagella motility apparatus that altogether lead to a serious decrease in the percentage of motile cells and consequently the fertilizing ability.

V. Kholodnyy (✉) · S. Boryshpolets · B. Dzyuba · J. Cosson
University of South Bohemia in České Budějovice, Faculty of Fisheries and Protection of Waters, Research Institute of Fish Culture and Hydrobiology, Vodňany, Czech Republic

J. Betsy, S. Kumar (eds.), *Cryopreservation of Fish Gametes*,
https://doi.org/10.1007/978-981-15-4025-7_4

Keywords

Spermatozoa · Energy · Flagella · ATP · cAMP

Abbreviations

ADP	Adenosine diphosphate
ATP	Adenosine triphosphate
BF	Flagellar beat frequency
cAMP	Cyclic-adenosine monophosphate
CASA	Computer-assisted sperm analysis
CrP	Creatine phosphate
DMSO	Dimethyl sulfoxide
KCN	Potassium cyanide
K_m	Affinity between a substrate and an enzyme
NMR	Nuclear magnetic resonance
PCr	Phosphocreatine
ROS	Reactive oxygen species
SF	Seminal fluid

4.1 Introduction

In fish with external fertilization, spermatozoa rely, for their motility, entirely on the endogenous energy readily available and transportable by an appropriate intra-flagellar shuttle so as to cover the needs for a short period of highly motile activity of their flagellum. In these fish species, the primitive structure (limited mid-piece) of the spermatozoon and the lack of important endogenous metabolic substrates are compensated by a biosynthetic capacity; the reuse of various compounds as well as a limited role of some metabolic pathway also contribute to cover the basal metabolism of the spermatozoa during storage and before their release and activation by the external milieu. In fish, the external fertilization has been well documented by pioneer studies on rainbow trout and considered as representative of other fish species that include some variants in their modes of motility activation. Gametes activity is usually short after release in external water, mostly because spermatozoa are sensitive to osmotic shock and the cortical reaction rapidly prevents fertilization in eggs. In these conditions, the strategy adopted by fish spermatozoa is to move at a high velocity during a short period of time so as to enter the micropyle before it becomes incapacitated by hypoosmotic shock and before the cortical reaction occurs to the egg. The pre-accumulated amount of ATP is sufficient to cover the energetic demand for motility during this brief period. In most cases, the endogenous ATP is not sufficient because of the decline in mitochondrial respiration capacity due to a long period of in vitro or in vivo storage called aging or due to damages during cryopreservation or also possible deleterious conditions during husbandry in fish farms (anoxic conditions, inappropriate feeding regime of fish, etc.).

In case of fish species with external fertilization, a large number of spermatozoa is released during spawning; this probably compensates for the low individual energetic load in each spermatozoon and finally leads to a correct fertilization success in an open and hostile environment, that is either hypoosmotic (freshwater) or hyperosmotic (marine water) as compared to physiological osmolarity of the seminal fluid.

This chapter intends to provide a literature review on the knowledge about the energy production, storage, and utilization of fish spermatozoa. Here follow successively the description of energy production (ATP) by mitochondrial respiration, the storage of this energy in energetic molecules (ATP and creatine phosphate), the use of this energy to produce movement by transformation of chemical into mechanical energy, the exhaustion of this energy during the short motility period, the involvement of mechanical energy in the hydrodynamic and thermodynamic aspects of fish sperm swimming, and by the end the potential deleterious effects of cryopreservation at these different levels. For a full understanding of the energetic exchanges engendered by a fish spermatozoon, one should consider two aspects that are complementary: on one hand, the input of energy that is mostly represented by the rate at which the ATP stores are exhausted (chemical energy) by the motor part, the flagellar axoneme, and on the other hand, the output of energy (physical energy) developed by the spermatozoon to progress efficiently in the fluid so-called swimming solution. Those two aspects will be successively developed in this chapter and by the end a balance sheet between both aspects is considered. By the end, this review will discuss the prime role of energy evaluation in the frames of cryobanking of fish spermatozoa (Martínez-Páramo et al. 2017). More practical aspects related to energetics and physiology of teleost spermatozoa were described in the review paper of Coward et al. (2002).

4.2 Brief Features Specific to Fish Sperm Morphology

In most fish species, spermatozoa lack a specific nuclear capping structure called acrosome that is present in many other species: exceptions in fishes are, for instance, Acipenseridae.

Details about spermatogenesis can be found in the comprehensive review of Schulz et al. (2010). Since long, a fish sperm cell is designated as "primitive type" with carp spermatozoon as a model (Billard 1970), mostly composed of an ovoid head (2–3 μm diameter), a short mid-piece inserted in the posterior part of the nucleus with centrioles and few mitochondria as a source of energy (ATP) that serves to activate the propelling flagellum (ranging 50 μm length) ubiquitously composed of a 9 + 2 structure enveloped in a membrane that forms fin-like ridges (usually 2) that develop in most species in a plan corresponding to that of the central pair of the 9 + 2 structure and are present almost all the way along the flagellum (Linhartova et al. 2013). Some variants of the ubiquitous "9 + 2" model were described such as: (1) absence of flagella (example in Mormyriformes by Mattei et al. (1972)), (2) bi-flagellated (example in *Porichtys notatus* by Stanley (1965) and Mattei et al. (2019), or even 9 + 4 (Marchand 1977) −9 + 0 instead of 9 + 2 (in Anguilliformes (Billard et al. 1973)). In the latter, the additional absence of radial spokes and outer dynein arms does not prevent motility (Gibbons et al. 1985).

In almost all cases, waves travel from the base to the tip of the flagellum in fish as well as in other animal spermatozoa, with very few exceptions such as in the annelid Myzostomum sp. (Mattei and Marchand 1988; Ishijima et al. 1994) where it is opposite, or the Tephritid flies (Baccetti et al. 1989) that present a bidirectional wave propagation of their swimming spermatozoa. Mitochondria are of prime importance in fish sperm metabolism: their location in the fish spermatozoon presents little variants and readers can get more details in different fish species in review papers such as that by Ulloa-Rodríguez et al. (2017).

4.3 The Production of Energy by Mitochondrial Respiration

4.3.1 Energy Resources

Energy for flagellar motility and basic cell metabolism is generated by the breakdown of nutrients from external or internal origin occurring in presence of absence of oxygen. A large majority of fish species use external fertilization where sperm is shed in the external aqueous medium lacking metabolic substrate. Therefore, sperm cells have to rely only on their cellular reserves, for instance phospholipids as it was shown by Minassian and Terner (1966) for *Alosa pseudoharengus* (alewife) and *Salmo gairdneri* (*Oncorhynchus mykiss*) (rainbow trout).

The main model species for further investigations has been trout. Spermatozoa of trout possess the enzymatic capacity for glycolysis, triglyceride, and phospholipid catabolism and triglyceride synthesis (Lahnsteiner et al. 1993). Trout spermatozoa have the capacity to incorporate glucose and acetate (Minassian and Terner 1966). Because of the lack of glucosidase and glycogen deposits (Billard 1986), trout sperm as well as other externally fertilizing species are unable to use polysaccharides as energy source: this contrasts with internally fertilizing fish species that use glycolysis for sperm motility (Gardiner 1978). This author demonstrated that the sperm of the shiner surfperch (Embiotocidae) and the guppy (Poeciliidae) exhibit a rate of glycolytic metabolism comparable to that of mammalian sperm (Bucci et al. 2010). Sperm metabolism is probably supported by ovarian sugars during the period of prolonged viability within the ovary, a characteristic of the two species.

According to Lahnsteiner et al. (1993), during the brief period of motility (<30 s) of the rainbow trout *O. mykiss* spermatozoa, a small decrease of intracellular glucose occurs but it cannot contribute to compensate the very rapid and important decrease of ATP concentration. During the same period, no change in triglyceride content occurs, but a significant decrease appears during longer periods (5–20 min) after motility cessation. During this resting period, it was shown that intracellular ATP store of immotile sperm is partly reconstituted which leads to conclude that energy restocking occurs mostly at the expense of triglyceride degradation.

4.3.2 ATP Production Resulting from Mitochondrial Respiration

The role of respiration and mitochondrial activity in teleost spermatozoa was reviewed in details by Ulloa-Rodríguez et al. (2017). In many fish species, measurement of respiratory activity presents difficulties because of the low oxygen consumption of spermatozoa, in contrast to model species such as sea urchin (Christen et al. 1983; Billard et al. 1986); in addition, the low respiratory activity remains almost unchanged when fish spermatozoa are transferred into motility-activating solutions while it is about 50-fold increased when sea urchin sperm is transferred into seawater (Christen et al. 1983). A promising method could circumvent the limits to measure low rate of respiration of fish spermatozoa (Lo et al. 1996): O_2 concentration is determined from the phosphorescence decay rate of the palladium (II) complex (Pd–phosphor). This measurement is based on quenching the phosphorescence of Pd–phosphor by O_2.

Results of sperm respiration rate presented in literature use different units, not always easy to compare between fish species and also because of differences in the conditions of measurement (temperature or presence vs. absence of motility). Values ranging from 20 to 40 μL O_2 per 10^{10} cells per hour were evaluated for sperm of rainbow trout, Atlantic salmon and Atlantic cod (Terner and Korsh 1963; Mounib 1967). More recently, oxygen consumption rate published in literature and presented in a comparative way for about ten different fish species by Ingermann (2008) shows that values vary from 1.4 to 70 nmol O_2/min/10^9 spermatozoa depending on species. Only one fish species (*Clarias gariepinus*) shows a strong increase of respiration rate (from 13 to 280 nmol O_2/min/10^9 sp.) accompanying motility activation (Mansour et al. 2003). In other fish species such as scophthalmidae (*Psetta maxima*—Dreanno et al. 1997, 1999b) or cyprinidae (*Chalcalburnus chalcoides*—Lahnsteiner et al. 1999) sperm shows only a modest increase of respiration rate after motility activation. Efficient respiration needs to be coupled in mitochondria to ATP production via the ATP-synthase (Somlo et al. 1982). For estimation of the full respiratory capacity of mitochondria, it is useful to apply diffusible "uncouplers" such as CCCP or FCCP (carbonylcyanide-4-trifluromethoxy-phenylhydrazone): these compounds are diffusible through the membranes and allow full rate of electron transfer in the electron chain of mitochondria without restriction due to its control by ATP-synthase. The effects of respiratory inhibitors such as oligomycin (Somlo et al. 1982) or KCN and their relationship with ATP stores of fish sperm were studied in detail by Dreanno et al. (1999b). Mitochondrial inhibitors have little effect in case of trout (Terner and Korsh 1963; Christen et al. 1987; Ingermann et al. 2003) or turbot (Dreanno et al. 1999b) spermatozoa.

As a general rule, it is considered that ATP present in fish spermatozoa has been accumulated during or at the end of spermatogenesis and the motility period relies mainly on this energy store. Thus, it appears that respiration rate in quiescent spermatozoa needs to be only minimal so as to maintain this ATP level prior to ejaculation. Such low but substantial respiration is enough for basal metabolism to maintain ionic exchanges and balances across the plasma membrane (Christen et al. 1987; Dreanno et al. 1999b).

An important aspect of oxidative phosphorylation is the release of CO_2 due to respiration. It has been shown in several fish species that CO_2 is a potent source of control of motility (Dreanno et al. 1999b). In the group of fish globally called flat fish (turbot or halibut as examples) as well as in sturgeon (*Acipenser transmontanus*), CO_2 blocks completely the flagellar motility, as shown by Inaba et al. (2003). This by-product of respiration is present at high concentration in the testis or male ducts prior to ejaculation: the inhibition of motility by CO_2 probably constitutes an additional control to prevent any motility and therefore, any ATP consumption, during storage in the genital tract of fish where the acidifying effect of CO_2 may also contribute to prevent motility.

4.3.3 Situation of the Fish Sperm Energy Storage in Various Fish Species

The energy stored in fish sperm prior to and used during the motility period was evaluated in several fish species (Gosh 1989; Billard and Cosson 1990). This was described in turbot as example (Dreanno et al. 1997, 1999b; Suquet et al. 1998), sea bass (Dreanno et al. 1999a), perch (Boryshpolets et al. 2009), bluegill (Burness et al. 2005), trout (Robitaille et al. 1987; Christen et al. 1987; Saudrais et al. 1998), carp (Perchec et al. 1995; Perchec Poupard et al. 1997), sturgeon (Tsvetkova et al. 1996), and catfish (Ziętara et al. 2004). Values for spermatozoa of other fish species can also be found (Cosson 2010, 2012; Dzyuba et al. 2017a).

4.3.4 Exposure of Fish Spermatozoa to Oxygen Stress at Ejaculation in the External Milieu

Mitochondria have a specific relation with oxydo-reduction balance and oxidative stress within the spermatozoa: the electron transport chain has a main role in the production of different reactive oxygen species (ROS) which exceeds the sperm-limited antioxidant defenses, and an oxidative stress is induced, characterized by peroxidation damage to the sperm plasma membrane and to the DNA in different species, mammals as well as fish (De Iuliis et al. 2009; Lahnsteiner and Mansour 2010; Aitken et al. 2010; Guthrie and Welch 2012). The variety of reactive oxygen species (ROS) molecular species covers superoxide ($O^{\bullet 2-}$), hydroxyl radical ($^{\bullet}OH$), hydrogen peroxide (H_2O_2), some acidic compounds (HOX, with $X = Cl^-$, Br^-, I^-, or SCN^-), nitric oxide ($NO^{\bullet}$), and peroxy-nitrite ($ONOO^-$). These molecules are highly disruptive to cellular function, and an increase in ROS production usually contributes significantly to several diseases, with exceptions leading to opposite effects (de Lamirande et al. 1997; de Lamirande and O'Flaherty 2008).

Therefore, mitochondrial antioxidant enzyme activities, such as superoxide dismutase (SOD), catalase (CAT), glutathione peroxidase (GPx), and glutathione reductase (GR), have been evaluated regarding their role and the effects of their alterations on cell bioenergetics (Halliwell and Gutteridge 1985; Agarwal and Allamaneni 2004; Guthrie and Welch 2012). The negative effects of oxidative stress on teleost sperm motility have been studied in several species (Dzyuba et al. 2016), as well as the redox balance under cryopreservation conditions showing that antioxidant supplementation could improve the post-thawed sperm quality and performance (Shaliutina-Kolešová et al. 2015). Quantification of ROS and total antioxidant status are both used as complementary parameters to assess fish sperm quality, especially during cell-stressing situations such as cryopreservation (Cabrita et al. 2014).

Quite a lot of recent studies have also emphasized the deleterious effects of oxygen dissolved in water when fish sperm cells are released by males from the seminal fluid (SF) into the external milieu. Delivery of sperm to the external milieu is a necessary condition for motility activation of fish spermatozoa because it is basically the osmotic signal that provokes initiation of motility (Morisawa 1985; Cosson 2010). As long as SF remains in the fish ducts, it mainly constitutes an anaerobic environment for spermatozoa with low oxygen concentration, mainly due to the respiration of highly concentrated sperm. Right at ejaculation, fish spermatozoa become exposed to a highly oxidative fluid (fresh- or seawater) that generate several of the reactive oxygen species above-mentioned and rapidly penetrating through the sperm membranes. The axonemal part of the fish sperm flagellum is probably less susceptible than membranes as shown after application of cryopreservation techniques to de-membranated fish spermatozoa (Dzyuba and Cosson 2001).

While various antioxidant compounds are present in SF where they protect spermatozoa from ROS, the external water is devoid of any such protection (Ciereszko et al. 1999). In trout SF, citric acid is present in quite high concentration and Liu et al. (1997) have shown that ascorbic acid acts as antioxidant that decreases lipid peroxidation in trout spermatozoa. Ciereszko et al. (1999) reported high levels of another antioxidant, uric acid, in fish SF. In externally fertilizing fish species, the protection against ROS by SF becomes immediately inefficient because of the high dilution rate of SF in the external milieu at ejaculation.

The presence of relatively high concentrations of these anti-ROS compounds in the SF suggests that they protect fish spermatozoa against oxidative damage so as to maintain a high potential for motility at spawning. Recent studies on fish sperm oxidative stress applied to species such as carp and trout (Shaliutina-Kolešová et al. 2016), sturgeon (Dzyuba et al. 2014b, 2016; Shaliutina-Kolešová et al. 2015), and several other species (Shaliutina-Kolešová et al. 2013). In most cases, it is observed that prevention of oxidative stress during the freeze-thawing process applied for cryopreservation is beneficial to the survival of fish spermatozoa.

4.3.5 Other Potent Sources of Energy

Seminal fluid that surrounds spermatozoa is an important factor to protect sperm cells until ejaculation. Analysis of the chemical components present in SF shows that it is a complex medium, where potential substrates for sperm respiration are present at various levels in oviparous and viviparous fish species: glucose and other sugars plus lipids (Lahnsteiner et al. 1993) as well as proteins (Billard and Cosson 1990) but it is not clear whether the latter are possible nutrients as well as amino acids that are present at quite high concentration in carp and trout SF (Billard and Menezo 1984). In case of turbot, the sperm motility period is accompanied by a decrease in intracellular levels of amino acids suggesting that the metabolism of amino acids would contribute to the oxidative metabolism (Dreanno et al. 2000). In some viviparous fish species, there are evidences for a high level of utilization of glucose to sustain sperm motility (Gardiner 1978). In case of mammals, it was shown that, when stimulated, the glycolytic pathway maintains ATP levels and motility of bovine epididymal sperm mitochondrially uncoupled (Losano et al. 2017). Even though monosaccharides appear as a potential sperm nutrient in various fish species, the mechanism by which they are taken up by fish spermatozoa is not known. In a recent paper of Lahnsteiner et al. (2010), the *Sparus aurata* sperm metabolism, carbohydrates and lipids content was related to motility capacity.

Levels of triglycerides are quite high in the seminal plasma. However, triglycerides are too large to penetrate cell membranes (Geise 1979). Spermatozoa have lipases that presumably hydrolyze the triglycerides, resulting in free fatty acids that can be taken into sperm cells by passive diffusion. Results of Minassian and Terner (1966) confirm this point. Most of these possibilities remain to be tested in more details in case of fish spermatozoa. Lipid composition of common carp spermatozoa was recently correlated with cryoresistance in this species and the sperm population presents some heterogeneity relatively to this property (Horokhovatskyi et al. 2016a, b).

Citrate represents another possible source of nutrient: it is present at quite high concentration in the SF of several fish species, ranging from 50 to 450 mg/L depending on species (Piironen and Hyvärinen 1983) including turbot (Dreanno et al. 2000). Citrate may be incorporated by the spermatozoa and oxidized directly for energy via the tricarboxylic acid cycle but it is not clear whether and how citrate is taken up by fish sperm cells.

The numerous chemicals above mentioned are of primary importance for the adjustment of optimal cryopreservation conditions of fish spermatozoa. The beneficial presence and concentration remain to be tested individually for each species because of the lack of basic knowledge about their biological functions.

After delivery of spermatozoa from the SF to the external milieu, fish sperm cells can only rely on their internal energy store. The main source of energy for long-term ATP regeneration was shown to be endogenous triglycerides and phosphatidylcholine with a preference for phosphatidylcholine in case of sea urchin (Mita et al. 1994) and mainly triglycerides and glycolysis in trout spermatozoa (Lahnsteiner et al. 1993). Oxidative metabolism of saturated fatty acids generates 5.7 moles of ATP per mole of oxygen (Stumpf and Barber 1957). Sea urchin spermatozoa have a respiratory quotient of 0.7, consistent with the oxidative metabolism of endogenous phospholipid (Mohri and Horiuchi 1961).

In a study of the aerobic metabolism of bull spermatozoa at 35°C, Rikmenspoel (1978) found that the ratio between the energy expended by the spermatozoa in swimming against the viscous resistance of the medium, and their total oxygen consumption, was equivalent to 8.8 kcals per mole of O_2. Other results give a value for this ratio of 8.8–9.8 kcals per mole of O_2, depending on the estimate of the percentage of spermatozoa which were nonmotile. The bull spermatozoa have a normal oxygen consumption per spermatozoon approximately six times that of *Ciona* spermatozoa (Brokaw and Benedict 1968).

In viviparous fish species using external fertilization, energy mostly relies on mitochondrial metabolism, a feature that contrasts with mammalian sperm. In the latter, recent experiments demonstrated that glycolytic pathway can substitute mitochondrial source so as to maintain normal ATP level and sustain flagellar motility (Losano et al. 2017). In a few externally fertilizing fish species, an unusually long period of sperm motility was observed (example reported by Elofsson et al. 2003a, b), which would indicate that ovarian fluid is able in these species to provide diffusible substrates sustaining motility. Similarly, the contribution of ovarian fluid to sperm physiology was suggested by Butts et al. (2017).

In case of oviparous fish species that use internal fertilization, ejaculated sperm in spermatophores is not exposed to osmotic damage and copulation lasts for very short (around 2 s in guppy according to Clark and Aronson (1951)). Spermatozoa have to cover 1 cm distance in the ovarian cavity (Billard 1990) where they are in contact with various nutrients such as glucose and where they can swim for long periods, from 1 up to 18 h in guppy and *Cymatogaster aggregata,* respectively, according to Gardiner (1978). In addition, spermatozoa of viviparous fish species possess granular glycogen stores associated with mitochondria that participate to an endogenous store of energy as a potential source of glucose by glycolysis during the swimming period (Billard and Breton 1970).

4.4 The Storage of Energy

ATP content of sperm cells can be evaluated by several methods, recently including evaluation in a single spermatozoon cell (Chen et al. 2015). A full evaluation of the storage of energy in fish sperm cells needs the determination not only of the internal content of ATP but also that other energetic compounds that are able to exchange high energy phosphate bonds that can be transferred to ADP and allow to reconstitute ATP store. This was established in case of sturgeon sperm (Fedorov et al. 2015, 2017) by use of liquid chromatography combined to HPRS or in case of turbot or sea bass sperm where the adenine nucleotides energetic balance was determined by H^+-NMR and ^{31}P-NMR analysis (Dreanno et al. 1999b), as well as in trout by ^{31}P-NMR (Robitaille et al. 1987; Saudrais et al. 1998). All these results clearly point out to the fact that ATP level can be rescued by the Cr–P generated by the mitochondrial metabolism. This means that other phosphagen compounds are as important as ATP in the energy balance of fish sperm cells (Ellington and Kinsey 1998).

4.4.1 Quantity of Stored Energy and its Cell Management

From the Sect. 4.3, it appears that most of the energy stored in fish spermatozoa originates mostly from oxidative phosphorylation in the sperm mitochondria. It is known that most of the ATP results from mitochondrial activity itself resulting from the coupling of the electron chain with the mitochondria, from which ATP is exported into the cytoplasmic compartment by a specific ATP–ADP translocator (Kunji et al. 2016); then, ATP should be transported and distributed all along the flagellum so as to supply chemical energy to sustain the mechanical energy generated by the motor part of the flagellum called the axoneme. For some period, it was believed that simple diffusion of ATP along the inner part of the flagellum would be enough (Nevo and Rikmenspoel 1970), but some theoretical considerations lead to the necessity of a distribution system to ensure that a constant ATP concentration be present on any point along the axoneme (Tombes and Shapiro 1985, 1987; Tombes et al. 1987). Such shuttle system involves an additional high energy compound and assistance of enzymatic system that will be detailed below and was shown to be present in fish sperm cells (Saudrais et al. 1998).

Even though the ATP molecule is the most common high-energy compound used as a fuel for many cell functions, including motility (Cosson 2012), several other high energy molecules are present in living cells such as creatine phosphate or arginine phosphate. In fish, creatine phosphate (CrP) is the main high-energy compound that was characterized in spermatozoa of several fish species as a complement of ATP (Dreanno et al. 1999a, b, 2000). In the last decades, many studies have demonstrated the decrease of the ATP concentration inside the fish sperm cells during the motility period (Perchec et al. 1995; Saudrais et al. 1998; Poupard et al. 1998; Dreanno et al. 1999a, b; Billard et al. 1999; Dreanno et al. 2000; Fedorov et al. 2017). A more restricted number of studies have investigated the concentration of ATP related compound such as ADP, AMP, CrP, and others (Dreanno et al. 1999a, b; Fedorov et al. 2017). All these compounds are part of an intracellular network under control of different enzymes that are able to transfer high-energy phosphate bonds from one to another (see Figure in Chap. 1 in book edited by Cosson 2015) as example the equilibrium ATP $\leftrightarrow$ ADP + Pi is catalyzed by enzymes called ATPases. As another example, ADP + CrP $\leftrightarrow$ ATP + Cr is controlled by enzymes called creatine kinases. One creatine kinase is mitochondrial while a second one is in the flagellum and distributed all along the axoneme (Fig. 4.1). The mitochondrial creatine kinase delivers CrP that diffuses along the flagellum, both creatine kinases being present in trout sperm cells. The rate of diffusion of CrP molecules is higher than that of ATP (Tombes and Shapiro 1987; Tombes et al. 1987). Such an arrangement of catalytic activities and substrates constitute an intracellular network ensuring the correct production and distribution of energy in fish sperm cells (Fig. 4.1) and is called the “ATP shuttle.”

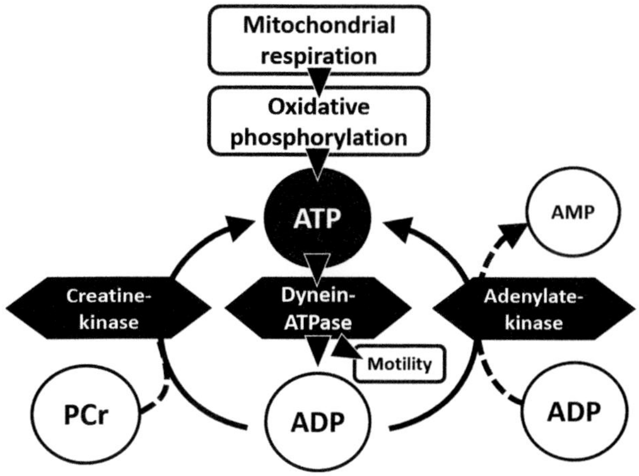

Fig. 4.1 Simplified schematic representation of the energetic exchanges in a fish sperm spermatozoon. ATP is produced by oxidative phosphorylation in the mitochondrion, then ATP diffuses along the flagellum and is transformed into ADP by hydrolysis so as movement is generated by the dynein–ATPase. The resulting ADP is recycled by two enzymatic activities, the creatine kinase (PCr + ADP $\geq$ ATP) and the adenylate kinase (2 ADP $\geq$ ATP + AMP). As a complement, the energy-rich molecule PCr, generated by the mitochondrion, constitutes an additional source of energy for movement via the ATP–PCr shuttle (see text for explanation)

A schematic but more detailed representation of the ATP–PCr shuttle can be found in the Fig. 1 in the paper of Dzyuba et al. (2016). The elements necessary for such ATP–CrP shuttle were shown to be present in fish sperm flagella such as trout (Saudrais et al. 1998).

The inhibition of creatine kinase by FDNB, a specific inhibitor of creatine kinase (Tombes et al. 1987), provokes the paralysis of the distal part of the flagellum in several fish species: in turbot and in trout spermatozoon, the distal part is devoid of bend and stiffened in presence of FDNB (Saudrais et al. 1998) but not in de-membranated sperm (Cosson, unpublished). This local axonemal paralysis indicates that the inhibition of the creatine kinase prevents the renewal of ATP and the production of ATP from Cr–Ph in this distal portion of the flagellum, while the proximal portion that is close to the mitochondrion, can use directly the ATP still produced by the latter. Stiffening of the distal flagellar tip was also observed in vivo (but not in reactivated sperm) of two sea urchin species (Brokaw 1965) and trout sperm (Cosson, unpublished) after application of thiourea, an inhibitor of respiratory phosphorylation.

Initiation of a bend after a high osmolarity shock (caused by high DMSO concentration) is blocked at the base (head–tail junction) of the flagellum (Prokopchuk et al. 2015). This blockage cannot be due to ATP accumulation because propagation immediately ensues with normal wave's amplitude (100 ms or less) after transfer in normal osmolarity medium that provokes motility activation.

In experiments using local delivery of ATP on de-membranated turbot flagellum, the bend activation can occur everywhere along the flagellum.

Other enzymatic activities present in the fish sperm flagellum also contribute to the renewal and maintenance of ATP, such as adenylate kinase that is responsible for the catalysis of the reaction 2 ADP ↔ ATP + AMP and is sensitive to a specific inhibitor called diAPP as was shown by Saudrais et al. (1998) in trout sperm.

4.5 Motility Activation: Energy Implications

4.5.1 The Maturation Step

At the end of spermatogenesis and before sperm ejaculation, it is necessary that fish sperm gets prepared for motility activation. In addition to energy storage, fish sperm cells must reach a mature state that ensures their ability to be activated by the external milieu. This maturation step is well described in case of mammals (de Souza et al. 2017) where maturation mostly occurs during the transit into the epididymis. This maturation process was also documented to a lower extend in several groups of fish species: salmonids (Morisawa et al. 1993), chondrostean (sturgeons, Dzyuba et al. 2014a, 2017b), cyprinids (Redondo-Müller et al. 1991), and occurs through quite different mechanisms as briefly summarized below. In salmon, maturation is mainly under control of the cAMP content and the pH of the water where fish are transiting during migration (Morisawa et al. 1993). In carp, results from Redondo-Müller et al. (1991) indicate that an ionic equilibration across the sperm membrane is the main factor responsible for maturation. In sturgeon, the contact of sperm with urine that occurs, in this group of species, prior to ejaculation is the way to reveal the potential of sperm to have its motility activated by contact with freshwater (Dzyuba et al. 2014a, 2017b). Thus, it appears that, just among those three groups of fish species, different strategies were adopted for controlling sperm maturation. In some cases, such as in sturgeon, it was shown that the maturation process is also dependent on the energetic store of spermatozoa (Fedorov et al. 2015) and their intracellular calcium level (Bondarenko et al. 2017). A proteomic analysis is useful to identify the proteins playing a major role in this process (Boccaletto et al. 2018) among which proteins involved in the energetic of sperm cells. This approach takes advantage of the analogy with the role of the epididymis in sperm quality and male fertility of mammals (de Souza et al. 2017). In some species, sperm maturation was shown to be partially under control of a posttranslational modification (polyglycylation) of axonemal tubulin (Bré et al. 1996).

4.5.2 Role of Osmolarity and Specific Ions at the Motility Activation Step

The role of osmolarity of the medium surrounding fish spermatozoa was reviewed by Alavi and Cosson (2006) and was demonstrated as having primordial role in many species such as in carp by Perchec et al. (1996) and by Perchec Poupard et al. (1997); Poupard et al. (1998), in sea bass by Fauvel et al. (1999), in paddlefish by Linhart et al. (2003), in Persian sturgeon by Alavi et al. (2008), in hake by Groison et al. (2008, 2010a, b) and Cosson et al. (2008b), in pike by Alavi et al. (2009), in Brycon by Orfão et al. (2011), in pink cusk eel (*Genypterus blacodes*) by Dumorné et al. (2018), in cod by Cosson et al. (2008b) or in tilapia, an euryhaline species, by Takei et al. (2015) and Legendre et al. (2016). This set of results shows that osmolarity is the key factor controlling sperm motility either for marine fish species where motility is triggered by a large increase of osmolarity (Suquet et al. 1994; Cosson et al. 2008a, b), or in freshwater species that activate sperm motility when in contact with very low osmolarity (Morisawa 1985; Alavi and Cosson 2006).

In sperm of several species, osmolarity plays a synergic role with specific ions (Takai and Morisawa 1995). This is the case for K^+ and osmolarity in sturgeon (Prokopchuk et al. 2016), in paddlefish (Linhart et al. 2002, 2003), and also for K^+ and Ca^{2+} in trout (Tanimoto and Morisawa 1988; Cosson et al. 1989; Boitano and Omoto 1991; Bondarenko et al. 2014) or in carp (Marian et al. 1993; Krasznai et al. 1995, 2000; Linhart et al. 2003).

Because osmolarity is the main activating factor, some results attempted to describe the changes in sperm volume at initiation of fish sperm motility (Bondarenko et al. 2013) but such change does not occur in all species and is a slow process compared to the time needed for the activation step (Prokopchuk et al. 2016). Therefore, the role of aquaporins was also investigated in the activation step as they are supposed to accelerate the water exchanges through the cell membrane (Chauvigné et al. 2011, 2013). This possible link between aquaporins and osmoregulation of fish sperm was investigated in several fish species such as sea bream (*S. aurata*) by Zilli et al. (2009) or tilapia by Takei et al. (2015).

Altogether, it appears that, because of the time period between application of osmotic signal and flagellar response is very brief, ranging less than 100 ms according to Prokopchuk et al. (2015), any change in the internal osmotic pressure cannot play a direct signaling role for flagellar activation because of its too long time period (several second or more) to get established in the cell and even its absence in some fish species (Bondarenko et al. 2013). Therefore, additional signaling pathways were explored to understand the nature of the activating signal. One is the membrane potential that was measured in trout spermatozoa (Gatti et al. 1990; Boitano and Omoto 1991) and recognized as an important and fast enough factor at motility initiation (in carp, see Krasznai et al. (2000)). In conjunction with membrane potential, transient potential receptors (TRP) represent potent candidates for the signal transduction at the membrane level. TRPs constitute a large family of membrane protein that are present in many cell types (McKemy 2007). As example, TRPV4 was identified as the temperature-sensitive ion channel of human sperm (Mundt et al. 2018) and some of those TRPV, such as TRPV1, were identified in sperm of the fish species *Labeo rohita* (Majhi et al. 2013) and shown to be present in several other fish species such as trout. In the latter study, authors could show that TRPV1 is involved in several sperm functions such as response to temperature, while TRPV1 is also known to respond to osmolarity signal. Stress-activated channels represent alternative candidates for fish sperm activation as proposed by Krasznai et al. (2003). It remains important in the future to identify clearly in many fish species, the link between the transduced membrane signal and its reception at the axoneme (Dzyuba and Cosson 2014).

4.5.3 cAMP, Its Non-energetic but Rather Regulatory Role

Cell signaling is important in the regulation of energetic and physiology of fish spermatozoa (Dzyuba and Cosson 2014); among intracellular signals is the cyclic-adenosine monophoshate (cAMP). The compound is another adenosine nucleotide present in fish spermatozoa but it has no high energy bond and thus is not directly involved as an energetic compound. Since the work of Morisawa's group (Morisawa et al. 1993), it is known that intracellular cAMP concentration is an important signal leading to flagellar activation of trout spermatozoa. This was also shown in other species such as sturgeon (Dzyuba and Cosson 2014; Dzyuba et al. 2016). A rise of the internal concentration of cAMP occurs right after the contact of trout sperm with freshwater at spawning (Cosson et al. 1991a; Morisawa et al. 1993; Cosson et al. 1995). This rise occurs concomitantly with the activation of adenylate cyclase by the intracellular Ca^{2+} concentration rise engendered by the decrease of K^{+} ions in the extracellular milieu (Cosson et al. 1989; Morisawa et al. 1993). This cAMP rise induces the activation of a cAMP-dependent protein kinase that activates a tyrosine kinase that finally phosphorylates a tyrosine residue of the 15 kDa MIPP (motility initiating phosphoprotein (Jin et al. 1994), subsequently identified at the basal region of the axoneme (Morisawa 1994)).

Several evidences argue against the absolute requirement of cAMP in this activation process. In trout spermatozoa, the presence of cAMP is not needed if the intracellular concentration of ATP does not exceed 25 μM, but it is needed if ATP is more than 100 μM (Cosson et al. 1995). The addition of phosphodiesterase in both situations has no effect. Additional results also show that the intracellular pH greatly influences the control of motility by the intracellular ATP concentration. In case of de-membranated sea urchin spermatozoa assayed for motility at pH 7.0–7.2, only 10% moving cells are observed at 1 mM ATP, while at least 60% motility occurs at 20 μM ATP (Yoshimura et al. 2007). The same authors show that the inhibitory effect of high ATP concentration is released by addition of ADP. In the case of trout spermatozoa, similar effects were observed independently of the cAMP role (Cosson, unpublished). In both species, no motility remains at ATP 20 μM after application of a protein phosphatase but reappears after ADP addition. According to Inaba (2002), a type 2A protein phosphatase is responsible of the dephosphorylation of TcTex-2-related dynein light chain. This whole set of results shows the complicated interplay that occurs between pH, ATP, ADP, and cAMP in the control of motility of trout sperm. Especially regarding pH effects, the relationship between intra- and extracellular pH of trout sperm was measured by two methods (Christen et al. 1983) and shows the internal pH, pH_i is about 1 unit higher than the external pH, pH_e (Woolsey and Ingermann 2003). The same authors show that axonemal dynein ATPase activity increased approximately 3.5-fold between pH 7.0 and 7.6, which represent approximate pH_i values, whereas the potential for trout sperm to become motile upon water activation is particularly sensitive to the pH_e in the range of 7.4–8.0. This emphasizes an additional role of pH_i: that is the control of maturation of salmonid sperm.

The cAMP rise occurring at motility initiation is of very brief duration: especially when measured at low temperature, about 5–6°C which is physiological for trout reproduction, it is remarkable that this cAMP rise occurs much later after the whole sperm population was activated, which opens the question of the chronological role of cAMP.

The involvement of cAMP is presented as a schematic drawing in Fig. 4.5 in the paper of Cosson (2012) that illustrates the cascade of events occurring during the trout sperm motility period where ATP or phosphagen-related molecules or signaling processes are implied.

The cAMP control of fish sperm motility was also demonstrated in different situations in various salmonid species as well as in chondrostean, mostly sturgeons (Dzyuba and Cosson 2014) but it does not occur neither in cyprinids such as carp nor in euryhaline species such as tilapia and nor in most marine species such as turbot or sea bass.

4.5.4 The Activation Step Per Se

As long as fish spermatozoa remain in testis and duct, their flagella are fully inactive and it is only when sperm is ejaculated that the contact with the external fresh- or marine water that motility starts. This waves activation is a very fast process as illustrated in Fig. 4.2.

4.5.4.1 Activation of Fish Sperm Motility: Role of Osmolarity

The composition of external milieu completely differs depending on fish reproductive strategy and conditions: either comprising a high salt concentration in seawater or a quasi-nil salt concentration in freshwater. This huge difference can also be expressed in terms of another physical parameter, the osmolality that is the total concentration of soluble molecules (including ions) in a solution. A large majority of fish spermatozoa (those with external fertilization) have their motility activated either by a high osmolarity environment ranging that of seawater (marine species) or by a very low osmolarity close to pure water (freshwater species). Osmolarity values expressed as mOsmol/kg are considered as high (around 1200 for seawater) or low (10–20) by comparison to that of SF (ranging 300), whatever the species marine or freshwater.

4.5.4.2 The Role of Specific Ions in the Activation of Fish Sperm Motility

In a restricted number of freshwater species (salmonid and chondrostean fish), it was shown that, on top of the osmolarity, specific ions (mostly K^+) control the motility initiation. In this case, the swimming medium should fulfill a double condition: low

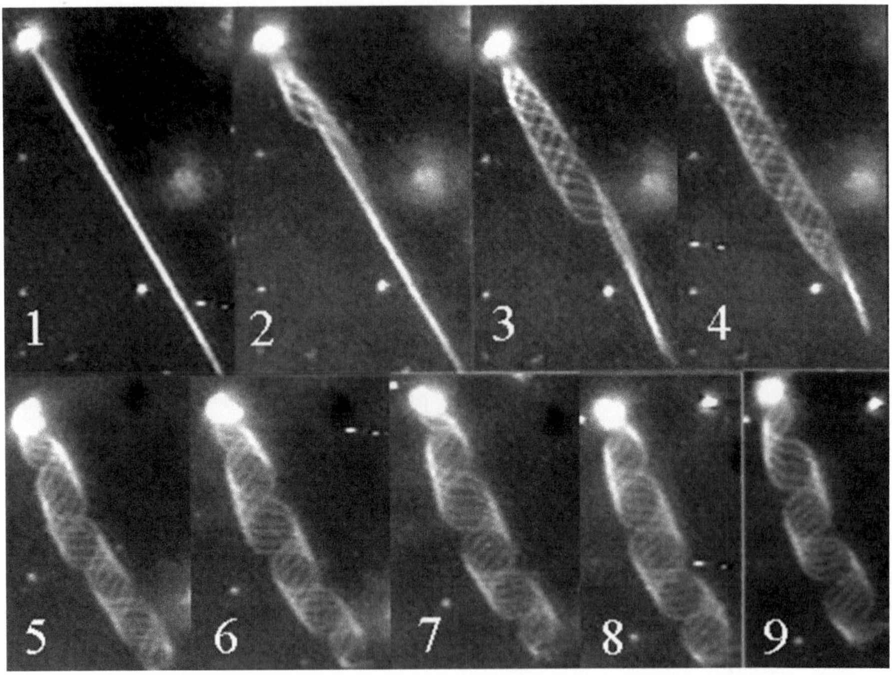

Fig. 4.2 Wave initiation of a turbot spermatozoon. Panels 1–9 correspond to images recorded every 20 ms, each panel corresponds to six successive positions of the same sperm flagellum (about 3 ms between two positions). Panel 1 shows an immotile flagellum, then panel 2 shows the appearance of the very first waves, then panels 3–9 show the gradual propagation of the waves along the flagellum. The first wave takes 50–100 ms to fully develop (see text below and Prokopchuk et al. (2015))

osmolarity and low K^+ concentration. In trout and sturgeon, the K^+ signal can be bypassed if strong osmolarity shock is applied prior to transfer in swimming medium (Takei et al. 2015; Prokopchuk et al. 2016).

4.5.4.3 Activation of Fish Sperm Motility by Ions: Role in the Control of Membrane Potential

Membrane potential reflects the difference on ions concentration between both sides of a membrane. In mammal spermatozoa, it was shown that the membrane potential, evaluated by a fluorescent technique, controls progressive sperm motility (Moscatelli et al. 2017). The spermatozoon membrane potential is under control of KSper, a pH-sensitive K^+ controller (Navarro et al. 2007) and of CatSper, a Ca^{2+}-sensitive current controller (Espinosa and Darszon 1995). In fish sperm, as already mentioned, the membrane potential was also described as controlling motility activation (Gatti et al. 1990; Kho et al. 2005) and Boitano and Omoto (1991) developed a model where the control of membrane potential by external K^+ ions concentration is the primary signal that activates trout sperm motility. Other authors (Kho et al. 2005) completed this model by including a cAMP concentration rise following the membrane potential signal. These authors leave unsolved an important point that was already emphasized by Boitano and Omoto (1991): the chronology of events contradicts part of the model, as flagella activation is a very fast event taking less than 1 s as shown by Prokopchuk et al. (2015) while membrane potential change and its subsequent effect on cAMP rise take a much longer time period.

4.5.4.4 Activation of Fish Sperm Motility: Role of Specific Inhibitor or Activator of Motility

It was shown that, in some fish species (mostly flatfish e.g., turbot), activation of flagella motility is prevented due to the inhibitory effects of CO_2 present in the testis and sperm ducts and the high content of carbonic anhydrase in the spermatozoa of this group of species (Inaba et al. 2003). Intracellular CO_2 can not only control flagella according to an on/off mechanism but also modulate the wave shape of flagella (Cosson et al. 1999b). A few examples of published results show that some specific compounds can diffuse from the egg and activate the motility of the spermatozoa of the same species (e.g., in herring Cherr et al. 2008). Alternatively, it was shown that in tilapia, a glycoprotein, naturally present in the semen represses the flagellar activity but becomes permissive for motility by simple dilution of the semen (Mochida et al. 1999). Those two examples are exceptional among the large number on investigated fish species.

4.6 The Flagella Motility Apparatus and Amount of Energy Consumed During the Motility Period

Details about the structure and function of the axoneme of fish spermatozoa can be found in reviews of Cosson (1996), and Boryshpolets et al. (2018) and in the book chapters by Cosson (2008a, b, 2015). Since one of the first and pioneering reports shows the presence of an ATPase activity-sustaining movement of fish spermatozoa by Tibbs (1959) in perch sperm flagella, abundant information has been obtained in many other species. Several authors emphasize the direct relationship between the flagellar beat frequency (BF) of fish spermatozoa and the rate of utilization of energy, mostly intra-flagellar ATP concentration (Billard 1986). The flagellum BF was evaluated by inelastic light scattering techniques (Bergé et al. 1967) or, in further studies, by stroboscopic illumination of the microscopy slide (Cosson et al. 1985). Figure 4.3 shows the wave propagation of a fish spermatozoon at the early phase of its motility period when load of intracellular ATP is maximal.

4.6.1 ATP Hydrolysis by Dynein and Its Coupling for Conversion to Mechanical Energy

The use of ATP in the axoneme of a flagellum is due to the mechanochemical enzyme called “dynein”: its name comes from the Greek “dyne” that means force (Gibbons 2012). Dynein molecules are proteins of high molecular weight that are

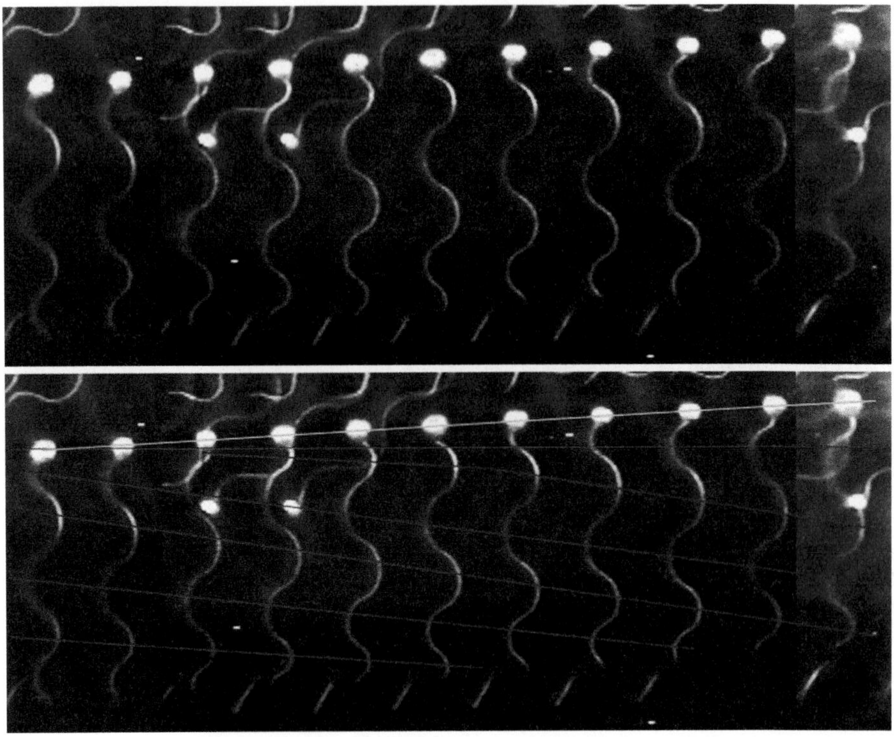

Fig. 4.3 Normal waves propagation during the motility period of a turbot spermatozoon. From left to right are images every millisecond of a turbot spermatozoon swimming in seawater. In the bottom panel, the red horizontal line is used as reference, the blue line shows the temporal and spatial propagation of the alternation of the crests or valleys of each wave curvature, and the green line shows the corresponding distance covered by the translated sperm head (upward). Dark field microscopy (×100 immersion lens) combined with stroboscopic illumination

localized all along the flagellar microtubules and constitute a major part of the proteins of the axoneme. Details about the function and regulation of dynein can be found in the recent review of Gibbons (2012).

In fish spermatozoa, the consumption of the stored energy during the motility period (Cosson 2008b, 2015; Prokopchuk and Cosson 2017) is gradual in most fish species, but is such that *the decrease of ATP concentration is not to zero*. The intracellular concentration of ATP remaining at the end of the motility period is predicted to be above the K_m for ATP of the dynein–ATPase. Therefore, this low ATP level does not completely explain why motility of fish sperm flagella does stop, as discussed by Cosson et al. (1999a). Nevertheless, when the Ca^{2+} ions concentration is high in the trout swimming medium, the motility period is prolonged (Billard 1990; Billard and Cosson 1992).

The phenomenon called wave dampening describes the decrease of amplitude versus distance in fish sperm flagella, as direct consequences it entails a decrease of the swimming efficiency of the waves. The energy (ATP) becomes exhausted in the distal part of the flagellum because it is far from the production site (mitochondrion) located close to the proximal flagellum.

The same dampening effect is observed in conditions of inhibition of Cr–K by FDNB that provokes a paralysis (rigor) of the distal part of the flagellum (see Sect. 4.4.1); it is related to the lack of energy in the distal part of the flagellum, which prevents wave generation in the distal segment.

When the dampening of waves is extreme, it leads to full arrest of motility and the axoneme adopts a rigid and linear shape called "rigor". It is observed in sea urchin (Gibbons 2012) and in fish (Cosson, unpublished) de-membranated spermatozoa when the ATP concentration is suddenly dropped to very low value, but it is reversible after re-addition of ATP. The rigor state (see lower panel in Fig. 4.4) is due to the ADP (generated by ATP hydrolysis) that remains bound to the ATPase active site of dynein. It was shown by Gray (1955) and further studies including fish sperm flagella that the number and the amplitude of waves along a flagellum relate to sperm velocity according to a linear relationship. This result from the fact that amplitude of waves as well as their beat frequency are under dependence of the intra-flagellar ATP concentration, a feature explaining the interest of many studies for the determination of ATP concentration present in fish sperm flagella.

It should be reminded that it is the earliest part of the sperm motility period which is the most important for the success of fertilization of fish egg because it is during this short period that the distance covered by the moving spermatozoon is larger (Cosson 2008b).

There is no relationship between sperm longevity and initial velocity, but there is a positive correlation between total distance covered by a fish spermatozoon and the motility period duration (Cosson 2010, 2016), which in particular was established for zebrafish sperm (Alavioon et al. 2017). In contrast, there is a clear decrease of several parameters that are linked together during the motility period: the ATP consumption, the beat frequency, and the percentage of swimming fish spermatozoa (Cosson 2010). This is illustrated as an example in carp spermatozoa in Fig. 4.5.

4.6.2 The Concentration of the Different Energetic Molecules During the Motility Period

Phosphagens are defined as any of several organic phosphate compounds (e.g., phosphor–creatine, PCr, or phosphoarginine) releasing energy on hydrolysis of the phosphate. The phosphagens are energy storage compounds, also known as "high-energy" phosphate chemicals, associated with ATP homeostasis. They allow a high-energy phosphate pool to be maintained in a concentration range, which, if it all were ATP, would create problems due to the ATP-consuming reactions in these tissues. In case of sudden demands for lots of energy, these phosphagen compounds can maintain a reserve of high-energy phosphates that can be used to provide the energy that could not be immediately supplied by glycolysis or oxidative phosphorylation. Most phosphagens cover immediate need of energy but in a limited way, as

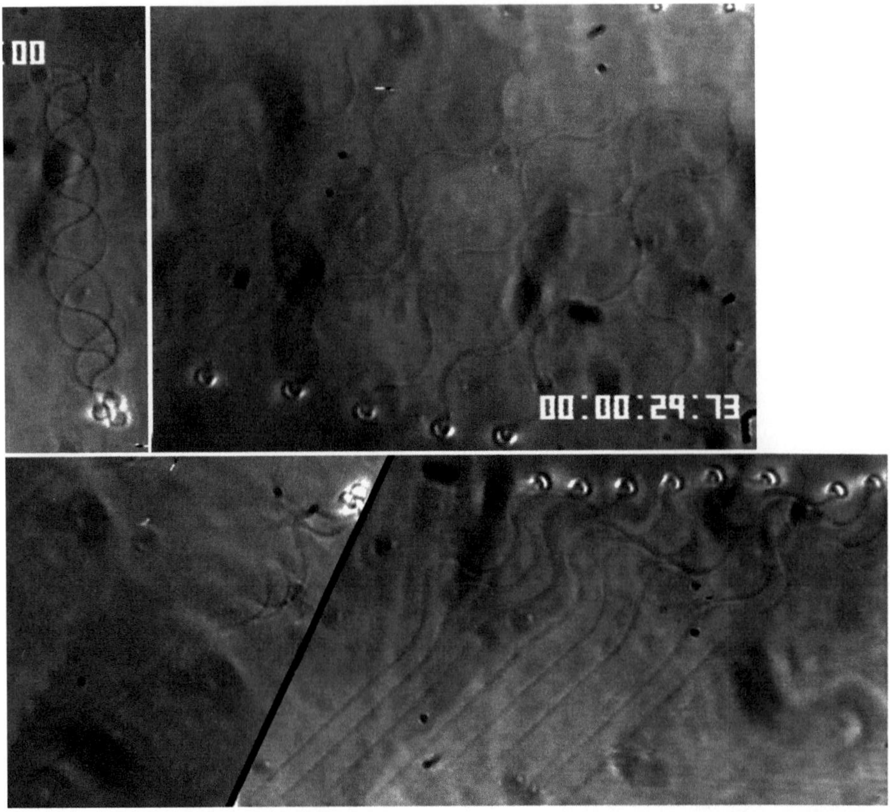

Fig. 4.4 Wave dampening appearing when a fish sperm cell is close to the arrest of the motility period. In this illustration, a turbot spermatozoon was video-recorded by microscopy (100× phase contrast and oil immersion objective lens) and stroboscopic illumination (three flashes per video frame). The upper panel shows on the left side a turbot spermatozoon about 20 s post-activation in seawater, the right part of this panel shows the same sperm cell recorded, while the microscope stage is translated to visualize individual positions of the propagated flagellar waves every 6 ms. The lower panel shows a similar spermatozoon close to the end of the motility period and visualizes the absence of waves in the distal part of the flagellum (rigid segment), while the proximal portion develops regular and propagated waves. This wavy segment being proximal to the head is favorably provided in ATP from the mitochondrion located at the head/tail junction

compared to ATP. In fish spermatozoa, nucleotide phosphate compounds other than the adenine derivatives such as GTP or UTP are not used because they are very poor substrates for the dynein-ATPase.

The AEC (adenylate energy charge) in a cell was defined by Atkinson (1968) as:

$$\mathrm{AEC}(\%) = 100 \times (([\mathrm{ADP}] + 2[\mathrm{ATP}])/2([\mathrm{AMP}] + [\mathrm{ADP}] + [\mathrm{ATP}]))$$

where [AMP], [ADP], and [ATP] are the intracellular concentration of each phosphagen or phospho-nucleotide. The value of AEC and the concentration of

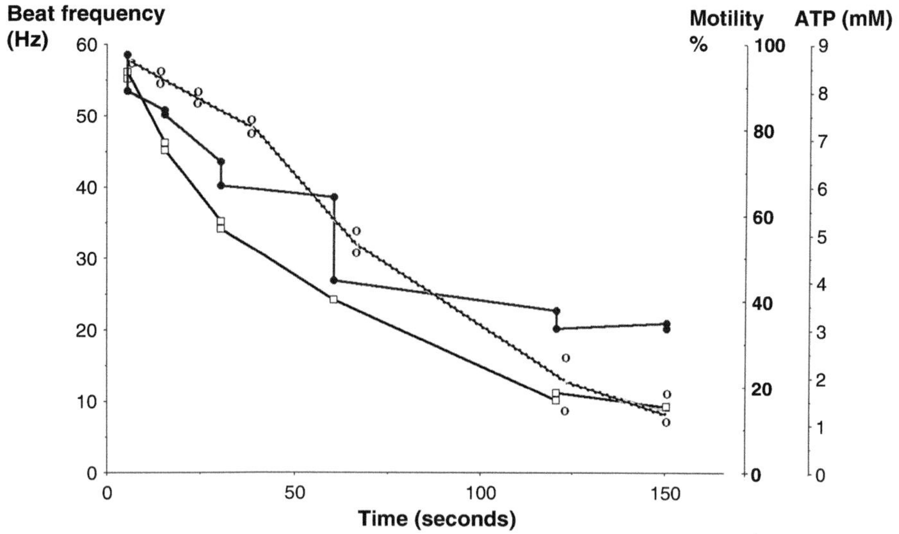

Fig. 4.5 Parallelism of the decrease during the motility period of three characteristics of fish spermatozoa (carp sperm in this example): intracellular ATP concentration, beat frequency, and percentage of motile cells. Motility % = full circles; Beat frequency = open circles; ATP concentration = squares. The abscissae represents the time elapsed during the motility period

ATP are considered as indicators of the quality of fish spermatozoa (Ingermann et al. 2003; Ziętara et al. 2004; Cosson 2012).

It appears in more and more publications that the energetic balance in cells is under control of AMP kinase (AMP-k). AMP-k function is that of a metabolic sensor as it takes in account the intracellular content of ATP, ADP, and AMP. As an example, the AMP-k is activated when the intracellular AMP and ATP concentrations increase. In its AMP-activated form and following its phosphorylation via the cAMP-dependent AMP-k-kinase, the AMP-k activates the metabolic ways producing ATP (glycolysis, fatty acids oxidation) while it inhibits the metabolic ways consuming ATP (cholesterol and triglycerides synthesis, lipolysis of adipocytes; reviewed by (Hardie et al. 2012).

These aspects demonstrate the importance of the evaluation of all the energetic phosphate compounds present in a cell such as a spermatozoon, ATP, ADP and AMP, as well as cAMP. There are examples in the literature of the role of AMP-k in vertebrate spermatozoa: in human (Hurtado de Llera et al. 2012), in goat (Zhu et al. 2018), in chicken (Nguyen et al. 2014), or in fish (Xu et al. 2016). In other fish species (but not their spermatozoa), there are not many publications in the literature about the presence of AMP-k, with the exception of Jibb and Richards (2008) in goldfish.

Several methodologies have been used to evaluate the concentration of the different metabolites involved in energetic balance: chemical methods involving chemo-luminescence of luciferin have historically been the most popular because of its cheapness and its high degree of sensitivity but its application is quite limited to ATP, with possible extension to ADP evaluation. More expensive and sophisticated are some biophysical techniques involving specialized technologies like detection of ^{31}P by RMN and that allows to measure not only ATP or ADP but also P_i as well as other compounds such as PCr; this was applied to trout (Robitaille et al. 1987), sea bass (Dreanno et al. 1999a), or turbot sperm (Dreanno et al. 2000). More recently, a new technique was developed that employs liquid chromatography coupled with high resolution mass spectrometry, having enough sensitivity to detect very low amounts of phospho-nucleotides and energy compounds present in fish spermatozoa (Fedorov et al. 2015, 2017).

These methods have led to evaluate energetic compounds in spermatozoa of different fish species, either prior to sperm motility activation or all along the motility period. In African catfish sperm, the presence of glycine, pyruvate plus lactate, or glucose allows a partial prevention of the decrease in ATP concentration and the AEC during in vitro storage for periods of one to several days (Ziętara et al. 2004). Regarding the situation in *Silurus glanis* spermatozoa during the motility period, some data were published by Billard et al. (1997).

In fish spermatozoa, most of ATP content is devoted to motility (Cosson 2010; Cosson et al. 2010), thus only a minor part of the energy is spent to maintain ionic gradient concentration across membrane and other "house-keeping" activity.

From results published in literature, it can be concluded that there is a close relationship between ATP content and sperm motility (as example, see Zilli et al. (2004)). Nevertheless, Dziewulska et al. (2012) did not observe a statistically significant correlation between the percentage of motile sperm and ATP concentration in case of the fresh sperm of wild Atlantic salmon (*Salmo salar* L.) before motility activation. In fish species, the total amount of ATP stored prior to activation (ATP per cell at activation) appears to be the factor limiting the duration of the motility period (at least in the studied species, e.g., six marine species reported by Cosson et al. (2008a)). This is also true considering the ATP store remaining present in fish sperm halfway in the motility period. In both cases, the lower this ATP store, the shorter is the motility duration (Cosson et al. 2008a; Cosson 2010). In addition, it was observed that during the process of turbot sperm aging, a decrease of the ATP content is occurring that correlates with a decrease of motility parameters (Suquet et al. 1998). Nevertheless, some exceptions are reported, in addition to the mentioned above Atlantic salmon and the other example is the cyprinid *Alburnus alburnus* in which no relationship was established between ATP and sperm motility parameters, including longevity (Lahnsteiner and Mansour 2010).

4.6.3 Concentration of Energetic Compounds in Relation with the Internal Volume of a Fish Spermatozoon

Most data on the energetic molecules in fish spermatozoa refer to quantity per a certain number of spermatozoa. For correct expression and understanding of its significance, one must refer quantity to volume if possible volume of one spermatozoon. The difficulty comes from the estimation of the value of the sperm cell volume, especially the water volume accessible inside the cell. Estimation of fish sperm cell volume can use microscopy images, either by high magnification optical microscopy or by electron microscopy. Both techniques lead to artifacts such as fixation damage and in any case, lead only to estimation of the volume by measuring the contour of cell, not the water volume per se. Several other methods have been used, including isotopic equilibration as example. In case of trout, cell volume was estimated as 0.16 μL per 10^7 sperm, that is, 1.6×10^{-6} μL per one sperm cell (Christen et al. 1987). In zebrafish, according to the results of Hagedorn et al. (2009), a simple calculation leads to a volume of $1.7–2.1 \times 10^{-6}$ μL per sperm cell. These volume estimations are in the same range as that of sea urchin spermatozoon or that of the human sperm (15–17 μm^3) (Laufer et al. 1977). One should remark that from the total water amount present in a cell, it is estimated that only 37% of water is osmotically active.

In addition, due to osmotic reaction of the fish sperm cell, the internal water volume is subjected to large variations. This is the case for carp spermatozoon as example where large changes of cell volume were measured (Perchec Poupard et al. 1997; Bondarenko et al. 2013). Consequently, a two- or threefold increase of the volume of water in a sperm cell will contribute to a corresponding decrease of ATP concentration by a dilution effect. These considerations allowed to revisit some previously published data. One should take into account the very small volume of cytoplasm surrounding the axoneme, because of the close contact of flagellar membrane with the periphery of the axonemal cylinder and very limited internal space inside the axoneme. This question is even complicated when considering the presence of bilateral folds (fins) of the flagellum membrane of fish sperm (Gillies et al. 2013). The osmolarity-driven reaction occurring at contact with external medium, such as freshwater as example, leads to water entry, and to the significant increase in volume of sperm cells in some fish species, which nevertheless could be tolerated, thanks to the membrane folds which allow to buffer the osmotic damage during the motility period (Perchec Poupard et al. 1997; Cosson 2012; Bondarenko et al. 2013). Solute molecules such as ATP will consequently have their concentration reduced independently of the metabolism occurring during motility. An opposite effect probably occurs in the conditions where the energy restocking is possible as explained in the next section, leading to an additional increase of the ATP concentration.

The osmotic effects on the ATP usage were explored in a publication of Perchec et al. (1996) where authors compared the ATP consumption of carp spermatozoa either in very low osmotic environment (pure water) or in a saline solution (around 150 mOsmol/kg). In pure water, they observe that the number of waves developed along the flagellum is lower, while the ATP consumption is lower and the motility duration is shorter due to the curling of the flagellum, a handicap observed in spermatozoa of other fish species at the end of the motility period (Cosson 2010). When exposed to drastic osmotic condition such as pure water, some abnormal blebs appear at the flagellar membrane level, with characteristics similar to damage observed for fish spermatozoa having underwent a freeze-thawing process. These features probably also contribute to affect the intracellular ATP stores.

Another aspect of the published evaluations of ATP (or other energetic compounds) concentration during the motility period of fish spermatozoa is that results take in account the total population of spermatozoa while is well known that during the fish sperm motility period, the percentage of motile sperm cells is gradually decreasing (see Fig. 4.4 and Cosson 2010, 2012). Therefore, the ATP concentration should be referred to the active (moving) population of spermatozoa. An example of such correction of the results is presented in Fig. 4.5, where ATP concentration was reestimated only by reference to the active and thus ATP-consuming portion of moving cells leads to a quite constant concentration of ATP all along the motility period. Similar observations hold for most other fish species mentioned above (Cosson, unpublished).

4.7 The Possible Restocking of Energy After the End of the Motility Period

There are many examples of the ability for fish spermatozoa to accomplish a second period of motility after they stopped at the end of the first motility period. As examples, reviving of sperm motility was described in trout (Christen et al. 1987)), in carp (Linhart et al. 2008a, b), in turbot (Dreanno et al. 1999b), or in sturgeon (Gatti et al. 1991; Dzyuba et al. 2013a).

Fish spermatozoa can sustain several rounds of activity provided they are exposed to a non-swimming solution after stopping their motility accomplished during a first motility period. A reloading of their ATP content occurs through respiration without ATP consumption because motility is not permitted in the non-swimming solution used for reloading conditions (Perchec Poupard et al. 1997). The publication of Linhart et al. (2008b) shows that several rounds of motility of carp spermatozoa can also be induced and at each round, the revived sperm are able to accomplish egg fertilization.

The possibility that some energetic compounds would be accessible to sperm in fish ovarian fluid in the vicinity of eggs was proposed by Turner and Montgomerie (2002) and by Elofsson et al. (2003b). This represents another form of restocking that would prevent a too fast exhaustion of the ATP stores as it occurs in free water (marine or freshwater).

Oppositely, the possibility of storage for very long period in female tracts was explained by Holt (2011) by the presence in the female fluid of some compound preventing the motility and in this way preserving the ATP stores.

4.8 The Hydrodynamic Aspects of Energy Consumption

The way micro-swimmers such as sperm cells move in water has been studied from a physical point of view since the pioneering studies of Sir Gray (1955) followed by many other specialists of hydrodynamics. A simple overview of this approach can be found in paper from Yates (1986) and in book edited by Cosson (2015). Movement needs for the spermatozoon to spend energy against viscosity and in this section we intend to compare by several aspects how biochemical energy in the form of ATP is converted in physical translation in the form of mechanical energy and tries to evaluate the yield of such transformation.

4.8.1 Motility Evaluation at the Sperm Population Level by CASA and at the Individual Sperm Flagellum Level

The translation of a sperm cell results from the flagellum activity; the evaluation of the translation efficiency comes from two groups of methods: either a follow-up of the head position as a function of time, mostly using successive video images (Inaba and Shiba 2018) or an analysis of the flagellum images obtained at high resolution and high frequency (for details see recent paper of Boryshpolets et al. 2018).

For evaluation of the flagellum efficiency in its production of the velocity of sperm translation, methods mostly rely on the early observations (Gray 1955) and the "Resistance Force Theory" (RFT) model of motility developed by Gray and Hancock (1955) and the prediction about the relationship between velocity and mechanical energy published in the pioneer papers of Holwill (1966, 1969, 1974).

The main power of the RFT is that the velocity of translation of a moving spermatozoon can be expressed by a formula using a restricted number of parameters accessible to experimental evaluation:

$$V_{\text{calc}} = 2f\pi^2b^2/\lambda\left\{1/1 + 4\pi^2b^2/\lambda^2 - \left(1 + 2\pi^2b^2/\lambda^2\right)^{1/2} \cdot 3a/n\lambda[(\text{lnd}/2\lambda)) + 1]\right\} \tag{1}$$

In Eq. 1, V_{calc} is the calculated velocity of propulsion of the spermatozoon, b is the amplitude of wave, λ is the length of wave, n is the number of simultaneous waves, f is the beat frequency of waves, a is the radius of head, and d is the radius of flagellum.

In a way similar to that leading to calculation of the velocity, V_{calc} in Eq. 1, from values of flagellar parameters, it is possible to evaluate the value of power (P) needed to sustain steady-state motility by Eq. 2 as it has been proposed by Holwill (1977) and Rikmenspoel (1966). It is remarkable that Eq. 2 for evaluation of P shows proportionality to the square of the frequency value, f^2 as well as proportionality to the square of the wave amplitude value, b^2.

$$P = 4\pi^3\mu f^2b^2L/\{0.62 - \ln(2\pi d/\lambda)\} \tag{2}$$

4.8.2 Energy Consumed by Flagella During Forwardly Swimming Fish Spermatozoa

Such evaluation mostly uses hydrodynamic laws of physics (Holwill 1982) and takes into account the properties of the environment of the swimming sperm: viscosity of the swimming solution or vicinity of the surface close to which sperm cell is swimming (Cosson et al. 2003; Boryshpolets et al. 2013) especially during the sperm egg encountering event (Ishimoto et al. 2016). Evaluation also takes account of fish sperm structural specificities: as an example, the results of Gillies et al. (2013) predict that a single fish spermatozoon develops a power of 0.0168 pW (picoWatt) when swimming close to a glass surface and 0.0145 pW far from any surface. For comparison, it is worth to remind that when 1 mole of ATP is transformed into ADP by hydrolysis, it releases 30.5 kJ, knowing that power (in Watts) is the ratio of energy (in Joules) to time (in seconds). In case of a swimming sturgeon spermatozoon as example, the ATP consumption is ranging 0.8×10^{-5} pmol per sperm per second. For comparison, an individual human spermatozoon swimming close to a surface has a power consumption about 32% higher (Gillies et al. 2009).

Regarding fertilization and the energetics of a fish sperm swimming in the vicinity of an egg, it was shown that the approach to the egg micropyle is needing less energy because sperm cells obey to guidance mechanisms (chemoattraction among others) rather than following a random trajectory (Cosson 2015).

4.8.3 Values of the Chemical and Physical Power Developed in Fish Sperm Flagella

Thermodynamic laws predict that 1 mole of ATP transformed into ADP by hydrolysis of the energetic bond of phosphate releases 30.5 kJ. For further calculation, let us take 1 pmol ATP $= 30.5 \times 10^{-12}$ kJ.

In case of a sturgeon spermatozoon (*Acipenser baeri*), the ATP consumption is 4 nmol/10^8 sperm/6 s or 0.8×10^{-5} pmol/sperm/s (Tsvetkova et al. 1996).

Knowing that:

$$\text{Power(in Watts)} = \text{Energy(in Joules)/Time(in seconds)} \tag{3}$$

the chemical power resulting from ATP hydrolysis is $0.8 \times 10^{-5} \times 30.5 \times 10^{-12}$ per s or 0.024 pW. This value is to compare the value of the physical power (0.0168–0.0145 pW) developed by a human sperm (Gillies et al. 2009), in addition, it was shown that within a single beat cycle, the punctual energy consumption varies up to tenfold. The power developed by a swimming fish spermatozoon, according to Gillies et al. (2009), can be calculated knowing the value of several parameters; the formula giving the cycle averaged power is:

$$P = \left(\mu f^2 L^3\right) p \tag{4}$$

where μ is viscosity, *f* is the beat frequency (BF) of the flagellum, *L* is its length and $0.4 < p < 0.6$ is depending on morphology. In Gillies et al. (2009), a viscosity of μ = 0.001 Pa·s was used in the simulations. At BF of 15.8 Hz, the physical power *P* spent by the sperm cell ranges 0.0102–0.0154 pW.

For a fish spermatozoon such as that of the sturgeon *Acipenser baeri* (Billard et al. 1999; Gillies et al. 2013), at a maximum BF of 55 Hz (or 45 Hz in Tsvetkova et al. (1996), the physical power *P* is ranging from 0.15 to 0.19 pW leading to a maximum velocity of 180 μm/s. The corresponding rate of ATP consumption for the same species in the same conditions leads to a value of 3.5 nmol/10^8 sperm per 4 s corresponding to an ATP rate of consumption of 1×10^{-5} pmol/spermatozoon/s, or 1.8×10^{-19} mol/spermatozoon/beat, which corresponds to energy and thus a chemical power of $P = 0.24$ pW. Values, according to Gillies et al. (2013) for *Acipenser baeri* spermatozoa, are over a period of the three beat cycles, of 0.19 pW if considering the presence of flagellar fins, while, for a flagellum devoid of fins, the power consumption is 0.15 pW, which is equivalent to 34×10^{-19} kJ/beat/spermatozoon.

In these conditions, the yield of energy conversion (ratio between the physical power output and the ATP input) for this fish sperm species is 18.8, a value that ranges the value that was estimated for sea urchin spermatozoon (Brokaw and Gibbons 1975).

For the sterlet sturgeon, *Acipencer ruthenus* (Tsvetkova et al. 1996; Fedorov et al. 2015) at the maximum BF (45 Hz) during the initial part of the motility period, the maximum measured velocity is 180 μm/s (175 μm/s in Alavi et al. (2008) and 150 μm/s in Prokopchuk et al. (2016)). In these conditions, the ATP hydrolysis rate is about 12 nmol/10^9 sperm/10 s, which represents an ATP consumption rate of 12×10^{-5} pmol/sperm/s. A similar calculation for shovel–nose sturgeon (Cosson et al. 2000), with a maximum BF of 48 Hz and a maximum velocity of 190 μm/s, the physical *P* ranges 0.114–0.144 pW. For paddle fish (Cosson et al. 2000), with a maximum BF of 45 Hz and a maximum velocity of 140 μm/s, a spermatozoon develops a power of 0.100–0.127 pW; the rate of ATP consumption was not evaluated in the two latter studies.

In the case of trout, evaluation from Robitaille et al. (1987), Cosson et al. (1991b) at 20 °C and Christen et al. (1987) at 15 °C lead to a maximum BF of 55 Hz, thus the corresponding physical power of 0.150–0.190 pW at maximum velocity of 220 μm/s; in these conditions, the ATP consumption rate was evaluated to 2.5–6 mM ATP/sperm/20 s (assuming a volume of 0.16 μL/10^7 sperm cells), corresponding to an ATP consumption rate of 4×10^{-5} pmol/s/spermatozoon or 7.2×10^{-19} mol/spermatozoon/beat. It is of interest to notice that, for the de-membranated spermatozoa of trout, the ATPase activity is ranging 2.27×10^{-5} pmol/spermatozoon/s or 1.25×10^{-19} mol/spermatozoon/beat, indicating that the axonemal flagellum is responsible of a major part of the ATP hydrolysis during motility of flagella.

For carp spermatozoa (Perchec et al. 1995), the maximum BF is 52 Hz and the maximum velocity is 135 μm/s, thus power developed by the flagellum is 0.133–0.169 pW, while the ATP hydrolysis rate is 8.5 nmol/10^8 spermatozoa/120 s corresponding to an ATP hydrolysis rate of $0.1 \cdot 10^{-5}$ pmol/s/spermatozoon, that is 1.9×10^{-19} mol/sperm/beat.

In the case of *S. glanis* (Billard et al. 1997), where spermatozoa are beating at a maximum BF of 38 Hz with a maximum velocity of 130 μm/s, the estimated power ranges 0.071–0.091 pW while the ATP consumption rate is 7.5 nmol/10^8 spermatozoa/20 s, corresponding to an ATP consumption rate of 0.375×10^{-5} pmol/s/spermatozoon or 1.0×10^{-19} mol/spermatozoon/beat.

For sea bass spermatozoa (Dreanno et al. 1999a), the maximum BF is 60 Hz and the maximum velocity is 110 μm/s; the power evaluation leads to 0.178–0.226 pW. In these conditions, the ATP consumption rate is 8 nmol/10^8 spermatozoa/10 s corresponding to an ATP consumption rate of 0.84×10^{-5} pmol/s/spermatozoon or 1.5×10^{-19} mol/spermatozoon/beat.

In case of turbot (Dreanno et al. 1999b), at a maximum BF of 57 Hz and a maximum velocity of 210 μm/s, the power developed by the flagellum is 0.161–0.204 pW, while in these conditions, the ATP consumption rate is 11 nmol/10^8 spermatozoa/10 s corresponding to an ATP hydrolysis rate of 1.1×10^{-5} pmol/s/spermatozoon or 0.616 pmol/spermatozoon/beat.

For hake, see Cosson et al. (2010), where authors develop considerations about ΔG_0 (see below) and ATP (see also (Groison et al. 2008, 2010a, b). For bluegill (Burness et al. 2005), the motility duration is 2 min, the value of BF is not mentioned but the maximum velocity is 120 μm/s. It is noticed that in this species, the ATP level remains very high for the 20 first seconds. During the first part of the motility period, the ATP consumption rate is 20 pmol/10^6 spermatozoa/10 s or 20×10^{-5} pmol/s/sperm.

For perch, Boryshpolets et al. (2009) have shown that the intracellular ATP concentration drops from 43.9 to 35 nmol/10^9 spermatozoa during the first 10 s of the motility period, meaning a rate of ATP decrease of about 0.9 nmol/s/10^9 spermatozoa. Given a flagellum beat frequency of 45 Hz (Alavi et al. 2015), this leads to an ATP consumption of 2×10^{-19} mol/spermatozoon/beat. During this early period, sperm velocity ranges 220 μm/s. Authors show that spermatozoa can sustain several episodes of motility with decreasing stepwise the external osmolarity. These results also show that ATP remains at higher concentration when activation occurs at higher osmolarity. This could indicate that ATP restocking occurs during the period the end of the first motility period and the initiation of the second motility period (see below).

For marine fish species, an evaluation of energetic aspects and motility characteristics was conducted comparatively in a broad diversity (35) of marine fish species by Ishijima (2012). Among other results, the author show that the power output is significantly related to the flagellar beat frequency and that swimming velocity is not related to the flagellum length. But this study does not provide information about the ATP input that sustains motility. The relationship between intracellular energetics and swimming capacities is also presented comparatively for spermatozoa of various fish species by Cosson et al. (2010). The review paper of Dzyuba et al. (2017a) also amply discusses the implications of energetic aspects in the motility and physiology of fish spermatozoa.

A tentative comparison of some of the above data about fish spermatozoa (in the species where necessary information is accessible) is presented in Table 4.1.

The above estimations were obtained from individual sperm where the swimming properties do not interfere between two neighbor sperm cells. In contrast, cooperative swimming where two sperm cells swim together was observed in many situations for spermatozoa of different fish species: such cooperative swimming mode was shown to save mechanical energy (Elfring and Lauga 2011).

4.9 The Thermodynamic Aspects of Fish Sperm Swimming at Different Temperatures

The knowledge of the effect of temperature on a biological function helps to predict the nature of the rate-limiting step in the series of reactions that control this function. In case of energetic utilization by fish spermatozoa, it is important to document the involvement of temperature in the motility process, especially considering that fish is a poikilotherm group of animals whose body temperature is adjusted by that of water where it lives and reproduces.

4.9.1 The Few Studies on Temperature Effects

Since the initial observations of Lindroth (1947), several studies have investigated the effect of the temperature of the swimming medium on the motility parameters of spermatozoa of different fish species either in the review papers of Alavi and Cosson (2005) and Dadras et al. (2017) or in different individual species by Lahnsteiner and Mansour (2012), as well as in carp by Dadras et al. (2016), in tilapia by Dzyuba et al. (2019), in hake compared to trout by Cosson et al. (2010) in which case difference between species was observed in relation to ATP store. In most species, the results show a positive relationship between motility parameters such as velocity or beat frequency and temperature: this is the case in carp (Dadras et al. 2016), in trout (Billard and Cosson 1992), in salmon and trout (Vladiĉ and Jätrvi 1997), in sole (Diogo 2010), in stickleback (Mehlis and Bakker 2014), in hake (Cosson 2008b, 2010), in cod (Cosson et al. 2008b), but this correlation was not observed in perch (Lahnsteiner 2011) or in sea bream (Lahnsteiner and Caberlotto 2012). In case of burbot sperm, results from Lahnsteiner et al. (1997) show, in this particular case, that temperature is a main factor controlling motility activation.

In case of trout sperm, the initial flagellar BF at 5°C is almost twice lower when compared to 25°C (Cosson et al. 1985); the former is physiological temperature of spawning for this species; the lower BF entails lower ATP consumption, i.e. its concentration remains high for longer period of time at low temperature (Cosson et al. 1995), as well as the concentration of cAMP remains elevated through whole

Table 4.1 Energetic and motility characteristics of a selected number of fish sperm species

Species	Wave number	Flagellar length, μm	Maximum velocity, μm/s	Max. beat frequency, Hz	ATP consumption rate, 10^{-19} mol/spermatozoon/beat	Physical power, pW
Sturgeon *Acipenser baeri*	4	49	180	45–55	1.8	0.15–0.19
Trout *Oncorhynchus mykiss*	4	43	220	55	7.2	0.17
Carp *Cyprinus carpio*	4	45	135	52	1.9	0.133–0.169
Turbot *Psetta maxima*	5	50	210	57	1.9	0.161–0.204
Wels *Silurus glanis*	4	55	130	38	1.0	0.071–0.091
Sea bass *Dicentrarchus labrax*	4.5	42	110	60	1.5	0.178–0.226

Note: Wave number in various fish species was evaluated at the earliest period of the motility period. Wave number represents the number of curvatures present along an active flagellum. The maximum velocity is measured during the earliest part of the motility period by CASA. The measurement of beat frequency is obtained by stroboscopic illumination. The ATP consumption rate is evaluated as the slope of the ATP hydrolysis versus time plot for sperm cells of each species during the earliest part of their motility period. The value of the physical power developed by the corresponding flagellum is predicted according to the calculation presented in the above paragraph motility period. Interestingly, the latter report shows that cAMP concentration rise occurs later than the onset of motility, while it was earlier believed to be a trigger of motility (Morisawa 1985) (as discussed above in this chapter).

In the studies of Cosson et al. (2008b, 2010), authors consider the results obtained regarding the relationship between temperature and intracellular ATP and made a comparison between trout and hake spermatozoa. Another study by Perchec et al. (1995) investigated the effect of temperature on carp sperm ATP content and show that ATP is consumed much faster at 20 °C that at 2 °C.

In fish, temperature can affect sperm motility at the membrane level as shown by the presence of thermosensitive ion channels (Majhi et al. 2013) as previously discussed in this chapter.

The thermodynamic aspects of the dynein ATPase were studied in many publications that are not detailed in this chapter showing that dynein obeys to enzymatic laws where the temperature parameter is largely contributing. This mainly explains susceptibility of flagellar swimming to temperature, even though membrane sensitivity contributes lightly to energy utilization in fish spermatozoa; in this respect, one can also consult the discussion part in the paper of Emri et al. (1998).

4.9.2 Temperature Effects

As mentioned, the general rule in many fish species is the lower the temperature, the longer the sperm motility period. This leads to an increase of the duration of sperm/egg approaching and thus rise the chances of fertilization when decreasing the temperature in the physiological limits of temperature.

In any thermodynamic system, including a living cell such as a spermatozoon, one defines a "state function" of its energetic situation that is formally expressed as ΔG (Gibbs energy), the "free" energy in the system, leading to the fundamental expression:

$$\Delta G_0 = \Delta H + T\Delta S$$

where ΔH is the enthalpy, T is the absolute temperature, and ΔS is the entropy. The relationship between ΔG_0 and ATP concentration was investigated by Cosson et al. (2010) in fish spermatozoa and which results are to compare with sea urchin spermatozoa (Silvester and Holwill 1965; Holwill 1969).

Another aspect, as discussed by Cosson et al. (2010), is that the ATP stored in a sperm cell is obviously related to the possible energy expanse of the same cell, expressed in terms of flagellum beat frequency (BF). When the BF was measured at different temperatures, the BF values were decreasing with decrease of temperature, this phenomenon is exemplified in the Fig. 5 of Cosson et al. (2010) that relates BF to temperature in two fish species, *O. mykiss* and *M. merluccius*. The data are presented in the graph as ln f/T versus T^{-1}, where T is the absolute temperature. Such graph is also known as an Arrhenius plot; it allows getting two important indices, the free energy (ΔH) and enthalpy (ΔS) of the ATP hydrolysis reaction

(Holwill and Silvester 1967), which controls the BF of sperm flagella (Gibbons 1981). Values of ΔH and ΔS can be deduced from an Arrhenius plot by measuring the slope and the intercept with the x-axis, leading to values of 10.945 kcal·mol^{-1} and 13.735 e.u. (entropy units), respectively, for *M. merluccius* spermatozoa; and similarly values of $\Delta H = 8.978$ kcal·mol^{-1} and $\Delta S = 12.940$ e.u. for *O. mykiss* spermatozoa. Knowing that $\Delta G = \Delta H - T{\cdot}\Delta S$, as above mentioned, one obtains ΔG values of 6.921 kcal mol^{-1} for *M. merluccius* and 5.760 kcal·mol^{-1} for *O. mykiss*. These values represent the minimal free energy necessary to accomplish the limiting reaction for optimal flagellar movement and which controls the beat frequency (BF).

From independent thermodynamic data, it is usually estimated that the total amount of free energy delivered by the reaction of hydrolysis of ATP has a value of $\Delta G = -31$ kJ·mol^{-1} or $= -7.4$ kcal·mol^{-1}. It is worth comparing the energy needed for movement ($\Delta G = 6.921$ and 5.760 kcal·mol^{-1} in *M. merluccius* and *O. mykiss*, respectively) with the total available energy for movement which corresponds to $\Delta G = -7.4$ kcal·mol^{-1} (theoretical energy content from ATP). Therefore, it appears that most of this available energy (89% and 78% in *M. merluccius* and *O. mykiss*, respectively) must be devoted to movement, which may severely impair other ATP demanding functions such as membrane ionic pumps; consequently, this ATP restriction probably results in an intracellular Ca^{2+} ions accumulation, leading to and possibly explaining the flagellar asymmetry and tracks circling as discussed elsewhere, or even premature arrest of motility of trout sperm. Ca^{2+} control of fish sperm circling while swimming close to a surface was also observed in other species such as sea urchin (Gray 1955; Gibbons 1962).

4.10 The Balance Sheet Total Between Respiration, ATP Storage, General Cell Usage of ATP, and Mechanical Power Developed for Swimming

4.10.1 Physical Aspects

According to physical principles applied to a swimming spermatozoon, the moments due to external viscous forces (M_f), internal shear (M_s), and internal elastic (M_e) should balance each other in such a way that: $M_f + M_s + M_e = 0$ (Rikmenspoel 1978). This relates to the balance of energy in the same spermatozoon so as the energy generated by internal forces (motor) should balance the viscous force offered by the medium that resist the displacement.

The forces developed by a flagellum that interacts with the fluid environment are mostly viscous while the inertial forces are negligible. This results from consideration of the Reynolds number that provides a rough evaluation of the ratio between viscous to inertial forces: $R_n = l{\cdot}v{\cdot}d/n$ where l is the length of the spermatozoon, v denotes its velocity, d is its density, and n the viscosity of the surrounding medium. These notions are basic for the development of the RFT theory initially elaborated by Gray and Hancock (1955) which allows to evaluate the swimming velocity and the physical energy required by a sperm flagellum when efficiently moving.

On another hand, the total number of dynein molecules per total length of a ciona sperm flagellum (47.5 μm long) was estimated by Brokaw and Benedict (1968), i.e. about 0.67×10^{-19} mol/spermatozoon, a value similar to that a fish sperm flagellum of the same length. The same authors estimated that the movement-coupled oxidative metabolism could generate 1.6–2.5 molecules ATP/dynein/beat. Knowing that respiration generates 6 moles ATP per mole of O_2 this leads to 1.6×10^{-19} mol ATP/beat/spermatozoon.

For comparison, as an example in turbot sperm (Dreanno et al. 1999b), the flagellum length is 43 μm and, at the earliest part of the motility period, the beat frequency ranges 60 Hz, while the respiration rate is 125 nmol $O_2/10^9$ spermatozoa/min or 0.35×10^{-19} mol O_2/beat/spermatozoon, which generates 2.1 mol ATP/beat/spermatozoon.

According to Brokaw and Benedict (1968), the power developed by a swimming sea urchin (*Lytechinus*) spermatozoon is 3×10^{-14} J/s/spermatozoon. At a beat frequency of 30 Hz, if $W = 14 \times 10^{-19}$ kJ/spermatozoon/beat, this means $W = 14 \times 10^{-16}$ J/spermatozoon. When expressed per second, this means 420×10^{-16} J/s/spermatozoon or 4.2×10^{-14} J/s/spermatozoon as it was evaluated by Brokaw and Benedict (1968).

4.11 Importance of Sperm Energetics at the Fertilization Step

As discussed by Burness et al. (2005), the fertilization success is positively related to sperm swimming speed (Kime et al. 2001) in case of fish and the swimming speed is depending on the amount of energy available (Jeulin and Soufir 1992) in human sperm. Because both swimming fast and for long periods of time require energy, a trade-off between these sperm traits is expected (Ball and Parker 1996; Levitan 2000). Sperm swimming speed was also demonstrated to be positively related to sperm flagellum length (Gomendio et al. 1991), while an increased energy is required to overcome the hydrostatic drag of a long flagellum further compromises sperm longevity (Stockley et al. 1997) because of a higher amount of consumed ATP.

Fertilization is a rapid event in fish: as this example in bluegill, the majority of a female's eggs are fertilized within 5–10 s of ejaculation (Burness et al. 2005). Such a rapid fertilization process probably places a priority on high initial sperm swimming speeds, rather than on motility longevity.

In externally fertilizing fish, a positive relationship between fertilization success and sperm swimming speed has been predicted theoretically (Ball and Parker 1996) and demonstrated empirically (Lahnsteiner et al. 1998). Theoretically, a longer flagellum generates greater forces and concomitantly higher swimming speeds than a short flagellum (Katz and Drobnis 1990; Gomendio et al. 1991). The increased hydrostatic drag of a long flagellum may, however, result in a decreased sperm longevity (Stockley et al. 1997), presumably due to the faster depletion of energy stores. A recent study on Atlantic salmon has shown that sperm with long flagella had a decreased longevity but no relationship between sperm tail length and swimming speed could be detected (Gage et al. 2002).

According to Burness et al. (2005), the single best predictor of variation in sperm ATP levels is the CPK (Creatine phosphokinase) activity; individuals with high CPK activity possess higher levels of ATP at 20 s post activation. It was shown that sperm motility depends on both ATP and PCr (Kamp et al. 1996; Saudrais et al. 1998, in trout sperm). PCr is acting in synergy as a shuttle of enzymatic activities, transporting energy-rich phosphate molecules from the mitochondria, where ATP is produced, to the dynein ATPase where ATP is consumed (Kamp et al. 1996), as discussed above in this chapter.

In addition to CPK, bluegill sperm with high levels of ATP at 20 s post activation also had high activities of a mitochondrial marker enzyme. This suggests that aerobically produced ATP may play a role in sperm motility (Burness et al. 2005).

For efficient fish egg fertilization, an important factor is the distance covered by sperm cells to reach the egg micropyle, for which task, the sperm should spend most of its previously stored energy (energy cost). From sperm velocity data and motility duration, a gross estimation of the distance covered by a fish spermatozoon is 2.3 mm in sea bass (Dreanno et al. 1999c), 12 mm in turbot (Chauvaud et al. 1995), 14 mm in cod (Cosson et al. 2008b), 11 mm in hake (Cosson et al. 2008b), 10 mm in tuna, and 9 mm in halibut (Billard et al. 1993). Egg size of the mentioned marine fish ranges 1–3 mm diameter, which value mostly leads to similar values in terms of the distance covered by a fish spermatozoon. Thus, any fish spermatozoon should be delivered in the close proximity of the oocyte so as to reach the corresponding egg micropyle and get a chance to fertilize.

4.12 The Deleterious Effects of Cryopreservation Affect the Energetic Levels

Defects appearing in sperm cells after cryotreatments are well documented in literature and the present book shows the diversity of such effects in case of fish spermatozoa. As an indication, the appearance of structurally abnormal membranous blebs can be detected by dynamic morphology (Billard et al. 2000) assessment of sturgeon spermatozoa after freeze-thawing treatment. As a consequence, energetic metabolism, among other functions, is clearly affected in freeze-thawed fish spermatozoa. Results of Dzyuba et al. (2010) show that a spontaneous activation of sperm motility appears during the freezing process in carp and perch and among five different fish species: mostly velocity of spermatozoa is affected due to the contact of sperm cells with areas of "pure" water appearing transiently during the melting of ice crystals in the surrounding freezing medium. In the stated paper, even though authors did not provide indication about the decrease of the internal ATP store, it is suspected that osmotic perturbations led to a partial exhaustion of the energetic stores, due to ATP and other energetic substrates leaking through the damaged spermatozoon membrane.

Osmotic effects occurring during cryopreservation were also described by Dzyuba et al. (2013b) showing that pre-exposure of carp sperm to a medium of moderate osmolarity (200 mOsm/kg) prevented partially the freezing damages and improved fertilizing ability: this could be due to induced volume changes such as shrinking of sperm cells but there is no internal ATP concentration measurement in this study.

According to a recent study (Horokhovatskyi et al. 2018), the fraction of sturgeon sperm that survived the freezing process can be selected by centrifugation on a Percoll gradient and demonstrate no change in their protein composition, probably reflecting their intactness. Even if these results provide no indication on the ATP content of selected spermatozoa, motility metrics predict that no change of energetic content occurs in the selected fraction of "native" spermatozoa. Nevertheless, recent results of Xin et al. (2018) show that in sturgeon (*A. ruthenus*) sperm there are proteome alterations after cryopreservation: authors detect several differences in protein profiles of fresh and cryopreserved sturgeon sperm, some affecting crucial energy metabolism enzymes such as the mitochondrial ATP synthase. Similar observations were published by Li et al. (2010) showing impact of cryodamage on proteome of carp spermatozoa. Membrane damage caused by cryotreatment was also visualized by video-microscopy of sturgeon spermatozoa (Tsvetkova et al. 1996) and probably participates to the energy (ATP and PCr) leaking out of the fish sperm cells that limits their swimming ability. In this respect, it is worth to mention results published on cryopreservation of halibut sperm (Billard et al. 1993) showing that the presence of external ATP in the freezing medium significantly improves the quality of cryopreserved sperm, leading to suspect that during the freeze-thawing process membrane permeability to ATP is enhanced.

4.13 Conclusion

By the end, this review emphasizes the prime role of energy evaluation in the frame of cryobanking of fish spermatozoa (Martínez-Páramo et al. 2017) and more generally in the understanding of the physiology of the male fish gametes that are intended to be used for artificial propagation (Fauvel et al. 2010). Nowadays, the parameters used for quantitative estimation of the fish sperm quality before and after application of cryotechnologies are more and more varied and sophisticated (Cosson 2016), among which measurement of sperm energetics resources emerges as crucial. More attention should be devoted to the sperm energetic factors in future works dealing with improvement of cryotechniques.

References

Agarwal A, Allamaneni SS (2004) Role of free radicals in female reproductive diseases and assisted reproduction. Reprod Biomed Online 9:338–347. https://doi.org/10.1016/S1472-6483(10)62151-7

Aitken RJ, De Iuliis GN, Finnie JM et al (2010) Analysis of the relationships between oxidative stress, DNA damage and sperm vitality in a patient population: development of diagnostic criteria. Hum Reprod 25:2415–2426. https://doi.org/10.1093/humrep/deq214

Alavi SMH, Cosson J (2005) Sperm motility in fishes. I. Effects of temperature and pH: a review. Cell Biol Int 29:101–110. https://doi.org/10.1016/j.cellbi.2004.11.021

Alavi SMH, Cosson J (2006) Sperm motility in fishes. (II) Effects of ions and osmolality: a review. Cell Biol Int 30:1–14. https://doi.org/10.1016/j.cellbi.2005.06.004

Alavi SMH, Rodina M, Cosson J et al (2008) Roles of extracellular Ca2+ and pH on motility and flagellar waveform parameters in sturgeon spermatozoa. Cybium 32:124–126

Alavi SMMH, Rodina M, Viveiros ATM et al (2009) Effects of osmolality on sperm morphology, motility and flagellar wave parameters in northern pike (Esox lucius L.). Theriogenology 72:32–43

Alavi SMH, Ciereszko A, Hatef A et al (2015) Sperm morphology, physiology, motility, and cryopreservation in Percidae. In: Biology and culture of Percid fishes. Springer, Dordrecht, pp 163–191

Alavioon G, Hotzy C, Nakhro K et al (2017) Haploid selection within a single ejaculate increases offspring fitness. Proc Natl Acad Sci U S A 114:8053–8058. https://doi.org/10.1073/pnas.1705601114

Atkinson DE (1968) Energy charge of the adenylate pool as a regulatory parameter. Interaction with feedback modifiers. Biochemistry 7:4030–4034. https://doi.org/10.1021/bi00851a033

Baccetti B, Gibbons BH, Gibbons IR (1989) Bidirectional swimming in spermatozoa of tephritid flies. J Submicrosc Cytol Pathol 21:619–625

Ball MA, Parker GA (1996) Sperm competition games: external fertilization and "adaptive" infertility. J Theor Biol 180:141–150. https://doi.org/10.1006/JTBI.1996.0090

Bergé P, Volochine B, Billard R, Hamelin A (1967) [Demonstration of the characteristic movement of living microorganisms owing to the study of the inelastic diffusion of light]. C R Acad Sci Hebd Seances Acad Sci D 265:889–92

Billard R (1970) Ultrastructure comparée de spermatozoïdes de quelques poissons téleostéens. In: Comparative spermatology. Academia Nazionale dei Lincei

Billard R (1986) Spermatogenesis and spermatology of some teleost fish species. Reprod Nutr Dév 26:877–920. https://doi.org/10.1051/rnd:19860601

Billard R (1990) La motilité du spermatozoide de poisson: aspects énergétiques. In: 5a Jornadas Internacionales de Reproducción animal. Zaragoza, España, pp 163–186

Billard R, Breton B (1970) Modifications ultrastructurales et cytochimiques des spermatozoïdes après dilution chez les poissons d'eau douce. In: Soc Fr Micros Electron, VIIe Congr Inter Micros Electron Grenoble, pp 632–637

Billard R, Cosson MP (1990) The energetics of fish sperm motility. In: Gagnon C (ed) Controls of sperm motility: biological and clinical aspects. CRC Press, Boca Raton, pp 153–173

Billard R, Cosson MP (1992) Some problems related to the assessment of sperm motility in freshwater fish. J Exp Zool 261:122–131. https://doi.org/10.1002/jez.1402610203

Billard R, Menezo Y (1984) The amino acid composition of rainbow trout (Salmo gairdneri) seminal fluid and blood plasma: a comparison with carp (Cyprinus carpio). Aquaculture 41:255–258. https://doi.org/10.1016/0044-8486(84)90288-6

Billard R, Ginsburg AS, Escaffre AM (1973) La spermiogenèse et le spermatozoïde d'Anguilla anguilla L. Étude ultrastructurale. Ann Biol Anim Biochim Biophys 13:523–534

Billard R, Christen R, Cosson MP et al (1986) Biology of the gametes of some teleost species. Fish Physiol Biochem 2:115–120. https://doi.org/10.1007/BF02264079

Billard R, Cosson J, Crim LW (1993) Motility of fresh and aged halibut sperm. Aquat Living Resour 6:67–75. https://doi.org/10.1051/alr:1993008

Billard R, Linhart O, Fierville F, Cosson J (1997) Motility of European catfish Silurus glanis spermatozoa in testes and in milt. Pol Arch Hydrobiol 44:115–122

Billard R, Cosson J, Fierville F et al (1999) Motility analysis and energetics of the Siberian sturgeon Acipenser baerii spermatozoa. J Appl Ichthyol 15:199–203. https://doi.org/10.1111/j.1439-0426.1999.tb00234.x

Billard R, Cosson J, Linhart O (2000) Changes in the flagellum morphology of intact and frozen/thawed Siberian sturgeon Acipenser baerii (Brandt) sperm during motility. Aquac Res 31:283–287. https://doi.org/10.1046/j.1365-2109.2000.00423.x

Boccaletto P, Siddique MAM, Cosson J (2018) Proteomics: a valuable approach to elucidate spermatozoa post-testicular maturation in the endangered Acipenseridae family. Anim Reprod Sci 192:18–27. https://doi.org/10.1016/J.ANIREPROSCI.2018.03.033

Boitano S, Omoto CK (1991) Membrane hyperpolarization activates trout sperm without an increase in intracellular pH. J Cell Sci 98(Pt 3):343–349

Bondarenko O, Dzyuba B, Cosson J et al (2013) Volume changes during the motility period of fish spermatozoa: interspecies differences. Theriogenology 79:872–881. https://doi.org/10.1016/J.THERIOGENOLOGY.2013.01.005

Bondarenko O, Dzyuba B, Cosson J et al (2014) The role of Ca2+ and Na+ membrane transport in brook trout (Salvelinus fontinalis) spermatozoa motility. Fish Physiol Biochem 40:1417–1421. https://doi.org/10.1007/s10695-014-9936-5

Bondarenko O, Dzyuba B, Rodina M, Cosson J (2017) Role of Ca2+ in the IVM of spermatozoa from the sterlet *Acipenser ruthenus*. Reprod Fertil Dev 29:1319–1328

Boryshpolets S, Dzyuba B, Stejskal V, Linhart O (2009) Dynamics of ATP and movement in Eurasian perch (Perca fluviatilis L.) sperm in conditions of decreasing osmolality. Theriogenology 72:851–859

Boryshpolets S, Cosson J, Bondarenko V et al (2013) Different swimming behaviors of sterlet (Acipenser ruthenus) spermatozoa close to solid and free surfaces. Theriogenology 79:81–86. https://doi.org/10.1016/j.theriogenology.2012.09.011

Boryshpolets S, Kholodnyy V, Cosson J, Dzyuba B (2018) Fish sperm motility analysis: the central role of the flagellum. Reprod Fertil Dev 30:833–841. https://doi.org/10.1071/RD17478

Bré MH, Redeker V, Quibell M et al (1996) Axonemal tubulin polyglycylation probed with two monoclonal antibodies: widespread evolutionary distribution, appearance during spermatozoan maturation and possible function in motility. J Cell Sci 109(Pt 4):727–738

Brokaw CJ (1965) Non-sinusoidal bending waves of sperm flagella. J Exp Biol 43:155–169

Brokaw CJ, Benedict B (1968) Mechanochemical coupling in flagella. II Effects of viscosity and thiourea on metabolism and motility of Ciona spermatozoa. J Gen Physiol 52:283–299. https://doi.org/10.1085/JGP.52.2.283

Brokaw CJ, Gibbons IR (1975) Mechanisms of movement in flagella and cilia. In: Wu TY, Brokaw CJ, Brennen CE (eds) Swimming and flying in nature. Plenum Press, New York, pp 89–126

Bucci D, Isani G, Spinaci M et al (2010) Comparative immunolocalization of GLUTs 1, 2, 3 and 5 in boar, stallion and dog spermatozoa. Reprod Domest Anim 45:315–322. https://doi.org/10.1111/j.1439-0531.2008.01307.x

Burness G, Moyes CD, Montgomerie R (2005) Motility, ATP levels and metabolic enzyme activity of sperm from bluegill (Lepomis macrochirus). Comp Biochem Physiol A Mol Integr Physiol 140:11–17. https://doi.org/10.1016/J.CBPB.2004.09.021

Butts IAE, Prokopchuk G, Kašpar V et al (2017) Ovarian fluid impacts flagellar beating and biomechanical metrics of sperm between alternative reproductive tactics. J Exp Biol 220:2210–2217. https://doi.org/10.1242/jeb.154195

Cabrita E, Martínez-Páramo S, Gavaia PJ et al (2014) Factors enhancing fish sperm quality and emerging tools for sperm analysis. Aquaculture 432:389–401. https://doi.org/10.1016/j.aquaculture.2014.04.034

Chauvaud L, Cosson J, Suquet M, Billard R (1995) Sperm motility in turbot, Scophthalmus marimus: initiation of movement and changes with time of swimming characteristics. Environ Biol Fish 43:341–349. https://doi.org/10.1007/BF00001167

Chauvigné F, Zapater C, Cerda J (2011) Role of aquaporins during teleost gametogenesis and early embryogenesis. Front Physiol 2:66. https://doi.org/10.3389/fphys.2011.00066

Chauvigné F, Boj M, Vilella S et al (2013) Subcellular localization of selectively permeable aquaporins in the male germ line of a marine teleost reveals spatial redistribution in activated spermatozoa. Biol Reprod 89(2):37. https://doi.org/10.1095/biolreprod.113.110783

Chen DTN, Heymann M, Fraden S et al (2015) ATP consumption of eukaryotic flagella measured at a single-cell level. Biophys J 109:2562–2573. https://doi.org/10.1016/J.BPJ.2015.11.003

Cherr GN, Morisawa M, Vines CA et al (2008) Two egg-derived molecules in sperm motility initiation and fertilization in the Pacific herring (Clupea pallasi). Int J Dev Biol 52:743–752. https://doi.org/10.1387/ijdb.072566gc

Christen R, Schackmann RW, Shapiro BM (1983) Metabolism of sea urchin sperm. Interrelationships between intracellular pH, ATPase activity, and mitochondrial respiration. J Biol Chem 258:5392–5399

Christen R, Gatti J-L, Billard R (1987) Trout sperm motility. The transient movement of trout sperm is related to changes in the concentration of ATP following the activation of the flagellar movement. Eur J Biochem 166:667–671. https://doi.org/10.1111/j.1432-1033.1987.tb13565.x

Ciereszko A, Dabrowski K, Kucharczyk D et al (1999) The presence of uric acid, an antioxidantive substance, in fish seminal plasma. Fish Physiol Biochem 21:313–315. https://doi.org/10.1023/A:1007886121663

Clark EM, Aronson CR (1951) Sexual behavior in the guppy Lebisteus reticulatus (Peters). Zoologica 36:49–66

Cosson J (1996) A moving image of flagella: news and views on the mechanisms involved in axonemal beating. Cell Biol Int 20(2):83–94. https://doi.org/10.1006/cbir.1996.0012

Cosson J (2008a) The motility apparatus of fish spermatozoa. In: Alavi SMH, Cosson J, Coward K, Rafiee G (eds) Fish spermatology. Alpha Science, Oxford, pp 281–316

Cosson J (2008b) Methods to analyse the movements of fish spermatozoa and their flagella. In: Alavi SMH, Cosson J, Coward K, Rafiee G (eds) Fish spermatology. Alpha Science International Ltd, Oxford, pp 63–102

Cosson J (2010) Frenetic activation of fish spermatozoa flagella entails short-term motility, portending their precocious decadence. J Fish Biol 76:240–279. https://doi.org/10.1111/j.1095-8649.2009.02504.x

Cosson JJ (2012) ATP, the sperm movement energizer. In: Kuestler E, Traugott G (eds) Adenosine triphosphate: chemical properties, byosinthesis and functions in cells. Nova Publisher Inc., Hauppauge, pp 1–46

Cosson JJ (ed) (2015) Flagellar mechanics and sperm guidance. Bentham Science Publishers, Sharjah

Cosson J (2016) Flagella parameters used as descriptors of fish spermatozoa motility. Anim Reprod Sci 169:128–129. https://doi.org/10.1016/j.anireprosci.2016.03.078

Cosson M-P, Billard R, Gatti JL, Christen R (1985) Rapid and quantitative assessment of trout spermatozoa motility using stroboscopy. Aquaculture 46:71–75. https://doi.org/10.1016/0044-8486(85)90178-4

Cosson MP, Billard R, Letellier L (1989) Rise of internal Ca2+ accompanies the initiation of trout sperm motility. Cell Motil Cytoskeleton 14:424–434. https://doi.org/10.1002/cm.970140312

Cosson M-P, Cosson J, Billard R (1991a) Synchronous triggering of trout sperm is followed by an invariable set sequence of movement parameters whatever the incubation medium. Cell Motil Cytoskeleton 20:55–68. https://doi.org/10.1002/cm.970200107

Cosson MP, Cosson J, Billard R (1991b) cAMP dependence of movement initiation in intact and demembranated trout spermatozoa. Bull Inst Zool Acad Sin 16:263–266

Cosson M-P, Cosson J, Andrè F, Billard R (1995) cAMP/ATP relationship in the activation of trout sperm motility: their interaction in membrane-deprived models and in live spermatozoa. Cell Motil Cytoskeleton 31:159–176. https://doi.org/10.1002/cm.970310208

Cosson J, Billard R, Gibert C et al (1999a) Ionic factors regulating the motility of fish sperm. In: Gagnon C (ed) The male gamete: from basic to clinical applications. Cache River Press, Vienna, pp 161–186

Cosson J, Dreanno C, Billard R et al (1999b) Regulation of axonemal wave parameters of fish spermatozoa by ionic factors. In: Gagnon C (ed) The male gamete: from basic knowledge to clinical applications. Cache River Press, Vienna, pp 161–186

Cosson J, Linhart O, Mims SD et al (2000) Analysis of motility parameters from paddlefish and shovelnose sturgeon spermatozoa. J Fish Biol 56:1348–1367. https://doi.org/10.1111/j.1095-8649.2000.tb02148.x

Cosson J, Huitorel P, Gagnon C (2003) How spermatozoa come to be confined to surfaces. Cell Motil Cytoskeleton 54:56–63. https://doi.org/10.1002/cm.10085

Cosson J, Groison A-L, Suquet M et al (2008a) Studying sperm motility in marine fish: an overview on the state of the art. J Appl Ichthyol 24:460–486. https://doi.org/10.1111/j.1439-0426.2008.01151.x

Cosson J, Groison AL, Suquet M, Fauvel C (2008b) Motility characteristics of spermatozoa in cod (Gadus morhua) and hake (Merluccius merluccius). Cybium 32:176–177

Cosson J, Groison A-L, Fauvel C, Suquet M (2010) Description of hake (Merlucius merlucius) spermatozoa: flagellar wave characteristics and motility parameters in various situations. J Appl Ichthyol 26:644–652. https://doi.org/10.1111/j.1439-0426.2010.01563.x

Coward K, Bromage NR, Hibbitt O, Parrington J (2002) Gamete physiology, fertilization and egg activation in teleost fish. Rev Fish Biol Fish 12:33–58. https://doi.org/10.1023/A:1022613404123

Dadras H, Dzyuba V, Cosson J et al (2016) The in vitro effect of temperature on motility and antioxidant response of common carp Cyprinus carpio spermatozoa. J Therm Biol 59:64–68. https://doi.org/10.1016/j.jtherbio.2016.05.003

Dadras H, Dzyuba B, Cosson J et al (2017) Effect of water temperature on the physiology of fish spermatozoon function: a brief review. Aquac Res 48:729–740. https://doi.org/10.1111/are.13049

De Iuliis GN, Newey RJ, King BV, Aitken RJ (2009) Mobile phone radiation induces reactive oxygen species production and DNA damage in human spermatozoa in vitro. PLoS One 4:e6446. https://doi.org/10.1371/journal.pone.0006446

de Lamirande E, O'Flaherty C (2008) Sperm activation: role of reactive oxygen species and kinases. Biochim Biophys Acta 1784:106–115. https://doi.org/10.1016/J.BBAPAP.2007.08.024

de Lamirande E, Leclerc P, Gagnon C (1997) Capacitation as a regulatory event that primes spermatozoa for the acrosome reaction and fertilization. Mol Hum Reprod 3:175–194. https://doi.org/10.1093/molehr/3.3.175

de Souza APB, Schorr-Lenz ÂM, Lucca F, Bustamante IC (2017) The epididymis and its role on sperm quality and male fertility. Anim Reprod 14:1234–1244. https://doi.org/10.21451/1984-3143-AR955

Diogo P (2010) Sperm motility in Solea senegalensis: effect of temperature, salinity, pH and ovarian fluid

Dreanno C, Suquet M, Quemener L et al (1997) Cryopreservation of turbot (Scophthalmus maximus) spermatozoa. Theriogenology 48:589–603. https://doi.org/10.1016/S0093-691X(97)00276-8

Dreanno C, Cosson J, Suquet M et al (1999a) Effects of osmolality, morphology perturbations and intracellular nucleotide content during the movement of sea bass (Dicentrarchus labrax) spermatozoa. Reproduction 116:113–125. https://doi.org/10.1530/jrf.0.1160113

Dreanno C, Cosson J, Suquet M et al (1999b) Nucleotide content, oxydative phosphorylation, morphology, and fertilizing capacity of turbot (Psetta maxima) spermatozoa during the motility period. Mol Reprod Dev 53:230–243. https://doi.org/10.1002/(SICI)1098-2795(199906)53:2<230::AID-MRD12>3.0.CO;2-H

Dreanno C, Suquet M, Fauvel C et al (1999c) Effect of the aging process on the quality of sea bass (Dicentrarchus labrax) semen. J Appl Ichthyol 15:176–180

Dreanno C, Seguin F, Cosson J et al (2000) 1H-NMR and31P-NMR analysis of energy metabolism of quiescent and motile turbot (Psetta maxima) spermatozoa. J Exp Zool 286:513–522. https://doi.org/10.1002/(SICI)1097-010X(20000401)286:5<513::AID-JEZ9>3.0.CO;2-5

Dumorné K, Valdebenito I, Contreras P et al (2018) Effect of pH, osmolality and temperature on sperm motility of pink cusk-eel (Genypterus blacodes, (Forster, 1801)). Aquacult Rep 11:42–46

Dziewulska K, Rzemieniecki A, Baranowska-Bosiacka I et al (2012) The parameters of energetic status of the wild Atlantic salmon (Salmo salar L.) spermatozoa. Aquac Res 43:139–148. https://doi.org/10.1111/j.1365-2109.2011.02833.x

Dzyuba B, Cosson J (2001) Cryopreservation of demembranated carp spermatozoa. Cryobiology 43:338–339

Dzyuba V, Cosson J (2014) Motility of fish spermatozoa: from external signaling to flagella response. Reprod Biol 14:165–175. https://doi.org/10.1016/j.repbio.2013.12.005

Dzyuba B, Boryshpolets S, Rodina M et al (2010) Spontaneous activation of spermatozoa motility by routine freeze-thawing in different fish species. J Appl Ichthyol 26:720–725. https://doi.org/10.1111/j.1439-0426.2010.01553.x

Dzyuba B, Cosson J, Boryshpolets S et al (2013a) Motility of sturgeon spermatozoa can sustain successive activations episodes. Anim Reprod Sci 138:305–313. https://doi.org/10.1016/J.ANIREPROSCI.2013.02.011

Dzyuba B, Cosson J, Yamaner G et al (2013b) Hypotonic treatment prior to freezing improves cryoresistance of common carp (Cyprinus carpio L.) spermatozoa. Cryobiology 66:192–194. https://doi.org/10.1016/j.cryobiol.2012.12.003

Dzyuba B, Cosson J, Boryshpolets S et al (2014a) In vitro sperm maturation in sterlet, Acipenser ruthenus. Reprod Biol 14:160–163. https://doi.org/10.1016/j.repbio.2014.01.003

Dzyuba V, Dzyuba B, Cosson J et al (2014b) The antioxidant system of sterlet seminal fluid in testes and Wolffian ducts. Fish Physiol Biochem 40:1731–1739. https://doi.org/10.1007/s10695-014-9963-2

Dzyuba V, Dzyuba B, Cosson J, Rodina M (2016) Enzyme activity in energy supply of spermatozoon motility in two taxonomically distant fish species (sterlet Acipenser ruthenus, Acipenseriformes and common carp Cyprinus carpio, Cypriniformes). Theriogenology 85:567–574. https://doi.org/10.1016/j.theriogenology.2015.09.040

Dzyuba B, Bondarenko O, Fedorov P et al (2017a) Energetics of fish spermatozoa: the proven and the possible. Aquaculture 472:60–72. https://doi.org/10.1016/j.aquaculture.2016.05.038

Dzyuba B, Cosson J, Dzyuba V et al (2017b) Sperm maturation in sturgeon (Actinopterygii, Acipenseriformes): a review. Theriogenology 97:134–138. https://doi.org/10.1016/J.THERIOGENOLOGY.2017.04.034

Dzyuba B, Legendre M, Baroiller JF, Cosson J (2019) Sperm motility of the Nile tilapia (Oreochromis niloticus): effects of temperature on the swimming characteristics. Anim Reprod Sci 202:65–72. https://doi.org/10.1016/j.anireprosci.2019.01.010

Elfring GJ, Lauga E (2011) Passive hydrodynamic synchronization of two-dimensional swimming cells. Phys Fluids 23:011902. https://doi.org/10.1063/1.3532954

Ellington WR, Kinsey ST (1998) Functional and evolutionary implications of the distribution of phosphagens in primitive-type spermatozoa. Biol Bull 195:264–272. https://doi.org/10.2307/1543138

Elofsson H, Mcallister BG, Kime DE et al (2003a) Long lasting stickleback sperm; is ovarian fluid a key to success in fresh water? J Fish Biol 63:240–253. https://doi.org/10.1046/j.1095-8649.2003.00153.x

Elofsson H, Van Look K, Borg B, Mayer I (2003b) Influence of salinity and ovarian fluid on sperm motility in the fifteen-spined stickleback. J Fish Biol 63:1429–1438. https://doi.org/10.1111/j.1095-8649.2003.00256.x

Emri M, Márián T, Trón L et al (1998) Temperature adaptation changes ion concentrations in spermatozoa and seminal plasma of common carp without affecting sperm motility. Aquaculture 167:85–94. https://doi.org/10.1016/S0044-8486(98)00309-3

Espinosa F, Darszon A (1995) Mouse sperm membrane potential: changes induced by Ca^{2+}. FEBS Lett 372:119–125. https://doi.org/10.1016/0014-5793(95)00962-9

Fauvel C, Savoye O, Dreanno C et al (1999) Characteristics of sperm of captive seabass in relation to its fertilization potential. J Fish Biol 54:356–369. https://doi.org/10.1111/j.1095-8649.1999.tb00835.x

Fauvel C, Suquet M, Cosson J (2010) Evaluation of fish sperm quality. J Appl Ichthyol 26 (5):636–643. https://doi.org/10.1111/j.1439-0426.2010.01529.x

Fedorov P, Dzyuba B, Fedorova G et al (2015) Quantification of adenosine triphosphate, adenosine diphosphate, and creatine phosphate in sterlet Acipenser ruthenus spermatozoa during maturation. J Anim Sci 93:5214–5221. https://doi.org/10.2527/jas.2015-9144

Fedorov P, Grabic R, Fedorova G et al (2017) Development and application of LC/HRPS for quantification of adenine nucleotides, creatine phosphate, and creatine in sturgeon spermatozoa. Czech J Anim Sci 62:67–74. https://doi.org/10.17221/4/2016-CJAS

Gage MJG, Macfarlane C, Yeates S et al (2002) Relationships between sperm morphometry and sperm motility in the Atlantic salmon. J Fish Biol 61:1528–1539. https://doi.org/10.1111/j.1095-8649.2002.tb02495.x

Gardiner DM (1978) Utilization of extracellular glucose by spermatozoa of two viviparous fishes. Comp Biochem Physiol A Physiol 59:165–168. https://doi.org/10.1016/0300-9629(78)90200-1

Gatti J-L, Billard R, Christen R (1990) Ionic regulation of the plasma membrane potential of rainbow trout (Salmo gairdneri) spermatozoa: role in the initiation of sperm motility. J Cell Physiol 143:546–554. https://doi.org/10.1002/jcp.1041430320

Gatti J-L, Fedrigo E, Jatteau P et al (1991) Siberian sturgeon, Acipenser baeri, spermatozoa: effects of osmotic pressure, sodium and potassium ions on motility. In: Williot P (ed) Acipenser. Cemagref Publication, Bordeaux, pp 143–151

Geise AC (1979) Cell physiology. W.B. Saunders Co., Philadelphia

Gibbons IR (1962) Observations on the circular movement of sperm at an interface. In: Proceedings of the 5th international congress for electron microscopy, vol. 2, p M-2

Gibbons IR (1981) Cilia and flagella of eukaryotes. J Cell Biol 91(3 Pt 2):107s–124s. https://doi.org/10.1083/jcb.91.3.107s

Gibbons IR (2012) Discovery of dynein and its properties: a personal account. In: King SM (ed) Dyneins. Academic, Boston, pp 2–87. https://doi.org/10.1016/B978-0-12-382004-4.10001-9

Gibbons BH, Baccetti B, Gibbons IR (1985) Live and reactivated motility in the 9 + 0 flagellum of Anguilla sperm. Cell Motil 5:333–350. https://doi.org/10.1002/cm.970050406

Gillies EA, Cannon RM, Green RB, Pacey AA (2009) Hydrodynamic propulsion of human sperm. J Fluid Mech 625:445. https://doi.org/10.1017/S0022112008005685

Gillies EA, Bondarenko V, Cosson J, Pacey AA (2013) Fins improve the swimming performance of fish sperm: a hydrodynamic analysis of the Siberian sturgeon *Acipenser baerii*. Cytoskeleton 70:85–100. https://doi.org/10.1002/cm.21093

Gomendio M, Roldan ERS, Bateson PPG (1991) Sperm competition influences sperm size in mammals. Proc R Soc Lond Ser B Biol Sci 243:181–185. https://doi.org/10.1098/rspb.1991.0029

Gosh RI (1989) Energy metabolism of fish spermatozoa, a survey. Gidrobiol Zhurnal 25:61–71

Gray J (1955) The movement of sea-urchin spermatozoa. J Exp Biol 32:775–801

Gray J, Hancock GJ (1955) The propulsion of sea-urchin spermatozoa. J Exp Biol 32:802–814

Groison A-L, Suquet M, Cosson J et al (2008) Biological characteristics of European hake (Merluccius merluccius) sperm. Cybium 32:178

Groison A-L, Fauvel C, Suquet M et al (2010a) Some characteristics of sperm motility in European hake (Merluccius merluccius, L., 1758). J Appl Ichthyol 26:682–689. https://doi.org/10.1111/j.1439-0426.2010.01541.x

Groison A-L, Suquet M, Cosson J et al (2010b) Sperm motility in European hake, Merluccius merluccius, and characterization of its spermatozoa concentration and volume, spermatocrit, osmolality and pH. Aquaculture 301:31–36. https://doi.org/10.1016/J.AQUACULTURE.2010.01.026

Guthrie HD, Welch GR (2012) Effects of reactive oxygen species on sperm function. Theriogenology 78:1700–1708. https://doi.org/10.1016/j.theriogenology.2012.05.002

Hagedorn M, Ricker J, McCarthy M et al (2009) Biophysics of zebrafish (Danio rerio) sperm. Cryobiology 58:12–19. https://doi.org/10.1016/j.cryobiol.2008.09.013

Halliwell B, Gutteridge JMC (1985) Free radicals in biology and medicine. J Free Radic Biol Med 1:331–332. https://doi.org/10.1016/0748-5514(85)90140-0

Hardie DG, Ross FA, Hawley SA (2012) AMPK: a nutrient and energy sensor that maintains energy homeostasis. Nat Rev Mol Cell Biol 13:251–262. https://doi.org/10.1038/nrm3311

Holt WV (2011) Mechanisms of sperm storage in the female reproductive tract: an interspecies comparison. Reprod Domest Anim 46:68–74. https://doi.org/10.1111/j.1439-0531.2011.01862.x

Holwill MEJ (1966) Physical aspects of flagellar movement. Physiol Rev 46:696–785. https://doi.org/10.1152/physrev.1966.46.4.696

Holwill ME (1969) Kinetic studies of the flagellar movement of sea-urchin spermatozoa. J Exp Biol 50:203–222

Holwill ME (1974) Some physical aspects of the motility of ciliated and flagellated microorganisms. Sci Prog 61:63–80

Holwill ME (1977) Some biophysical aspects of ciliary and flagellar motility. Adv Microb Physiol 16:1–48

Holwill ME (1982) Dynamics of eukaryotic flagellar movement. Symp Soc Exp Biol 35:289–312

Holwill ME, Silvester NR (1967) Thermodynamic aspects of flagellar activity. J Exp Biol 47:249–265

Horokhovatskyi Y, Sampels S, Cosson J et al (2016a) Lipid composition in common carp (Cyprinus carpio) sperm possessing different cryoresistance. Cryobiology 73:282–285. https://doi.org/10.1016/J.CRYOBIOL.2016.08.005

Horokhovatskyi Y, Sampels S, Dzyuba B et al (2016b) Heterogeneity of cryoresistance in common carp sperm. Anim Reprod Sci 169:114–115. https://doi.org/10.1016/J.ANIREPROSCI.2016.03.050

Horokhovatskyi Y, Dietrich MA, Lebeda I et al (2018) Cryopreservation effects on a viable sperm sterlet (Acipenser ruthenus) subpopulation obtained by a Percoll density gradient method. PLoS One 13:e0202514. https://doi.org/10.1371/journal.pone.0202514

Hurtado de Llera A, Martin-Hidalgo D, Gil MC et al (2012) AMP-activated kinase AMPK is expressed in boar spermatozoa and regulates motility. PLoS One 7:e38840. https://doi.org/10.1371/journal.pone.0038840

Inaba K (2002) Dephosphorylation of Tctex2-related dynein light chain by type 2A protein phosphatase. Biochem Biophys Res Commun 297:800–805. https://doi.org/10.1016/S0006-291X(02)02303-3

Inaba K, Shiba K (2018) Microscopic analysis of sperm movement: links to mechanisms and protein components. Microscopy 67:144–155. https://doi.org/10.1093/jmicro/dfy021

Inaba K, Dréanno C, Cosson J et al (2003) Control of flatfish sperm motility by CO2 and carbonic anhydrase. Cell Motil Cytoskeleton 55:174–187. https://doi.org/10.1002/cm.10119

Ingermann RL (2008) Energy metabolism and respiration in fish spermatozoa. In: Alavi SMH, Cosson JJ, Coward K, Rafiee G (eds) Fish spermatology. Alpha Science International Ltd., Oxford, pp 241–266

Ingermann RL, Robinson ML, Cloud JG (2003) Respiration of steelhead trout sperm: sensitivity to pH and carbon dioxide. J Fish Biol 62:13–23. https://doi.org/10.1046/j.1095-8649.2003.00003.x

Ishijima S (2012) Comparative analysis of movement characteristics of lancelet and fish spermatozoa having different morphologies. Biol Bull 222:214–221. https://doi.org/10.1086/BBLv222n3p214

Ishijima S, Ishijima SA, Afzelius BA (1994) Movement of Myzostomum spermatozoa: calcium ion regulation of swimming direction. Cell Motil Cytoskeleton 28:135–142. https://doi.org/10.1002/cm.970280205

Ishimoto K, Cosson J, Gaffney EA (2016) A simulation study of sperm motility hydrodynamics near fish eggs and spheres. J Theor Biol 389:187–197. https://doi.org/10.1016/J.JTBI.2015.10.013

Jeulin C, Soufir J-C (1992) Reversible intracellular ATP changes in intact rat spermatozoa and effects on flagellar sperm movement. Cell Motil Cytoskeleton 21:210–222. https://doi.org/10.1002/cm.970210305

Jibb LA, Richards JG (2008) AMP-activated protein kinase activity during metabolic rate depression in the hypoxic goldfish, Carassius auratus. J Exp Biol 211:3111–3122. https://doi.org/10.1242/jeb.019117

Jin ZX, Inaba K, Manaka K et al (1994) Monoclonal antibodies against the protein complex that contains the flagellar movement-initiating phosphoprotein of Oncorhynchus keta. J Biochem 115:885–890

Kamp G, Büsselmann G, Lauterwein J (1996) Spermatozoa: models for studying regulatory aspects of energy metabolism. Experientia 52:487–494. https://doi.org/10.1007/BF01919321

Katz DF, Drobnis EZ (1990) Analysis and interpretation of the forces generated by spermatozoa. In: Bavister BD, Cummins J, Roldan ERS (eds) Fertilization in mammals. Sereno Symposia, Nowell, pp 125–137

Kho KH, Morisawa M, Choi KS (2005) Cell signaling mechanisms of sperm motility in aquatic species. J Microbiol Biotechnol 15:665–671. https://doi.org/10.1111/j.1365-2516.2004.00947.x

Kime D, Van Look KJ, McAllister B et al (2001) Computer-assisted sperm analysis (CASA) as a tool for monitoring sperm quality in fish. Comp Biochem Physiol Part C Toxicol Pharmacol 130:425–433. https://doi.org/10.1016/S1532-0456(01)00270-8

Krasznai Z, Márián T, Balkay L et al (1995) Potassium channels regulate hypo-osmotic shock-induced motility of common carp (Cyprinus carpio) sperm. Aquaculture 129:123–128. https://doi.org/10.1016/0044-8486(94)00234-F

Krasznai Z, Marian T, Izumi H et al (2000) Membrane hyperpolarization removes inactivation of Ca2+ channels, leading to Ca2+ influx and subsequent initiation of sperm motility in the common carp. Proc Natl Acad Sci U S A 97:2052–2057. https://doi.org/10.1073/pnas.040558097

Krasznai Z, Morisawa M, Krasznai ZT et al (2003) Gadolinium, a mechano-sensitive channel blocker, inhibits osmosis-initiated motility of sea- and freshwater fish sperm, but does not affect human or ascidian sperm motility. Cell Motil Cytoskeleton 55:232–243. https://doi.org/10.1002/cm.10125

Kunji ERS, Aleksandrova A, King MS et al (2016) The transport mechanism of the mitochondrial ADP/ATP carrier. Biochim Biophys Acta Mol Cell Res 1863:2379–2393. https://doi.org/10.1016/j.bbamcr.2016.03.015

Lahnsteiner F (2011) Spermatozoa of the teleost fish Perca fluviatilis (perch) have the ability to swim for more than two hours in saline solutions. Aquaculture 314:221–224. https://doi.org/10.1016/J.AQUACULTURE.2011.02.024

Lahnsteiner F, Caberlotto S (2012) Motility of gilthead seabream Sparus aurata spermatozoa and its relation to temperature, energy metabolism and oxidative stress. Aquaculture 370–371:76–83. https://doi.org/10.1016/J.AQUACULTURE.2012.09.034

Lahnsteiner F, Mansour N (2010) A comparative study on antioxidant systems in semen of species of the Percidae, Salmonidae, Cyprinidae, and Lotidae for improving semen storage techniques. Aquaculture 307:130–140. https://doi.org/10.1016/J.AQUACULTURE.2010.07.011

Lahnsteiner F, Mansour N (2012) The effect of temperature on sperm motility and enzymatic activity in brown trout Salmo trutta, burbot Lota lota and grayling Thymallus thymallus. J Fish Biol 81:197–209. https://doi.org/10.1111/j.1095-8649.2012.03323.x

Lahnsteiner F, Patzner RA, Weismann T (1993) Energy resources of spermatozoa of the rainbow trout Oncorhynchus mykiss (Pisces, Teleostei). Reprod Nutr Dev 33:349–360

Lahnsteiner F, Berger B, Weismann T, Patzner R (1997) Sperm motility and seminal fluid composition in the burbot, Lota lota. J Appl Ichthyol 13:113–119. https://doi.org/10.1111/j.1439-0426.1997.tb00110.x

Lahnsteiner F, Berger B, Weismann T, Patzner RA (1998) Determination of semen quality of the rainbow trout, Oncorhynchus mykiss, by sperm motility, seminal plasma parameters, and spermatozoal metabolism. Aquaculture 163:163–181. https://doi.org/10.1016/S0044-8486(98)00243-9

Lahnsteiner F, Berger B, Weismann T (1999) Sperm metabolism of the teleost fishes Chalcalburnus chalcoides and Oncorhynchus mykiss and its relation to motility and viability. J Exp Zool 284:454–465. https://doi.org/10.1002/(SICI)1097-010X(19990901)284:4<454::AID-JEZ12>3.0.CO;2-O

Lahnsteiner F, Mansour N, Caberlotto S (2010) Composition and metabolism of carbohydrates and lipids in Sparus aurata semen and its relation to viability expressed as sperm motility when activated. Comp Biochem Physiol B Biochem Mol Biol 157:39–45. https://doi.org/10.1016/J.CBPB.2010.04.016

Laufer N, Segal S, Yaffe H et al (1977) Volume and shape of normal human spermatozoa. Fertil Steril 28:456–458

Legendre M, Alavi SMH, Dzyuba B et al (2016) Adaptations of semen characteristics and sperm motility to harsh salinity: extreme situations encountered by the euryhaline tilapia Sarotherodon melanotheron heudelotii (Dumeril, 1859). Theriogenology 86:1251–1267. https://doi.org/10.1016/J.THERIOGENOLOGY.2016.04.066

Levitan DR (2000) Sperm velocity and longevity trade off each other and influence fertilization in the sea urchin *Lytechinus variegatus*. Proc R Soc Lond Ser B Biol Sci 267:531–534. https://doi.org/10.1098/rspb.2000.1032

Li P, Hulak M, Koubek P et al (2010) Ice-age endurance: the effects of cryopreservation on proteins of sperm of common carp, Cyprinus carpio L. Theriogenology 74:413–423. https://doi.org/10.1016/J.THERIOGENOLOGY.2010.02.024

Lindroth A (1947) Time of activity of fresh water fish spermatozoa in relation to temperature. Zool Bide Uppsala:154–168

Linhart O, Cosson J, Mims SD et al (2002) Effects of ions on the motility of fresh and demembranated paddlefish (Polyodon spathula) spermatozoa. Reproduction 124:713–719

Linhart O, Cosson J, Mims SD et al (2003) Effects of ions on the motility of fresh and demembranate spermatozoa of common carp (Cyprinus carpio) and paddlefish (Polyodon spathula). Fish Physiol Biochem 28:203–205. https://doi.org/10.1023/B:FISH.0000030528.03098.6d

Linhart O, Alavi SMH, Rodina M et al (2008a) Comparison of sperm velocity, motility and fertilizing ability between firstly and secondly activated spermatozoa of common carp (*Cyprinus carpio*). J Appl Ichthyol 24:386–392. https://doi.org/10.1111/j.1439-0426.2008.01138.x

Linhart O, Alavi SMH, Rodina M et al (2008b) After finishing of motility, common carp (Cyprinus carpio) sperm is able to re-initiate a second motility period and to fertilize eggs. Cybium 32:187–188

Linhartova Z, Rodina M, Nebesarova J et al (2013) Morphology and ultrastructure of beluga (Huso huso) spermatozoa and a comparison with related sturgeons. Anim Reprod Sci 137:220–229. https://doi.org/10.1016/J.ANIREPROSCI.2013.01.003

Liu L, Ciereszko A, Czesny S, Dabrowski K (1997) Dietary ascorbyl monophosphate depresses lipid peroxidation in rainbow trout spermatozoa. J Aquat Anim Health 9:249. https://doi.org/10.1577/1548-8667(1997)009<0249:DAMDLP>2.3.CO;2

Lo L-W, Koch CJ, Wilson DF (1996) Calibration of oxygen-dependent quenching of the phosphorescence of Pd-meso-tetra (4-carboxyphenyl) porphine: a phosphor with general application for measuring oxygen concentration in biological systems. Anal Biochem 236:153–160. https://doi.org/10.1006/ABIO.1996.0144

Losano JDAA, Padín JF, Méndez-López I et al (2017) The stimulated glycolytic pathway is able to maintain ATP levels and kinetic patterns of bovine epididymal sperm subjected to mitochondrial uncoupling. Oxid Med Cell Longev 2017:1–8. https://doi.org/10.1155/2017/1682393

Majhi RK, Kumar A, Yadav M et al (2013) Thermosensitive ion channel TRPV1 is endogenously expressed in the sperm of a fresh water teleost fish (*Labeo rohita*) and regulates sperm motility. Channels 7:483–492. https://doi.org/10.4161/chan.25793

Mansour N, Lahnsteiner F, Berger B (2003) Metabolism of intratesticular spermatozoa of a tropical teleost fish (Clarias gariepinus). Comp Biochem Physiol B Biochem Mol Biol 135:285–296. https://doi.org/10.1016/S1096-4959(03)00083-6

Marchand B (1977) Un type nouveau de structure flagellaire. Type 9 + n. J Cell Biol 72:707–713. https://doi.org/10.1083/jcb.72.3.707

Marian T, Krasznai Z, Balkay L et al (1993) Hypo-osmotic shock induces an osmolality-dependent permeabilization and structural changes in the membrane of carp sperm. J Histochem Cytochem 41:291–297. https://doi.org/10.1177/41.2.8419464

Martínez-Páramo S, Horváth Á, Labbé C et al (2017) Cryobanking of aquatic species. Aquaculture 472:156–177. https://doi.org/10.1016/J.AQUACULTURE.2016.05.042

Mattei X, Marchand B (1988) La spermiogenèse de Myzostomum sp. (Procoelomata, Myzostomida). J Ultrastruct Mol Struct Res 100:75–85. https://doi.org/10.1016/0889-1605(88)90060-2

Mattei X, Mattei C, Reizer C, Chevalier JL (1972) Ultrastructure des spermatozoïdes aflagellés des mormyres (Poissons Téléostéens). J Microsc (Paris) 15:67–78

Mattei X, Marchand B, Quilichini Y (2019) A biflagellate spermatozoon in the African bonytongue *Heterotis niloticus* (Teleostei, Osteoglossidae). J Fish Biol 94(2):335–338. https://doi.org/10.1111/jfb.13894

McKemy DD (2007) Temperature sensing across species. Pflügers Arch 454:777–791. https://doi.org/10.1007/s00424-006-0199-6

Mehlis M, Bakker TCM (2014) The influence of ambient water temperature on sperm performance and fertilization success in three-spined sticklebacks (Gasterosteus aculeatus). Evol Ecol 28:655–667. https://doi.org/10.1007/s10682-014-9707-x

Minassian ES, Terner C (1966) Biosynthesis of lipids by human and fish spermatozoa. Am J Phys 210:615–618. https://doi.org/10.1152/ajplegacy.1966.210.3.615

Mita M, Oguchi A, Kikuyama S et al (1994) Endogenous substrates for energy metabolism in spermatozoa of the sea urchins Arbacia lixula and Paracentrotus lividus. Biol Bull 186:285–290. https://doi.org/10.2307/1542274

Mochida K, Kondo T, Matsubara T et al (1999) A high molecular weight glycoprotein in seminal plasma is a sperm immobilizing factor in the teleost Nile tilapia, Oreochromis niloticus. Dev Growth Differ 41:619–627. https://doi.org/10.1046/j.1440-169x.1999.00463.x

Mohri H, Horiuchi K (1961) Studies on the respiration of sea-urchin spermatozoa: III. Respiratory quotient. J Exp Biol 38:249–257

Morisawa M (1985) Initiation mechanism of sperm motility at spawning in teleosts. Zool Sci 2:605–615

Morisawa M (1994) Cell signaling mechanisms for sperm motility. Zool Sci 11:647–662

Morisawa S, Ishida K, Okuno M, Morisawa M (1993) Roles of pH and cyclic adenosine monophosphate in the acquisition of potential for sperm motility during migration from the sea to the river in chum salmon. Mol Reprod Dev 34:420–426. https://doi.org/10.1002/mrd.1080340411

Moscatelli N, Spagnolo B, Pisanello M et al (2017) Single-cell-based evaluation of sperm progressive motility via fluorescent assessment of mitochondria membrane potential. Sci Rep 7:17931. https://doi.org/10.1038/s41598-017-18123-1

Mounib MS (1967) Metabolism of pyruvate, acetate and glyoxylate by fish sperm. Comp Biochem Physiol 20:987–992. https://doi.org/10.1016/0010-406X(67)90068-0

Mundt N, Spehr M, Lishko PV (2018) TRPV4 is the temperature-sensitive ion channel of human sperm. elife 7. https://doi.org/10.7554/eLife.35853

Navarro B, Kirichok Y, Clapham DE (2007) KSper, a pH-sensitive K+ current that controls sperm membrane potential. Proc Natl Acad Sci U S A 104:7688–7692. https://doi.org/10.1073/pnas.0702018104

Nevo AC, Rikmenspoel R (1970) Diffusion of ATP in sperm flagella. J Theor Biol 26:11–18. https://doi.org/10.1016/S0022-5193(70)80027-3

Nguyen TMD, Alves S, Grasseau I et al (2014) Central role of 5′-AMP-activated protein kinase in chicken sperm functions. Biol Reprod 91:121. https://doi.org/10.1095/biolreprod.114.121855

Orfão LH, Nascimento AF, Corrêa FM et al (2011) Extender composition, osmolality and cryoprotectant effects on the motility of sperm in the Brazilian endangered species Brycon opalinus (Characiformes). Aquaculture 311:241–247. https://doi.org/10.1016/J.AQUACULTURE.2010.11.041

Perchec Poupard GP, Gatti JL, Cosson J et al (1997) Effects of extracellular environment on the osmotic signal transduction involved in activation of motility of carp spermatozoa. Reproduction 110:315–327. https://doi.org/10.1530/jrf.0.1100315

Perchec G, Jeulin C, Cosson J et al (1995) Relationship between sperm ATP content and motility of carp spermatozoa. J Cell Sci 108:747–753

Perchec G, Cosson MP, Cosson J et al (1996) Morphological and kinetic changes of carp (Cyprinus carpio) spermatozoa after initiation of motility in distilled water. Cell Motil Cytoskeleton 35:113–120. https://doi.org/10.1002/(SICI)1097-0169(1996)35:2<113::AID-CM4>3.0.CO;2-B

Piironen J, Hyvärinen H (1983) Composition of the milt of some teleost fishes. J Fish Biol 22:351–361. https://doi.org/10.1111/j.1095-8649.1983.tb04757.x

Poupard GP, Paxion C, Cosson J et al (1998) Initiation of carp spermatozoa motility and early ATP reduction after milt contamination by urine. Aquaculture 160:317–328. https://doi.org/10.1016/S0044-8486(97)00301-3

Prokopchuk G, Cosson J (2017) Biophysics of fish sperm flagellar movement: present knowledge and original directions. In: Jimenez-Lopez JC (ed) Cytoskeleton—structure, dynamics, function and disease. InTech, pp 127–150

Prokopchuk G, Dzyuba B, Bondarenko O et al (2015) Motility initiation of sterlet sturgeon (Acipenser ruthenus) spermatozoa: describing the propagation of the first flagellar waves. Theriogenology 84:51–61. https://doi.org/10.1016/j.theriogenology.2015.02.011

Prokopchuk G, Dzyuba B, Rodina M, Cosson J (2016) Control of sturgeon sperm motility: antagonism between K^+ ions concentration and osmolality. Anim Reprod Sci 164:82–89. https://doi.org/10.1016/j.anireprosci.2015.11.015

Redondo-Müller C, Cosson M-P, Cosson J, Billard R (1991) In vitro maturation of the potential for movement of carp spermatozoa. Mol Reprod Dev 29:259–270. https://doi.org/10.1002/mrd.1080290308

Rikmenspoel R (1966) Elastic properties of the sea urchin sperm flagellum. Biophys J 6:471–479. https://doi.org/10.1016/S0006-3495(66)86670-5

Rikmenspoel R (1978) Movement of sea urchin sperm flagella. J Cell Biol 76:310–322. https://doi.org/10.1083/jcb.76.2.310

Robitaille P-ML, Mumford KG, Brown GG (1987) 31 P nuclear magnetic resonance study of trout spermatozoa at rest, after motility, and during short-term storage. Biochem Cell Biol 65:474–485. https://doi.org/10.1139/o87-061

Saudrais C, Fierville F, Loir M et al (1998) The use of phosphocreatine plus ADP as energy source for motility of membrane-deprived trout spermatozoa. Cell Motil Cytoskeleton 41:91–106. https://doi.org/10.1002/(SICI)1097-0169(1998)41:2<91::AID-CM1>3.0.CO;2-I

Schulz RW, de França LR, LeGac F et al (2010) Spermatogenesis in fish. Gen Comp Endocrinol 165:390–411. https://doi.org/10.1016/J.YGCEN.2009.02.013

Shaliutina-Kolešová A, Gazo I, Cosson J et al (2013) Comparison of oxidant and antioxidant status of seminal plasma and spermatozoa of several fish species. Czech J Anim Sci 58:313–320. https://doi.org/10.17221/6861-CJAS

Shaliutina-Kolešová A, Cosson J, Lebeda I et al (2015) The influence of cryoprotectants on sturgeon (Acipenser ruthenus) sperm quality, DNA integrity, antioxidant responses, and resistance to oxidative stress. Anim Reprod Sci 159:66–76. https://doi.org/10.1016/j.anireprosci.2015.05.014

Shaliutina-Kolešová A, Kotas P, Štěrba J et al (2016) Protein profile of seminal plasma and functionality of spermatozoa during the reproductive season in the common carp (Cyprinus carpio) and rainbow trout (Oncorhynchus mykiss). Mol Reprod Dev 83:968–982. https://doi.org/10.1002/mrd.22737

Silvester NR, Holwill MEJ (1965) Molecular hypothesis of flagellar Activity. Nature 205 (4972):665–668

Somlo M, Cosson J, Clavilier L et al (1982) Identity problems concerning subunits of the membrane factor of the mitochondrial ATPase of Saccharomyces cerevisiae. Eur J Biochem 122:369–374. https://doi.org/10.1111/J.1432-1033.1982.TB05890.X

Stanley HP (1965) Electron microscopic observation of the biflafellate spermatid of the teleost fish Porichtys notatus. Anat Rec 151:477. https://doi.org/10.1002/ar.1091510312

Stockley P, Gage MJ, Parker GA, Møller AP (1997) Sperm competition in fishes: the evolution of testis size and ejaculate characteristics. Am Nat 149:933–954. https://doi.org/10.1086/286031

Stumpf PK, Barber GA (1957) Fat metabolism in higher plants IX. Enzymic synthesis of long chain fatty acids by avocado particles. J Biol Chem 227:407–418

Suquet M, Billard R, Cosson J et al (1994) Sperm features in turbot *(Scophthalmus maximus):* a comparison with other freshwater and marine fish species. Aquat Living Resour 7:283–294. https://doi.org/10.1051/alr:1994031

Suquet M, Dreanno C, Dorange G et al (1998) The ageing phenomenon of turbot spermatozoa: effects on morphology, motility and concentration, intracellular ATP content, fertilization, and storage capacities. J Fish Biol 52:31–41. https://doi.org/10.1006/jfbi.1997.0556

Takai H, Morisawa M (1995) Change in intracellular K+ concentration caused by external osmolality change regulates sperm motility of marine and freshwater teleosts. J Cell Sci 108:1175–1181

Takei GL, Mukai C, Okuno M (2015) Regulation of salmonid fish sperm motility by osmotic shock-induced water influx across the plasma membrane. Comp Biochem Physiol A Mol Integr Physiol 182:84–92. https://doi.org/10.1016/j.cbpa.2014.12.013

Tanimoto S, Morisawa M (1988) Roles for potassium and calcium channels in the initiation of sperm motility in rainbow trout. (K+/Ca2+/channel/sperm motility/rainbow trout). Dev Growth Differ 30:117–124. https://doi.org/10.1111/j.1440-169X.1988.00117.x

Terner C, Korsh G (1963) The oxidative metabolism of pyruvate, acetate and glucose in isolated fish spermatozoa. J Cell Comp Physiol 62:243–249. https://doi.org/10.1002/jcp.1030620303

Tibbs J (1959) The adenosine triphosphatase activity of perch sperm flagella. Biochim Biophys Acta 33:220–226. https://doi.org/10.1016/0006-3002(59)90517-7

Tombes RM, Shapiro BM (1985) Metabolite channeling: a phosphorylcreatine shuttle to mediate high energy phosphate transport between sperm mitochondrion and tail. Cell 41:325–334. https://doi.org/10.1016/0092-8674(85)90085-6

Tombes RM, Shapiro BM (1987) Enzyme termini of a phosphocreatine shuttle. J Biol Chem 262:16011–16019

Tombes RM, Brokaw CJ, Shapiro BM (1987) Creatine kinase-dependent energy transport in sea urchin spermatozoa. Flagellar wave attenuation and theoretical analysis of high energy phosphate diffusion. Biophys J 52:75–86. https://doi.org/10.1016/S0006-3495(87)83190-9

Tsvetkova LI, Cosson J, Linhart O, Billard R (1996) Motility and fertilizing capacity of fresh and frozen-thawed spermatozoa in sturgeons Acipenser Baeri and A. ruthenus. J Appl Ichthyol 12:107–112. https://doi.org/10.1111/j.1439-0426.1996.tb00071.x

Turner E, Montgomerie R (2002) Ovarian fluid enhances sperm movement in Arctic charr. J Fish Biol 60:1570–1579. https://doi.org/10.1006/jfbi.2002.2018

Ulloa-Rodríguez P, Figueroa E, Díaz R et al (2017) Mitochondria in teleost spermatozoa. Mitochondrion 34:49–55. https://doi.org/10.1016/J.MITO.2017.01.001

Vladiĉ T, Jätrvi T (1997) Sperm motility and fertilization time span in Atlantic salmon and brown trout-the effect of water temperature. J Fish Biol 50(5):1088–1093

Woolsey J, Ingermann RL (2003) Acquisition of the potential for sperm motility in steelhead (Oncorhynchus mykiss): effect of pH on dynein ATPase. Fish Physiol Biochem 29:47–56. https://doi.org/10.1023/B:FISH.0000035896.84954.f6

Xin M, Shaliutina-Kolesova A, Sterba J et al (2018) Impact of cryopreservation on sterlet, Acipenser ruthenus sperm motility and proteome. Anim Reprod Sci 192:280–289. https://doi.org/10.1016/J.ANIREPROSCI.2018.03.025

Xu Z, Li E, Xu C, Gan L, Qin JG, Chen L (2016) Response of AMP-activated protein kinase and energy metabolism to acute nitrite exposure in the Nile tilapia Oreochromis niloticus. Aquat Toxicol 177:86–97

Yates GT (1986) How microorganisms move through water: the hydrodynamics of ciliary and flagellar propulsion reveal how microorganisms overcome the extreme effect of the viscosity of water. Am Sci 74:358–365. https://doi.org/10.2307/27854249

Yoshimura A, Nakano I, Shingyoji C (2007) Inhibition by ATP and activation by ADP in the regulation of flagellar movement in sea urchin sperm. Cell Motil Cytoskeleton 64:777–793. https://doi.org/10.1002/cm.20222

Zhu Z, Li R, Ma G et al (2018) 5′-AMP-activated protein kinase regulates goat sperm functions via energy metabolism in vitro. Cell Physiol Biochem 47:2420–2431. https://doi.org/10.1159/000491616

Ziętara MS, Slominska E, Swierczynski J et al (2004) ATP content and adenine nucleotide catabolism in African catfish spermatozoa stored in various energetic substrates. Fish Physiol Biochem 30:119–127. https://doi.org/10.1007/s10695-005-2493-1

Zilli L, Schiavone R, Zonno V et al (2004) Adenosine triphosphate concentration and β-d-glucuronidase activity as indicators of sea bass semen quality. Biol Reprod 70:1679–1684. https://doi.org/10.1095/biolreprod.103.027177

Zilli L, Schiavone R, Chauvigné F et al (2009) Evidence for the involvement of aquaporins in sperm motility activation of the teleost Gilthead Sea bream (Sparus aurata). Biol Reprod 81:880–888. https://doi.org/10.1095/biolreprod.109.077933

Fish Sperm Quality Evaluation After Cryopreservation

5

Sergii Boryshpolets, Vitaliy Kholodnyy, Jacky Cosson, and Borys Dzyuba

Abstract

In this chapter, we will focus on specificity of fish spermatozoa and their cryopreservation. We will discuss common cryoinjury issues and their outcomes affecting sperm behavior, with the emphasis on the importance of evaluation methods, and highlight possible development of quality control for cryopreservation methods of fish spermatozoa.

Keywords

Sperm quality · Cryopreservation · DNA damage · Cryo damage · Spermatozoa

5.1 Introduction

Cryopreservation has become an important tool for the conservation of many endangered species or long-term preservation and shipping of gametes in commercially important species including fish. Mainly due to their small size and relatively high cryoresistance, the spermatozoa have become almost the only tissue which is used for cryopreservation in fish (Martínez-Páramo et al. 2017). Cryopreserved sperm samples can provide the hatcheries with stable and valuable male genetic information ready for the immediate mass larva production, selective breeding programs, or research needs. Nevertheless, most fish farmers are not inclined to practice cryopreservation, probably due to the relatively high expenditures and lower

S. Boryshpolets (✉) · V. Kholodnyy · J. Cosson · B. Dzyuba
Laboratory of Reproductive Physiology, Faculty of Fisheries and Protection of Waters, South Bohemian Research Center of Aquaculture and Biodiversity of Hydrocenoses, Research Institute of Fish Culture and Hydrobiology, University of South Bohemia in České Budějovice, Vodňany, Czech Republic
e-mail: sboryshpolets@frov.jcu.cz

J. Betsy, S. Kumar (eds.), *Cryopreservation of Fish Gametes*,
https://doi.org/10.1007/978-981-15-4025-7_5

commercial importance so far, as well as lower "male value" in fish livestock, compared to mammals (Tiersch 2008; Asturiano et al. 2017). However, many different cryopreservation protocols have been developed during several decades and currently cover a large variety of fish species (Tiersch and Mazik 2011). Contemporary methods and protocols are simplified and commonly involve ready-to-use commercial equipment and extenders (Morris 2000; Asturiano et al. 2017). As a consequence, most of the recent reports in fish sperm cryopreservation are dealing with some improvements of the widely used "recipes" such as concentration or type of cryoprotectant, equilibration time, introduction of different antifreeze proteins, etc. Moreover, an understanding of the basic principles of cryobiology is still essential for the correct application of cryopreservation methods in order to minimize cell damage and optimize the conditions to get the highest number of viable sperm after thawing (Morris 2000; Morris et al. 2012). This is even more apparent for fish sperm, because usually fertilization of an enormous amount of eggs is needed, and the higher the number of viable spermatozoa is, the lower is the required sample volume. Thus, methods for the cryopreservation of spermatozoa still need to be improved in many medical, veterinary, and conservation applications (Morris 2000; Tiersch and Mazik 2011; Morris et al. 2012) and the cryopreservation of fish sperm is not an exception. Many different protocols are used all over the world and sometimes the outcomes of their usage lead to varying and even contradictory results/conclusions (Tiersch and Mazik 2011; Horokhovatskyi et al. 2016b; Boryshpolets et al. 2017). This discrepancy could be partly explained by huge variations in gamete features among fish subpopulations even in the frames of the same species. This could be due to the different rearing or environmental conditions, which affect reproduction physiology and consequently cryoresistance of sperm (Boryshpolets et al. 2009a; Horokhovatskyi et al. 2016a, b; Blecha et al. 2018). It is still not clear which particular properties of sperm samples may be the best ones to predict their tolerance to the freezing/thawing process (Kopeika et al. 2003; Kopeika and Kopeika 2008). Moreover, the existing differences in some reported results for cells survival can be a consequent of wrong estimation of sperm quality after cryopreservation that can in turn impair the effect of application of cryopreserved samples, and in addition may lead to a misevaluation of cryopreservation methods per se (Asturiano et al. 2017). In this chapter, we will focus on specificity of fish spermatozoa and their cryopreservation. We will discuss common cryoinjury issues and their outcome on sperm behavior, aiming to demonstrate the importance of evaluation methods and highlight possible direction in the development of quality control and cryopreservation methods of fish spermatozoa.

5.2 Specificity of Fish Spermatozoa and Their Cryopreservation

5.2.1 Cryopreservation

During freezing and thawing, the cells undergo a variety of processes which may result in stress and damage. Cooling of aqueous solutions usually involves a supercooling phenomenon, when temperature of liquid firstly drops below the melting point before the appearance of ice nucleation, which results in accumulation of latent heat and its instantaneous release after nucleation onset. High supercooling is usually associated with massive cell damage due to the intracellular ice formation (crystallization) and can be avoided by an artificial initiation of extracellular ice nucleation, so-called seeding (Morris 2000). Another crucial point is associated with the choice of the cooling rate. It is generally accepted that, if the cooling rate is very high, the water inside the cell may freeze and form crystals that are lethal for the cell. Low cooling rate may prevent intracellular ice formation. However, during a slow cooling regime, the extracellular water crystallization will lead to increase of concentration of the substances dissolved in water that in turn will cause the outflow of some water from cells through the membranes so as to restore the osmotic balance between intracellular and surrounding media (Mazur et al. 1972). During this process, cells remain in unfrozen channels located in between ice crystals, where they are subjected to dehydration, and their membrane experiences phase transitions of lipid bilayer due to the temperature changes. Osmotic stress resulted from increased concentration of intracellular solutes may lead to irreversible changes of macromolecule conformation and plasma membrane damage by shear stress (Fuller and Paynter 2004; Morris et al. 2012). Such dehydration course depends on several factors, e.g. duration of cooling, ratio between liquid phase and ice, and some cell features like parameters of membrane permeability to water and solutes, amount of intracellular water, surface-to-volume ratio, etc. (Fuller and Paynter 2004). The above observations were used to develop a model commonly called "two-factor model of freezing injury," which can generally explain a cryoinjury of cells cooled to ultralow temperatures (Mazur et al. 1972). It considers a balance between two set of effects, i.e. damage due to high solutes concentration at slow cooling and damage caused by intracellular ice formation during fast cooling, forming a basis for the prediction of an optimum cooling rate for maximum cell survival.

A number of chemicals, known as cryoprotective additives (CPA), are widely used in cells cryopreservation due to their ability to inhibit or fully prevent the processes accompanying freezing/thawing that lead to cell damage (Elliott et al. 2017). In particular, these substances may increase the volume of unfrozen solution at low temperatures by binding free water molecules, thus reducing the ionic hyperconcentration, while some of the CPAs could promote crystal-free solidification of water (vitrification), etc. Most of CPAs easily permeate cell membranes during several minutes (or even seconds) following their application. As a general rule, cells survive a high concentration of CPA better than a high ionic concentration. Use of different CPAs at the same concentration often results in very similar cryopreservation outcome when applied in the same species, while different outcomes of spermatozoa survival could be often obtained with the same cryoprotectant in case of various cell types or species used. This phenomenon may be related to specific toxicity of different CPAs for different cell types (Morris et al. 2012; Elliott et al. 2017).

During thawing, the processes which occurred in the sample during freezing proceed in a reversed order and the cells experience an osmolality drop, due to non-uniform dilution of highly concentrated solution by water released from melting ice and appearance of areas with low (or zero) osmolality. During this process, water may enter the cell due to osmotic forces, leading to a possible formation of intracellular ice. To avoid this phenomenon, a rapid thawing is the most preferable for biological samples containing living cells. Following thawing, the cells will experience a high concentration of CPA, some of which may be cytotoxic in these conditions (Morris 2000; Morris et al. 2012).

Thus, the individual characteristics of cells, such as osmo-sensitivity and tolerance to CPA, predict their cryoresistance. In other words, the optimal cryopreservation regime will provide conditions, which protect cells from: (a) formation of intracellular ice; (b) mechanical pressure of growing extracellular ice; (c) long-term exposure to solutes, including potentially cytotoxic CPAs.

5.2.2 Specific Features of Spermatozoa in Relation to Cryopreservation

As it was mentioned above, the properties of cells, e.g. specific features of membrane permeability, could affect the outcome of freezing/thawing cycle, and these peculiarities should be considered during the development of cryopreservation methods. In general, spermatozoon is a highly simplified and specific cell. Many typical cell organelles disappear during sperm differentiation. The remaining "simple" structure is perfect to execute the main sperm function, i.e. to deliver male genetic information stored in spermatozoon nucleus inside its head. This is accomplished by means of propagation to the external environment due to flagella beating, using the energy produced by mitochondria inside midpiece (Cosson 2012; Dzyuba et al. 2017). The small size and the "simplicity" of sperm structure (absence of complex compartmentalization typical for somatic cells) as well as its reduced cytoplasm volume are a perfect background for successful cryopreservation (Martínez-Páramo et al. 2017). On the other hand, presence of the quite fragile flagellum and its high functional importance could be considered as an important challenge to be solved (Boryshpolets et al. 2018). In the majority of fish species, spermatozoa are immotile in the seminal fluid and are activated after changing the environment (Alavi and Cosson 2006; Browne et al. 2015). Since most fish species are external fertilizers, this environment is water: sea water or fresh water depending on the habitat area of particular species, correspondingly hyper- or hypotonic compared to the seminal fluid osmolality. In other words, fish spermatozoa motility activation usually requires changes in external osmolality. In some fish species, e.g. belonging to salmonids or acipenserids (Cosson et al. 1991; Takai and Morisawa 1995; Alavi and Cosson 2006), a decrease in external K^+ ions concentration is required for motility activation rather than osmolality changes. This means that the main physiological function of spermatozoa is highly dependent on the environmental osmolality and ion composition. For this reason, the above mentioned changes of the aqueous media and consequently osmolality during cryopreservation could have a significant effect on fish spermatozoa. In particular, ice crystal formation and growth result in an increase of ion concentration and osmolality that present a risk to trigger the spermatozoa motility activation in marine fishes, and vice versa and the decrease of surrounding media osmolality during thawing may activate freshwater fish spermatozoa (Boryshpolets et al. 2009b; Dzyuba et al. 2010). Interestingly, in case of marine fish species sperm, cryopreservation outcomes are actually more successful as compared to their freshwater counterparts (Suquet et al. 2000). One of the possible explanations is the sensitivity of freshwater fish spermatozoa to low external osmolality, playing the role of a trigger for their activation, and the absence of any tolerance to high osmolality. Considering the short duration of freshwater fish sperm motility (Cosson 2010), the possibility of motility activation during sample thawing could definitely affect the efficiency of the fertilization and consequently the outcome and applicability of the cryopreservation method. Thus, spermatozoa survival depends on mechanical damage due to ice crystal formation, osmotic stress, or macromolecules conformation due to solutes hyperconcentration like most cell types. Additionally, they also could lose their physiological value, due to simple activation of motility, even in the absence of any extremal change in media composition. However it should be mentioned that the osmotic component of cryoinjuries is considered to be the most important factor responsible of sperm cryodamage (Morris et al. 2012). In the next section, we will discuss in more detail the most common examples of damages occurring during sperm cryopreservation and their consequences for the sperm quality.

5.3 Damages of Spermatozoa During Cryopreservation

Proper development of cryopreservation methods is not possible without clear understanding of the reasons and consequences of damages that occur in the cells during the cryopreservation processes. Usually, one compares the quality of cells before and after freezing/thawing process. For a convenient description of the negative consequences of freezing/thawing procedure, all the cryodamages could be divided into several groups, associated mainly with the function and/or organelles affected by these damages: the nucleus and DNA integrity, midpiece/mitochondria and energy stores, flagellum and motility, plasma membrane, and ionic/osmotic balance.

5.3.1 Damages to the Plasma Membrane

The composition of the sperm plasma membrane differs significantly among species in terms of content of particular proteins and lipids (Li et al. 2010; Drokin et al. 1989). Spermatozoon plasma membrane plays key function in sperm motility through regulation of intracellular ion concentration involving different signaling pathways and membrane channels (Alavi and Cosson 2006). Hence its integrity is essential for sperm function and is widely used as a sperm quality parameter (Cabrita et al. 2014; Horokhovatskyi et al. 2016a). As already mentioned, fish spermatozoa plasma membrane is sensitive even to minor changes in osmolality of the external medium. In more labile freshwater fish spermatozoa, both the post-thaw sperm motility percentage and the membrane integrity are affected dramatically (Cabrita et al. 2001; Beirao et al. 2011). The majority of lethal damages to the plasma membrane will arise because of intracellular ice formation and osmotically driven swelling or shrinkage of sperm cells. Nevertheless, the membrane of spermatozoa which avoided the lethal damage may experience further negative influences that may affect its properties, e.g. appearance of reactive oxygen species which lead to peroxidation of membrane lipids (Chen et al. 2010; Li et al. 2010) and degradation of membrane proteins (Dietrich et al. 2015; Nynca et al. 2015), as well as leakage of intracellular elements related to cytoskeleton or metabolism (Zilli et al. 2005, 2008; Dietrich et al. 2015; Nynca et al. 2015; Westfalewicz et al. 2015). Integrity of sperm membrane is widely checked by using specific fluorescent or nonfluorescent dyes. Nowadays dual staining in which cells with undamaged membrane and cells with damaged one are stained by different fluorescent stains is commonly used. In particular, the popular live/dead fluorescent assay comprises two substances, one penetrates only cells with damaged membrane and binds nucleic acids with appearance of red fluorescence (propidium iodide), while a second membrane permeant dye enters all cells and after binding with nucleic acids fluoresces green (SYBR® 14). Also, assessment of changes in the composition of plasma membrane (lipids and/or proteins) or seminal plasma content before and after cryopreservation is the commonly used approach for the evaluation of cryodamage of spermatozoa (Cabrita et al. 2014).

5.3.2 Damages of the Cytoskeleton/Flagellum

Various events occurring during cryopreservation may also affect cytoskeletal proteins, such as tubulin, actin or dynein etc., which are the main components of the spermatozoon flagellum in many species (Cosson 2007). There are several reports about changes in flagellar protein profile following cryopreservation in mammalian and fish spermatozoa (Desrosiers et al. 2006; Chen et al. 2014; Dietrich et al. 2015; Nynca et al. 2015). During freezing/thawing spermatozoa may lose part of flagella due to mechanical damage or just small part of specific proteins due to degradation or other chemical reactions. Usually these changes in flagellum lead to different motility disorders such as lower velocity and/or compromised symmetry of wave, as well as progressive efficiency, that will finally affect fertilizing ability of spermatozoa. Such type of damages could be revealed directly by appearance of specific proteins in the medium after cryopreservation (indicating their leakage) or non-directly, by observation of spermatozoa/flagellar behavior, i.e. motility or shape (Correa et al. 2007; Gómez-Torres et al. 2017).

5.3.3 Damages of the Midpiece Zone and Mitochondria

The sperm flagellum represents a structure comprising small "biological motors" situated along the microtubules, i.e. the dyneins that require some stable level of ATP to fulfil their function (Cosson 2007, 2012). Mitochondria are the source of ATP in spermatozoa similar to other cell types. It is quite obvious to associate the drop of metabolic activity and content of ATP after cryopreservation with damages in mitochondria or whole midpiece. The latter may result in leakage of substrates and macro-ergic molecules and, all together, these damages may compromise the process of energy production, resulting in lack or absence of ATP (He and Woods 2004; Boryshpolets et al. 2009b; Madeddu et al. 2010; Chen et al. 2014; Dietrich et al. 2015; Nynca et al. 2015; Figueroa et al. 2017). Short lasting sperm motility of fish spermatozoa is highly dependent on the intracellular level of ATP pre-accumulated before motility activation, since these cells are not able to produce enough ATP to sustain it during motility period (Perchec et al. 1995; Dreanno et al. 1999; Cosson 2012). On the contrary, in case of long-time swimmers, e.g. mammalian sperm, lower initial level of ATP may be less crucial, provided that the energy producing pathways are working correctly. Generally, this type of damage explains a decrease in spermatozoa motility and fertilizing ability, while further studies are required to describe the exact mechanisms and relations between cryodamage, metabolic activity, motility, and fertilization ability of sperm.

5.3.4 DNA Damage

The main motivation for sperm cryopreservation is long-term storage of male DNA that underlines the need for preventing DNA damage during freezing/thawing. There are a lot of studies pointing out the consequences of DNA damage during assisted reproduction (Drevet 2016, 2017). Widespread sperm count and motility assays can reflect only overall sperm quality, while assessment of genetic quality (DNA integrity) is often not performed either for sperm or for egg because it requires more complicated and time-consuming assays. Nevertheless, the DNA damage could arise as a cryodamage and generally will not affect any other parameters of sperm quality. During the fertilization, fusion of parental DNAs will create the genome of an embryo and will determine the quality of the progeny. Some natural amount of errors in the sperm DNA is acceptable, because the oocyte has repair mechanisms. However, this correction of DNA damages may not be complete and some errors might be transferred to the progeny; and the amount of unfixed anomalies will depend on their initial number. Thus, the success of fertilization is directly related to the ability of egg to repair DNA damage and the initial amount of errors in spermatozoa (Kopeika et al. 2003; Kopeika and Kopeika 2008). At some point, sperm nuclei may accumulate more errors than egg can repair, thus resulting in the absence of development or a wrong embryonic development, ultimately leading to the transfer of damaged DNA to the progeny (Drevet 2016, 2017). Several methods of DNA damage assessment have been elaborated so far, having their pros and cons, they provide different information and involve different experimental approaches, while dealing either with sperm or egg (Agarwal et al. 2003). In case of fish, the possibility of using a large quantity of gametes and comparatively short duration of embryo development make a fertilization test the most direct way for the quick evaluation of DNA damage, which represents a simultaneous overall test for sperm DNA damage and egg repairing capacity.

DNA damages of fish spermatozoa are usually associated with DNA oxidation or fragmentation during the freezing/thawing process or CPA toxicity (Ciereszko et al. 2005; Dietrich et al. 2005; Martínez et al. 2012; Gosálvez et al. 2014). The DNA damage was supposed to lead to the decreased fertilization rate and increase in amount of abnormally developing larva in many freshwater fish species (Cabrita et al. 2001; Gwo et al. 2003; Zhou et al. 2006; Pérez-Cerezales et al. 2010, 2011). However, several other studies on fish spermatozoa cryopreservation found no difference in the amount of DNA damages before and after cryopreservation or non-significance of this damages regarding the fertilization success (Labbe et al. 2001; Zilli et al. 2003; Cabrita et al. 2005), suggesting that DNA damage may be also related to individual properties of egg/sperm cells in different species.

Nevertheless, changes in composition of plasma membrane or seminal plasma content as well as the DNA/membrane integrity before and after cryopreservation do not reveal the mechanism of damage itself (Martínez-Páramo et al. 2009), and furthermore, they do not distinguish lethal from nonlethal damages of sperm. This stresses the importance of process evaluation and correct interpretation of the existing data.

5.4 Indications and Consequences of Cryodamage on Fish Sperm Functions. Evaluation of Sperm Quality After Cryopreservation

At a first glance, the methods for evaluation of sperm quality are generally the same for fresh and cryopreserved spermatozoa: DNA and membrane integrity, protein and lipids profiles, ATP level or metabolic activity and, of course, motility and fertilization tests (Cabrita et al. 2014; Asturiano et al. 2017; Martínez-Páramo et al. 2017). However, it is known that during freezing and thawing, spermatozoa are affected by numerous physical and chemical factors which result in the appearance of viable, nonviable, and damaged sperm subpopulations in the thawed sperm. The presence of such subpopulations in the investigated sperm suspension obscures the real cryopreservation effects on the fraction of spermatozoa that survived the freeze-thawing process. Here below, we discuss the possible contribution of these non-uniformities on the evaluation process, with particular attention on subpopulations which survived after nonlethal damages and consider the consequences of these damages for spermatozoa quality.

After ice crystallization onset, the appearing crystals usually grow fast and part of spermatozoa will be caught by the ice front, and might be squashed and cut by the ice crystals or in some cases the growing ice crystal might provoke intracellular crystallization after contact with the cell (Morris et al. 2012). This kind of damages can be observed during direct plunging of sperm suspension into the liquid nitrogen. In this case, the spermatozoa viability in most fish species will be tending to zero. The content of destroyed cells will be released to the medium, as well as fragments of membranes and flagella. Even in case of using the conventional freezing/thawing protocol optimized to cryopreserve the cells, part of spermatozoa will well survive cryopreservation and keep their motile ability and potential to fertilize, while others will be completely destroyed. In this case, the indices of quality/survival should be correctly selected to avoid errors in the overall evaluation, e.g. overestimation due to the fact that destroyed spermatozoa may "disappear" from the analysis after freezing/thawing. That is why the proper control of sperm concentration before and after thawing may show amount of destroyed (or sometimes agglutinated into clumps (Horvath et al. 2003)) spermatozoa, which will be otherwise excluded from the evaluation. Simple visual inspection of sample and observation of agglutination or fragmentation of destroyed spermatozoa may also indicate the presence of absolutely broken cells. In an ideal case, most spermatozoa should keep their original shape, the post-thaw concentration of cells in the suspension should not be changed, and the sample should include alive spermatozoa (motile and with intact membrane) mixed with dead ones (immotile, damaged, and nonfunctional). The correct level of cryopreservation efficiency could be estimated by only evaluating the ratio between live/dead or motile/static cells etc. During assessment of the latter, one should pay a special attention to avoid any overestimation. Too high concentration of spermatozoa in the observation field may result in erroneous calculation of the ratio of survival (Boryshpolets et al. 2013b, 2018). For example, motile spermatozoa have the tendency to accumulate close to any surface, especially that of a glass slide (Cosson et al. 2003; Boryshpolets et al. 2013a) while immotile sperm may stay in the middle layer of the drop (in a different focal plane), thus not considered and therefore not included in the counting. Special counting chambers and the same dilution ratio (sperm concentration) with activating media help to avoid this problem. In most cases, the calculation of number of cells with lethal damage (or amount of sperm survived freezing/thawing) is believed to be sufficient to evaluate success rate of cryopreservation method. Thus, motility estimation and membrane integrity tests ("live/dead" staining) are the most widely used methods.

Nevertheless, conducting fertilization test can be considered as essential in some cases, e.g. screening tests for new CPAs, or conservation program of new species (i.e., without developed cryo-technology) etc., and as mentioned above, level of sperm DNA damage and ability of egg to repair it may be checked simultaneously. Herewith, a special attention should be payed to setting up a correct fertilization protocol. For instance, considering a limiting sperm per egg ratio and avoiding too high concentration of sperm in fertilization media are essential, because too high indices will mask the intermediate changes in sperm quality, and in any case, these values should be kept the same for fresh and cryopreserved samples. In order to detect if the decrease in percentage of motility or other parameters of sperm may be compensated by increasing of the sperm concentration during fertilization, additional experimental trials may be required. Not less important is to take into account potential toxic effect of CPA, present in the medium that is why fertilization tests for that purpose might be conducted as well.

It might happen that the fertilization outcome (percentage of fertilized eggs/ developed embryos) is not correlated with the percentage of motile sperm after cryopreservation. In some experiments, even no fertilization could be observed in case of samples containing significant amount of motile spermatozoa (Psenicka et al. 2008, 2011; Boryshpolets et al. 2011), and this phenomenon was observed particularly in case of sturgeon species. Therefore, authors suggested that DMSO may have a side effect by untimely triggering of the acrosomal reaction, thus preventing any further fertilization. In general, the explanations could lie in CPA toxicity rendered directly on eggs/sperm and compromising the sperm/egg interaction or embryo development, at the same time it may be associated with nonlethal damages (changes) in the survived population which do not affect the motility percentage of sperm cells, but decrease their fertilization abilities. Usually these changes are associated with DNA damage, as discussed above, or changes in sperm motility/ behavior or both.

More detailed analysis of post-thaw spermatozoa quality could be done by the assessment of motility traits using computer-assisted sperm analysis (CASA). Such studies were already performed for many freshwater and marine fish species (Martínez-Páramo et al. 2009, 2017). The possible changes in the parameters observed by CASA may be associated with osmotic changes during freezing/ thawing process. A decrease in spermatozoa velocity may be associated with lower concentration of ATP, due to the volume changes. On the other hand, a lower velocity of spermatozoa could be an indicator of flagellar apparatus damage, e.g. caused by the loss of some microtubular units. Damage of flagella could be accompanied by injuries in the spermatozoon membrane, while in some cases it still may remain intact. Slightly shorter flagellum may be observed in motile spermatozoa after thawing. This particular phenomenon is usually explained by breakage of flagellum tip that occurred during freezing/thawing, while nonlethally damaged membrane may be repaired, thus keeping functional activity of sperm (Cosson, unpublished).

Other specific changes in post-thaw sperm could be related to violation of the membrane signaling pathways, and may lead to the motility activation delay, i.e. sperm start to move in only several seconds after mixing with water. This phenomenon could be related, among others, to the presence of cryoprotectant, which may interfere with the water transport through membrane, thus affecting the activation process in osmo-sensitive spermatozoa (Prokopchuk et al. 2015). The other reason for this delay may be related to direct changes in the membrane signal sensing ability. The same significant delay in sperm activation was also observed in fresh sperm samples with different lipid composition (Blecha et al. 2018) or activated for a second round (Boryshpolets et al. 2009a), as well as during activation under different temperatures (Boryshpolets, unpublished). Sometimes, this behavior may lead to transient activation and stop of sperm motility, working in switch on/off manner (presence/absence of signal).

Additional negative effect of high concentration of solutes may lead to a decreased tolerance to the subsequent hypotonic shock occurring at the sperm motility activation step, resulting in the appearance of damage and much shorter duration of sperm motility. In particular cases, changes in osmolality during freezing/thawing may lead to an activation of fish sperm motility, i.e. "cryoactivation" (Boryshpolets et al. 2009b; Dzyuba et al. 2010); when immotile during the freezing process, spermatozoa become activated during thawing. While this process will not necessarily affect the fertilization success, it clearly indicates the changes and effects that spermatozoa are experiencing during the process of cryopreservation. This phenomenon, as well as cryoresistance in general, is related to specific sperm properties and obviously is species specific. Most probably, lipid/protein composition defines fluidity/rigidity properties of membrane and as a result may affect fish sperm sensitivity to external signals, including ion concentration or osmolality. Further research of fish sperm membrane properties, osmoregulation, and signaling in general will help to understand this specific sperm behavior occurring after sperm cryopreservation, thus potentially contributing to the development of more efficient methods of cryopreservation or fertilization.

In some cases, cryopreservation may act as a mechanism of selecting "healthy" and good swimming spermatozoa (Boryshpolets et al. 2011; Horokhovatskyi et al. 2017, 2018), which potentially did not undergo changes in membrane, metabolism, motility etc., as compared to fresh sperm cells. Interestingly, the overall assessment of post-thaw indices of spermatozoa (sperm/seminal fluid protein and lipid profiles, ATP level, metabolic activity etc.) usually shows dramatic changes in the mentioned parameters, as discussed above. It is not clear whether these changes occur in all cells present in the suspension, but in varying extent, or if survived cells differ substantially from the damaged cells. Selection of undamaged cells could help to clear up this issue, e.g. to understand specific sperm properties, predefining their cryoresistance, or at least will provide an indication of nonlethal changes in sperm, which in turn will help to improve methods of cryopreservation. Recent study of Horokhovatskyi et al. (2018) made a step forward toward a better understanding of these phenomena and demonstrated that changes in protein profile are associated only with damaged cells, while the survived cells do not reveal dramatic changes compared to fresh cells.

5.5 Conclusion

In this chapter, we described the most common cryoinjuries and specificity of fish sperm quality evaluation after cryopreservation. Correct estimation of sperm quality after cryopreservation is essential not only for a correct application of cryopreserved samples and evaluation of cryopreservation methods efficiency but also for understanding of spermatozoa changes occurring during freeze-thawing process and their possible effect on sperm fertilizing capacity. Separation methods, which allow to extract only spermatozoa survived cryopreservation from total population of all spermatozoa in post-thaw samples, open new possibilities to study cryoinjury of spermatozoa. In this respect, a new basic question could be addressed: if the survived cryopreservation population of spermatozoa possesses any unique parameters which determine their cryoresistance or this population is random. It is worth mentioning, that it could be species specific, thus suggesting two directions of fish sperm cryopreservation improvement. If survived spermatozoa are selected due to their specific properties, then cryopreservation works as a selection mechanism. This will require further research for better understanding of the impact of this selection on the forthcoming progeny. New methods for the assessment of sperm quality before or after cryopreservation focused on indication of these specific properties could be developed. On the other hand, if the survived population is random, then the methods of cryopreservation can be improved to maximize post-thaw survival (selection of optimal concentration of cells and cryoprotectant in suspension during freezing, regime of freezing/thawing, additional agents, etc.). Vitrification is believed to be the most perspective in this case, since during this process no ice crystals are formed at all (due to fast cooling speed and/or high concentration of cryoprotector (Morris 2006)). This technique has been applied for fish sperm in several species with variable success (Cuevas-Uribe et al. 2011, 2015; Figueroa et al. 2013; Abed-Elmdoust et al. 2015). Usually the decrease in survival rate and presence of damaged cells are associated with recrystallization during warming of vitrified sample (Morris 2000; Morris et al. 2012). The possible solution for this could be an optimization of warming the vitrified samples, e.g. using microwaves or gold nanoparticles etc., basically providing higher speed and more efficient warming than conventional water bath (Wang et al. 2014, 2016; Khosla et al. 2017). Nevertheless, the stable success in whole vitrification/heating procedure is still not reached and further studies are needed (Martínez-Páramo et al. 2017).

Acknowledgments Authors are acknowledging following projects: Ministry of Education, Youth and Sports of the Czech Republic—project „CENAKVA "(LM2018099), project "Biodiversity" (CZ.02.1.01./0.0/0.0/16_025/0007370 Reproductive and genetic procedures for preserving fish biodiversity and aquaculture); the Grant Agency of the University of South Bohemia in Ceske Budejovice (125/2016/Z); the Czech Science Foundation (No. 16-02407Y); European Union's Horizon 2020 research and innovation program under grant agreement No. 652831 (AQUAEXCEL2020).

References

Abed-Elmdoust A, Farahmand H, Mojazi-Amiri B et al (2015) Novel droplet vitrification combined with fish antifreeze protein type III enhances cryoprotection of semen in wild endangered Persian sturgeon Acipenser persicus (Borodin, 1897). Aquac Res 46(10):2392–2397. https://doi.org/10.1111/are.12397

Agarwal A, Sharma RK, Nelson DR (2003) New semen quality scores developed by principal component analysis of semen characteristics. J Androl 24:343–352

Alavi SMH, Cosson J (2006) Sperm motility in fishes. (II) Effects of ions and osmolality: a review. Cell Biol Int 30:1–14. https://doi.org/10.1016/j.cellbi.2005.06.004

Asturiano JF, Cabrita E, Horváth A (2017) Progress, challenges and perspectives on fish gamete cryopreservation: a mini-review. Gen Comp Endocrinol 245:69–76. https://doi.org/10.1016/j.ygcen.2016.06.019

Beirao J, Cabrita E, Perez-Cerezales S et al (2011) Effect of cryopreservation on fish sperm subpopulations. Cryobiology 62:22–31. https://doi.org/10.1016/j.cryobiol.2010.11.005

Blecha M, Dzyuba B, Boryshpolets S et al (2018) Spermatozoa quality and sperm lipid composition in intensively cultured and wild burbot (Lota lota). Anim Reprod Sci 198:129–136. https://doi.org/10.1016/J.ANIREPROSCI.2018.09.011

Boryshpolets S, Dzyuba B, Drokin S (2009a) Pre-spawning water temperature affects sperm respiration and reactivation parameters in male carps. Fish Physiol Biochem 35:661–668. https://doi.org/10.1007/s10695-008-9292-4

Boryshpolets S, Dzyuba B, Rodina M et al (2009b) Freeze-thawing as the factor of spontaneous activation of spermatozoa motility in common Carp (Cyprinus carpio L.). Cryobiology 59:291–296. https://doi.org/10.1016/j.cryobiol.2009.08.005

Boryshpolets S, Dzyuba B, Rodina M et al (2011) Cryopreservation of sterlet (Acipenser ruthenus) spermatozoa using different cryoprotectants. J Appl Ichthyol 27:1147. https://doi.org/10.1111/j.1439-0426.2011.01866.x

Boryshpolets S, Cosson J, Bondarenko V et al (2013a) Different swimming behaviors of sterlet (Acipenser ruthenus) spermatozoa close to solid and free surfaces. Theriogenology 79:81–86. https://doi.org/10.1016/j.theriogenology.2012.09.011

Boryshpolets S, Kowalski RKK, Dietrich GJJ et al (2013b) Different computer-assisted sperm analysis (CASA) systems highly influence sperm motility parameters. Theriogenology 80:758–765. https://doi.org/10.1016/j.theriogenology.2013.06.019

Boryshpolets S, Sochorová D, Rodina M et al (2017) Cryopreservation of arp (Cyprinus carpio L.) sperm: impact of seeding and freezing rates on post-thaw outputs. Biopreserv Biobank 15:234–240. https://doi.org/10.1089/bio.2016.0065

Boryshpolets S, Kholodnyy V, Cosson J, Dzyuba B (2018) Fish sperm motility analysis: the central role of the flagellum. Reprod Fertil Dev 30:833. https://doi.org/10.1071/RD17478

Browne RK, Kaurova SA, Uteshev VK et al (2015) Sperm motility of externally fertilizing fish and amphibians. Theriogenology 83:1–13. https://doi.org/10.1016/j.theriogenology.2014.09.018

Cabrita E, Anel L, Herraéz MP (2001) Effect of external cryoprotectants as membrane stabilizers on cryopreserved rainbow trout sperm. Theriogenology 56:623–635. https://doi.org/10.1016/S0093-691X(01)00594-5

Cabrita E, Robles V, Rebordinos L et al (2005) Evaluation of DNA damage in rainbow trout (Oncorhynchus mykiss) and gilthead sea bream (Sparus aurata) cryopreserved sperm. Cryobiology 50:144–153. https://doi.org/10.1016/j.cryobiol.2004.12.003

Cabrita E, Martínez-Páramo S, Gavaia PJ et al (2014) Factors enhancing fish sperm quality and emerging tools for sperm analysis. Aquaculture 432:389–401. https://doi.org/10.1016/j.aquaculture.2014.04.034

Chen YK, Liu QH, Li J et al (2010) Effect of long-term cryopreservation on physiological characteristics, antioxidant activities and lipid peroxidation of red seabream (Pagrus major) sperm. Cryobiology 61:189–193. https://doi.org/10.1016/j.cryobiol.2010.07.003

Chen X, Zhu H, Hu C et al (2014) Identification of differentially expressed proteins in fresh and frozen-thawed boar spermatozoa by iTRAQ-coupled 2D LC-MS/MS. Reproduction 147:321–330. https://doi.org/10.1530/REP-13-0313

Ciereszko A, Wolfe TD, Dabrowski K (2005) Analysis of DNA damage in sea lamprey (Petromyzon marinus) spermatozoa by UV, hydrogen peroxide, and the toxicant bisazir. Aquat Toxicol 73:128–138. https://doi.org/10.1016/j.aquatox.2005.03.003

Correa LM, Thomas A, Meyers SA (2007) The macaque sperm actin cytoskeleton reorganizes in response to osmotic stress and contributes to morphological defects and decreased motility. Biol Reprod 77:942–953. https://doi.org/10.1095/biolreprod.107.060533

Cosson JJ (2007) The motility apparatus of fish spermatozoa. In: Fish spermatology. Alpha Science, Oxford, pp 281–316

Cosson J (2010) Frenetic activation of fish spermatozoa flagella entails short-term motility, portending their precocious decadence. J Fish Biol 76:240–279. https://doi.org/10.1111/j.1095-8649.2009.02504.x

Cosson JJ (2012) ATP, the sperm movement energizer. In: Kuestler E, Traugott G (eds) Adenosine triphosphate: chemical properties, biosynthesis and functions in cells. Nova Publisher Inc, Hauppauge, pp 1–46

Cosson M-P, Cosson J, Billard R (1991) Synchronous triggering of trout sperm is followed by an invariable set sequence of movement parameters whatever the incubation medium. Cell Motil Cytoskeleton 20:55–68. https://doi.org/10.1002/cm.970200107

Cosson J, Huitorel P, Gagnon C (2003) How spermatozoa come to be confined to surfaces. Cell Motil Cytoskeleton 54:56–63. https://doi.org/10.1002/cm.10085

Cuevas-Uribe R, Leibo SP, Daly J, Tiersch TR (2011) Production of channel catfish with sperm cryopreserved by rapid non-equilibrium cooling. Cryobiology 63:186–197. https://doi.org/10.1016/J.CRYOBIOL.2011.06.004

Cuevas-Uribe R, Chesney EJ, Daly J, Tiersch TR (2015) Vitrification of sperm from marine fish: effect on motility and membrane integrity. Aquac Res 46:1770–1784. https://doi.org/10.1111/are.12337

Desrosiers P, Légaré C, Leclerc P, Sullivan R (2006) Membranous and structural damage that occur during cryopreservation of human sperm may be time-related events. Fertil Steril 85:1744–1752. https://doi.org/10.1016/J.FERTNSTERT.2005.11.046

Dietrich GJ, Szpyrka A, Wojtczak M et al (2005) Effects of UV irradiation and hydrogen peroxide on DNA fragmentation, motility and fertilizing ability of rainbow trout (Oncorhynchus mykiss) spermatozoa. Theriogenology 64:1809–1822. https://doi.org/10.1016/j.theriogenology.2005.04.010

Dietrich MA, Arnold GJ, Fröhlich T et al (2015) Proteomic analysis of extracellular medium of cryopreserved carp (Cyprinus carpio L.) semen. Comp Biochem Physiol Part D Genomics Proteomics 15:49–57. https://doi.org/10.1016/j.cbd.2015.05.003

Dreanno C, Cosson J, Suquet M et al (1999) Effects of osmolality, morphology perturbations and intracellular nucleotide content during the movement of sea bass (Dicentrarchus labrax) spermatozoa. J Reprod Fertil 116:113–125. https://doi.org/10.1530/jrf.0.1160113

Drevet JR (2016) Sperm DNA damage and assisted reproductive technologies: reasons to be cautious! Basic Clin Androl 26:11. https://doi.org/10.1186/s12610-016-0038-3

Drevet JR (2017) Sperm DNA integrity testing: a valuable addition to the tool box of infertility clinicians. Transl Androl Urol 6:S590–S591. https://doi.org/10.21037/tau.2017.05.08

Drokin SI, Kopeika EF, Grishchenko VI (1989) Differences in resistance to cryopreservation and specific features of lipid compositions of spermatozoa of sea and fresh-water species of fishes. Dokl Akad Nauk SSSR 304:1493–1496

Dzyuba B, Boryshpolets S, Rodina M et al (2010) Spontaneous activation of spermatozoa motility by routine freeze-thawing in different fish species. J Appl Ichthyol 26:720–725. https://doi.org/10.1111/j.1439-0426.2010.01553.x

Dzyuba B, Bondarenko O, Fedorov P et al (2017) Energetics of fish spermatozoa: the proven and the possible. Aquaculture 472:60–72. https://doi.org/10.1016/j.aquaculture.2016.05.038

Elliott GD, Wang S, Fuller BJ (2017) Cryoprotectants: a review of the actions and applications of cryoprotective solutes that modulate cell recovery from ultra-low temperatures. Cryobiology 76:74–91. https://doi.org/10.1016/j.cryobiol.2017.04.004

Figueroa E, Risopatrón J, Sánchez R et al (2013) Spermatozoa vitrification of sex-reversed rainbow trout (Oncorhynchus mykiss): effect of seminal plasma on physiological parameters. Aquaculture 372–375:119–126. https://doi.org/10.1016/J.AQUACULTURE.2012.10.019

Figueroa E, Valdebenito I, Zepeda AB et al (2017) Effects of cryopreservation on mitochondria of fish spermatozoa. Rev Aquac 9:76–87. https://doi.org/10.1111/raq.12105

Fuller B, Paynter S (2004) Fundamentals of cryobiology in reproductive medicine. Reprod Biomed Online 9:680–691. https://doi.org/10.1016/S1472-6483(10)61780-4

Gómez-Torres MJ, Medrano L, Romero A et al (2017) Effectiveness of human spermatozoa biomarkers as indicators of structural damage during cryopreservation. Cryobiology 78:90–94. https://doi.org/10.1016/J.CRYOBIOL.2017.06.008

Gosálvez J, López-Fernández C, Hermoso A et al (2014) Sperm DNA fragmentation in zebrafish (Danio rerio) and its impact on fertility and embryo viability—implications for fisheries and aquaculture. Aquaculture 433:173–182. https://doi.org/10.1016/j.aquaculture.2014.05.036

Gwo J, Wu C, Chang WP, Cheng H (2003) Evaluation of damage in Pacific oyster (Crassostrea Gigas) spermatozoa before and after cryopreservation using comet assay. Cryo Letters 24 (3):171–180

He S, Woods LC (2004) Effects of dimethyl sulfoxide and glycine on cryopreservation induced damage of plasma membranes and mitochondria to striped bass (Morone saxatilis) sperm. Cryobiology 48:254–262. https://doi.org/10.1016/j.cryobiol.2004.01.009

Horokhovatskyi Y, Sampels S, Cosson J et al (2016a) Lipid composition in common carp (Cyprinus carpio) sperm possessing different cryoresistance. Cryobiology 73:282–285. https://doi.org/10.1016/J.CRYOBIOL.2016.08.005

Horokhovatskyi Y, Sampels S, Dzyuba B et al (2016b) Heterogeneity of cryoresistance in common carp sperm. Anim Reprod Sci 169:114–115. https://doi.org/10.1016/j.anireprosci.2016.03.050

Horokhovatskyi Y, Rodina M, Asyabar HD et al (2017) Consequences of uncontrolled cooling during sterlet (Acipenser ruthenus) sperm cryopreservation on post-thaw motility and fertilizing ability. Theriogenology 95:89–95. https://doi.org/10.1016/j.theriogenology.2017.03.007

Horokhovatskyi Y, Dietrich MA, Lebeda I et al (2018) Cryopreservation effects on a viable sperm sterlet (Acipenser ruthenus) subpopulation obtained by a Percoll density gradient method. PLoS One 13(8):e0202514. https://doi.org/10.1371/journal.pone.0202514

Horvath A, Miskolczi E, Urbanyi B (2003) Cryopreservation of common carp sperm. Aquat Living Resour 16:457–460. https://doi.org/10.1016/s0990-7440(03)00084-6

Khosla K, Wang Y, Hagedorn M et al (2017) Gold nanorod induced warming of embryos from the cryogenic state enhances viability. ACS Nano 11:7869–7878. https://doi.org/10.1021/acsnano.7b02216

Kopeika E, Kopeika J (2008) Variability of sperm quality after cryopreservation in fish. In: Fish spermatology. Alpha Science, Oxford, pp 347–396

Kopeika J, Kopeika E, Zhang T et al (2003) Detrimental effects of cryopreservation of loach (Misgurnus fossilis) sperm on subsequent embryo development are reversed by incubating fertilised eggs in caffeine. Cryobiology 46:43–52. https://doi.org/10.1016/S0011-2240(02)00162-1

Labbe C, Martoriati A, Devaux A, Maisse G (2001) Effect of sperm cryopreservation on sperm DNA stability and progeny development in rainbow trout. Mol Reprod Dev 60:397–404. https://doi.org/10.1002/mrd.1102

Li P, Hulak M, Koubek P et al (2010) Ice-age endurance: the effects of cryopreservation on proteins of sperm of common carp, Cyprinus carpio L. Theriogenology 74:413–423. https://doi.org/10.1016/j.theriogenology.2010.02.024

Madeddu M, Berlinguer F, Pasciu V et al (2010) Differences in semen freezability and intracellular ATP content between the rooster (Gallus gallus domesticus) and the Barbary partridge

(Alectoris barbara). Theriogenology 74:1010–1018. https://doi.org/10.1016/j.theriogenology.2010.05.001

Martínez JG, García VA, Carrasco SP (2012) DNA fragmentation and membrane damage of bocachico Prochilodus magdalenae (Ostariophysi: Prochilodontidae) sperm following cryopreservation with dimethylsulfoxide and glucose. Neotrop Ichthyol 10:577–586. https://doi.org/10.1590/S1679-62252012005000018

Martínez-Páramo S, Pérez-Cerezales S, Gómez-Romano F et al (2009) Cryobanking as tool for conservation of biodiversity: effect of brown trout sperm cryopreservation on the male genetic potential. Theriogenology 71:594–604. https://doi.org/10.1016/j.theriogenology.2008.09.034

Martínez-Páramo S, Horváth Á, Labbé C et al (2017) Cryobanking of aquatic species. Aquaculture 472:156–177. https://doi.org/10.1016/j.aquaculture.2016.05.042

Mazur P, Leibo SP, Chu EHY (1972) A two-factor hypothesis of freezing injury: evidence from Chinese hamster tissue-culture cells. Exp Cell Res 71:345–355. https://doi.org/10.1016/0014-4827(72)90303-5

Morris J (2000) Asymptote cool guide to cryopreservation, vol 44. Asymptote Ltd, Cambridge, pp 1–42

Morris GJ (2006) Rapidly cooled human sperm: no evidence of intracellular ice formation. Hum Reprod 21:2075–2083. https://doi.org/10.1093/humrep/del116

Morris GJ, Acton E, Murray BJ et al (2012) Freezing injury: the special case of the sperm cell. Cryobiology 64:71–80. https://doi.org/10.1016/j.cryobiol.2011.12.002

Nynca J, Arnold GJ, Fröhlich T, Ciereszko A (2015) Cryopreservation-induced alterations in protein composition of rainbow trout semen. Proteomics 15:2643–2654. https://doi.org/10.1002/pmic.201400525

Perchec G, Jeulin C, Cosson J et al (1995) Relationship between sperm ATP content and motility of carp spermatozoa. J Cell Sci 108:747–753

Pérez-Cerezales S, Martínez-Páramo S, Beirão J, Herráez MP (2010) Fertilization capacity with rainbow trout DNA-damaged sperm and embryo developmental success. Reproduction 139:989–997. https://doi.org/10.1530/REP-10-0037

Pérez-Cerezales S, Gutiérrez-Adán A, Martínez-Páramo S et al (2011) Altered gene transcription and telomere length in trout embryo and larvae obtained with DNA cryodamaged sperm. Theriogenology 76:1234–1245. https://doi.org/10.1016/j.theriogenology.2011.05.028

Prokopchuk G, Dzyuba B, Bondarenko O et al (2015) Motility initiation of sterlet sturgeon (Acipenser ruthenus) spermatozoa: describing the propagation of the first flagellar waves. Theriogenology 84:51–61. https://doi.org/10.1016/j.theriogenology.2015.02.011

Psenicka M, Dietrich GJ, Wojtczak M et al (2008) Acrosome staining and motility characteristics of sterlet spermatozoa after cryopreservation with use of methanol and DMSO. Cryobiology 56:251–253. https://doi.org/10.1016/j.cryobiol.2008.03.006

Psenicka M, Kaspar V, Alavi SMH et al (2011) Potential role of the acrosome of sturgeon spermatozoa in the fertilization process. J Appl Ichthyol 27:678–682. https://doi.org/10.1111/j.1439-0426.2010.01642.x

Suquet M, Dreanno C, Fauvel C et al (2000) Cryopreservation of sperm in marine fish. Aquac Res 31:231–243. https://doi.org/10.1046/j.1365-2109.2000.00445.x

Takai H, Morisawa M (1995) Change in intracellular K+ concentration caused by external osmolality change regulates sperm motility of marine and freshwater teleosts. J Cell Sci 108:1175–1181

Tiersch TR (2008) Strategies for commercialization of cryopreserved fish semen. Rev Bras Zootec 37:15–19. https://doi.org/10.1590/S1516-35982008001300003

Tiersch TR, Mazik PM (2011) Cryopreservation in aquatic species, 2nd edn. World Aquaculture Society, Baton Rouge

Wang T, Zhao G, Liang XM et al (2014) Numerical simulation of the effect of superparamagnetic nanoparticles on microwave rewarming of cryopreserved tissues. Cryobiology 68:234–243. https://doi.org/10.1016/J.CRYOBIOL.2014.02.002

Wang J, Zhao G, Zhang Z et al (2016) Magnetic induction heating of superparamagnetic nanoparticles during rewarming augments the recovery of hUCM-MSCs cryopreserved by vitrification. Acta Biomater 33:264–274. https://doi.org/10.1016/J.ACTBIO.2016.01.026

Westfalewicz B, Dietrich MA, Ciereszko A (2015) Impact of cryopreservation on bull (Bos taurus) semen proteome1. J Anim Sci 93:5240–5253. https://doi.org/10.2527/jas.2015-9237

Zhou B, Liu W, Siu WHL et al (2006) Exposure of spermatozoa to duroquinone may impair reproduction of the common carp (Cyprinus carpio) through oxidative stress. Aquat Toxicol 77:136–142. https://doi.org/10.1016/j.aquatox.2005.11.006

Zilli L, Schiavone R, Zonno V et al (2003) Evaluation of DNA damage in Dicentrarchus labrax sperm following cryopreservation. Cryobiology 47:227–235. https://doi.org/10.1016/j.cryobiol.2003.10.002

Zilli L, Schiavone R, Zonno V et al (2005) Effect of cryopreservation on sea bass sperm proteins. Biol Reprod 72:1262–1267. https://doi.org/10.1095/biolreprod.104.036202

Zilli L, Schiavone R, Storelli C, Vilella S (2008) Effect of cryopreservation on phosphorylation state of proteins involved in sperm motility initiation in sea bream. Cryobiology 57:150–155. https://doi.org/10.1016/j.cryobiol.2008.07.006

Cryopreservation: History and Development

6

Judith Betsy and Stephen Kumar

Abstract

Cryopreservation is an age-old technology which was practised in different forms. It has evolved from preserving cells in snow to liquid nitrogen. As necessity is the mother of invention, the need for mankind invents new technologies. Cryopreservation is also such a technology which helped increasing the livestock production. It is a technology which made human sperm bank and stem cell storage possible. In this chapter, various historical developments of cryopreservation in all fields of life science with special emphasis on fisheries are discussed.

Keywords

Cryopreservation · History · Cryogenic · Cryonic · Fish · Human

6.1 Introduction

Cryopreservation is the process of preserving cells, tissues, organs, or any biological material at −196 °C in liquid nitrogen to maintain their viability. Preserving cells in such a low temperature arrests all biological activities, including biochemical reactions that lead to cell death and DNA degradation. Through cryopreservation, cells can be preserved for centuries which can be revived later upon thawing.

J. Betsy (✉)
Department of Aquaculture, Fisheries College and Research Institute, Tamil Nadu Dr. J. Jayalalithaa Fisheries University, Tuticorin, India

S. Kumar
Directorate of Sustainable Aquaculture, Tamil Nadu Dr. J. Jayalalithaa Fisheries University, Thanjavur, India

J. Betsy, S. Kumar (eds.), *Cryopreservation of Fish Gametes*,
https://doi.org/10.1007/978-981-15-4025-7_6

However, it is not always possible to revive all the cells after thawing because intracellular ice formation has to be controlled to keep the cell membrane intact and the cells alive. This can be accomplished by following optimum freezing rate and workable diluent. Diluent consists of extender and cryoprotectants which can range from simple to complex media. Selection of these diluents varies between different cells and different species which can be understood only if the spermatology of the species is known.

Research on spermatology and gamete preservation dates back to 1600s (Sherman 1964). These studies were mostly on short-term preservation with the gametes. The actual cryopreservation studies started only after the development of artificial insemination (AI) during the late 1950s and early 1960s.

Spermatology or studies of spermatozoa and their qualities started as early as 1776 with Italian Catholic priest, Lazzaro Spallanzani who did works on amphibian sperm. He identified, cooled, and heated amphibian sperm and measured the unitary dimension of sperm to fertilize egg. He further determined the ratio between "spermatic particle" and egg for amphibian to be about one to one billion. In the same year, he reported that when cooled by snow, human sperm could become motionless (Spallanzani 1776). It is also pointed out that the effects of cold temperatures on cells have been investigated from 1787 (Luyet and Gehenio 1940). But it is also true that efforts to actually freeze sperm did not begin until the mid-1800s.

It becomes important to know about cryobiology, cryogenics, and cryonics while we deal with cryopreservation. These terminologies have acquired very much importance in recent years.

6.1.1 Cryobiology

The word cryobiology is derived from the Greek words κρῦος (kryos) which means "cold," βίος (bios) meaning "life" and λόγος (logos) meaning "word" here called science. It was coined in the year 1921. It is the branch of biology that studies the effects of low temperatures on organisms.

6.1.2 Cryogenics

The term cryogenics was coined in the year 1899. Cryogenics is defined as the scientific study of materials and their behavior at extremely low temperatures. The term cryogenics was derived from the Greek *cryo*, which means "cold," and *genic*, which means "producing."

6.1.3 Cryonics

In the process of cryonics, animals and humans are cryopreserved so that they can be revived in future. The first corpse to be frozen was that of Dr. James Bedford in 1967 (Iserson 2001). According to reports, about 250 bodies were cryopreserved in the United States as of 2014 and 1500 people had made arrangements for cryopreservation after their legal death (Moen 2015).

The procedure for cryonics begins within minutes of death (Robert and David 2014) when they are legally dead. It uses cryoprotectants to prevent ice formation during cryopreservation (Best 2008). According to Hannah (2016), it is unlikely that a corpse could be reanimated after undergoing vitrification, which causes damage to the brain including its neural networks.

6.2 Milestones in Cryopreservation

1683	The first documented cryobiological study by Sir Robert Boyle who documented the effects of freezing temperatures on living animals in his monograph "New Experiments and Observations Touching Cold"
1949	Ernest John Christopher Polge discovered glycerol as a cryoprotectant
1953	Jerome K. Sherman founded the world's first sperm bank from which the first human birth from cryopreserved sperm was recorded
1953	Bunge and Sherman gave first birth from frozen ejaculated sperm in human
1959	Lovelock and Bishop discovered DMSO
1964	The Society for Cryobiology was founded to bring together those from the biological, medical, and physical sciences who had a common interest in the effect of low temperatures on biological systems. These scientists call themselves "cryogenicists"
1965	Peter Mazur demonstrated the relation between the speed of freezing and intracellular ice formation, and Stanley Leibo demonstrated that each cell type has their unique freezing curve
1972	D. Whittingham, S. Leibo, and P. Mazur applying a slow freezing process achieved the first embryo freezing in mouse
1982	The successful cryopreservation of mammalian embryos was first achieved
1983	Alan Trounson frozen early human embryos 1–3 days after fertilization and successfully achieved pregnancy
1983	First child birth from a cryopreserved human embryo
1985	Protocol for freezing human embryos using propanediol and sucrose was established
1986	First child birth from a cryopreserved human oocyte
1986	Christopher Chen successfully frozen and thawed human oocytes from which a twin pregnancy was achieved after insemination and replacement in utero
1987	It was found that morphologically normal embryos could be frozen more successfully than morphologically abnormal embryos
1987	Records stated that at least 63 babies had been born from cryopreserved human embryos to date
1988	The first attempt to freeze immature oocytes was made

(continued)

1995	Devrowy gave first birth from frozen human epididymal sperm
1996	Gil-Salmon and his team gave first birth from frozen testicular sperm
2004	First child birth from the cryopreserved ovarian tissue was reported

6.3 Fish Spermatozoa Preservation

Fish spermatozoa are unique in nature with distinctive characters. The quality and quantity of spermatozoa are species dependent and also individual dependent. In general, they are immotile and only very few species have motile spermatozoa. The immotile spermatozoa need an activation agent that is normally the surrounding water medium. They are very tiny and their size ranges from 45 to 55 μm for the fishes which have internal fertilization. Fishes which exhibit external fertilization have "aquasperms" that have ovoid or spherical nucleus that measure less than 5 μm in its maximal extension. Their overall length may vary from 35 to 45 μm which are smaller than the bacteria. Therefore, their density is too high ranging from 0.1×10^9 to 50×10^{12}cells/ml. Their life is also very short and hence, once collected they should be used for fertilization within short period or else they will be spoiled due to external contaminants like bacteria and fungi. Apart from contamination, they are more possibly influenced or affected by the temperature and evaporation of seminal plasma. Therefore, it is necessary to preserve them after proper processing. The spermatozoa can be preserved by either of the following two ways and maintained for future use during off-season.

- Short-term preservation
- Long-term preservation

6.3.1 Short-Term Preservation

The history of fish gamete preservation is believed to have started during 1850s. Preservation of spermatozoa of pike, carp, perch, and barbs was done during 1853 (de Quatrefages 1853). Following these investigations, experiments were conducted involving the storage of salmonid sperm at temperatures from 0 to 9 °C (Scheuring 1924; Bennigton 1936; Smith and Quistorff 1943; Barrett 1951).

In short-term preservation, the spermatozoa along with the seminal plasma are preserved for a short duration either in diluted or undiluted form. Short-term preservation or preservation of male gametes under chilled condition at around 5 °C is simpler and easily adoptable. Therefore, it has greater scope for adoption in the field. Sizable quantum of research was also done in this line and many researchers reported on the short-term preservation and its effect on the spermatological parameters for many species of fishes (Hara et al. 1982; Harvey and Kelley 1984a, b; Moore 1987; DiLauro et al. 1994; Wayman et al. 1998). Short-term preservation of spermatozoa is generally done with refrigerator or in an ice box or an insulated box. No special equipment is required and no big cost is involved. The duration of such preservation is limited to less than a week.

The attempts on short-term preservation are believed to be around 95 years old, as the oldest report is seemed to be during 1923. When stored at 5–6 °C, Atlantic salmon (*Salmo salar*) sperm maintained high level of fertility up to 2.5 days (Brofeldt 1923). Barrett (1951) later attempted the short-term preservation of sperm from chum salmon (*Oncorhynchus keta*) at 2.5–5.8 °C and came with the result of 50% fertilization after 4 days of storage. When lake trout (*Salvelinus namaycush*) sperm was stored at 34 °F, it resulted in 80% fertilization after 24 h (Gibson 1952). Henderson and Dewar (1959) tried short-term preservation of undiluted sperm of brook trout (*Salvelinus fontinalis*) and reported higher hatching yield when stored for 5 days under chilled condition. When the milt of same species was stored at 2–3 °C, fertility was possible up to 5 days (Truscott et al. 1968). Dushkina (1975) reported that when sperm of Pacific herring (*Clupea pallasi*) was stored at 0.8–1 °C, fertility can be retained for 3 weeks. Buyukhatipoglu and Holtz (1978) reported that undiluted milt of trout had motility for 8.3 ± 2.7 days when stored at a temperature of 4 °C. Similar short-term preservation was observed and reported for salmon also. In Fringed-lipped peninsula carp (*Labeo fimbriatus*), the spermatozoa stored at 4 °C and room temperature retained motility up to 84 and 24 h, respectively (Akash 2001). Thus, there have been reports specifying the positive results of short-term storage in low temperatures. Stoss et al. (1987) observed better survival rates of rainbow trout sperm until day 16 when kept undisturbed.

Quality changes in the milt after storage in low temperature ranged from 1 to 7 °C have been reported by some researchers (Buyukhatipoglu and Holtz 1978; Stoss et al. 1987). Harvey and Kelley (1984b) found decline in post activation motility of *Sarotherodon mossambicus* sperm within 60–120 h when stored at 5 °C. Henderson and Dewar (1959) reported higher hatching yield from undiluted brook trout (*S. fontinalis*) sperm, stored for 5 days under chilled condition than that of the diluted sperm fluid. Basavaraja and Hegde (2005) reported that motility of the spermatozoa of *Tor khudree* stored at 4 °C or room temperature showed a slow to moderate decrease and the fall in motility rate was much quicker at room temperature than at 4 °C. They also stated that spermatozoa of *T. khudree* can be preserved in a motile state for 4–5 days. Agarwal et al. (2013) reported that in undiluted snow trout (*Schizothoraichthys progastus*) semen, 50% of sperm remain motile up to 3 days of storage.

6.3.2 Long-Term Preservation

Long-term preservation is the storage of spermatozoa for an indefinite period. This is possible only by "cryopreservation" of the spermatozoa. Period between 1935 and 1945 was more potential in the investigations of storage of spermatozoa at very low temperatures. Both in mammals and animals, a number of scientists observed that sperm could survive freezing and storage temperatures as low as −321 °F. It is interesting to note that in 1939, investigators from several universities founded the American Society of Animal Production (ASAP) to discuss and develop protocols for semen collection, evaluation, and preservation from domestic animals, especially the bull (Foote 1998). However, successful cryopreservation attempts were not reported for many cells due to the poor results obtained in getting the viable cells back after cryopreservation.

The modern sperm cryobiology was established when Polge et al. (1949) discovered the use of glycerol as a cryoprotectant. The discovery of cryoprotectants paved way for new type of cryobiological study, because new compounds had been identified that were used in the optimization of cryopreservation protocols. They successfully cryopreserved fowl sperm and produced chicks using cryopreserved spermatozoa. Dr. Jerome K. Sherman, an American pioneer in sperm freezing further refined this process in 1953 by introducing a simple method of preserving human sperm using glycerol. He combined this with a slow cooling of sperm, and storage with solid carbon dioxide as a refrigerant. He also demonstrated for the first time that frozen sperm, when thawed, were able to fertilize an egg and induce its normal development.

In 1953, Blaxter cryopreserved spermatozoa of Herring and proceeded up to artificial fertilization with the cryopreserved spermatozoa at the Marine Laboratory, Aberdeen. This must have been a standard report that stimulated large scale research in this line.

It is interesting to note that in nature also, cryopreservation concept is found working in the organisms like Water Bears living in polar region which should live there at ultralow subzero temperature. They are found to survive in the freezing low temperatures by replacing their internal water with the natural sugar trehalose. This sugar is found to prevent water from crystallization.

Similar methodology is followed for the spermatozoa cryopreservation also. The spermatozoa collected are diluted with some chemicals that are similar to that of their seminal plasma in pH, osmolality, and other features. Cryoprotectants are added for protection and stored in a cryogen that can produce very low temperature due to its varied state in existence (e.g., liquid nitrogen at −196 °C). Thus, cryopreservation involves cooling of a cell and storage at a temperature where all metabolic processes are arrested. Under cryopreservation, cells can be stored for indefinite period if properly processed and cryopreserved.

Cryopreservation is done in three methods, viz., isochoric cryopreservation, isobaric cryopreservation, and hyperbaric cryopreservation.

6.3.2.1 Isochoric Cryopreservation

In Isochoric (constant volume) cryopreservation, mostly biological materials are cryopreserved at low temperatures and freezing is done in a constant volume chamber. During isochoric cryopreservation, an increase in intracellular ionic concentration during freezing can cause damage to cell components.

6.3.2.2 Isobaric Cryopreservation

Isobaric (constant pressure) cryopreservation is done by keeping the pressure constant at 1 atm. In this method, compounds that depress the freezing point of the solution are added to reduce chemical damages at high subzero or negative temperatures.

6.3.2.3 Hyperbaric Cryopreservation

This is the method followed for freezing cells. It reduces the ionic concentration at subzero temperatures. It also leads to elevated pressure which is followed by rapid freezing leading to reduction of ice crystal.

Although there had been deep interests among scientists in the sperm cryopreservation, this area of research has assumed a status only when there was a real practical application of cryopreserved spermatozoa in the dairy industry. There was a need for long-term storage methods for bull sperm to attempt artificial fertilization in the cattle. With the continued development of AI, the dairy industry in the United States started progressive breeding programs and began approaching the industry from a genetic perspective. During 1980s, research programs across the country began investigating sperm biology in terms of collection, use of equipment, evaluation of sperm morphology, and composition of extenders for both sperm biology and fundamental cryobiology (Foote 1998).

After 2000, it has been observed that like diary industry cryopreservation attempts have been done in aquaculture industry also. However, it is not well recognized due to lesser application. Therefore, it can be said that although there is a technology available for cryopreserving the spermatozoa of cultivable fishes, as it is observed in the large animals' gamete cryopreservation, in fishes also it can gain importance and attention only when the technology is applied in the field in true sense and results are visible. Until that time, cryopreservation of fish gametes will remain only in research. So research must continue to achieve the goal of large scale application of cryopreservation of fish gametes.

6.4 Cryopreservation in Aquaculture

Although cryopreservation is not a recent happening, in aquaculture it is found adopted only in recent years. Advent of artificial insemination has brought in revolutionary changes in the dairy industry. Cryopreservation of spermatozoa at subzero temperature has been considered as a viable technique for the success of organized breeding programs through artificial insemination. Introduction of this biotechnological approach in fish breeding program becomes necessary to ensure successful breeding and continuous seed production even during monsoon failures. While induced breeding helps in making the female brooders to evaluate and produce eggs, the cryopreserved male gametes can possibly help in the nonuse of male brooders. In recent years, this technology has gained attention for improving offspring quality and availability of fish fry for stocking throughout the year. Cryopreservation and application of cryopreserved spermatozoa for the breeding of fishes may be a possible answer to produce quality seeds and genetically improved varieties overcoming the adversely affecting natural monsoon conditions.

Cryopreservation of gametes increases the longevity of gametes for many years without affecting their efficiency or potency. In aquaculture, it principally focuses on crossbreeding of fishes within the populations and aids in saving endangered species by facilitating the storage of their gametes in the gene bank. Production of hybrids with desired characters is a possible major advantage in the application of cryopreservation in aquaculture. Through cryopreservation gamete availability of sexes can be synchronized, economy in sperm can be obtained, broodstock management can be simplified, gametes from different fish farms can be transported and germplasm storage can be done for genetic selection programs or conservation of species.

Sperm cryopreservation can lead to the germ storage of all transgenic lines. Since DNA damage may impair fertility or embryo development, studies on DNA have been made using approaches like comet assay (single cell gel electrophoresis), TUNEL (terminal deoxynucleotidyl transferase-nick-end-labeling), SCSA (sperm chromatin structure assay), and the analysis of specific DNA sequences using RT-PCR. In addition to being a simple technique for the preservation of gametes, cryopreservation can contribute immensely for the gene mapping studies in aquaculture and subsequent gene manipulation.

The fishes for which the cryopreservation was done successfully either in laboratory or in the field have been listed in the Tables 6.1 and 6.2.

Table 6.1 Successful cryopreservation of spermatozoa in fish and shrimps

Species	Scientific name	Stage	References
Herring	*Clupea harengus*	Testes	Blaxter (1953)
Atlantic cod	*Gadus morhua*	Spermatozoa	Mounib et al. (1968); Rideout et al. (2003); DeGraaf and Berlinsky (2004)
Grey mullet	*Mugil cephalus*	Spermatozoa	Chao et al. (1975)
Rainbow trout	*Oncorhynchus mykiss*	Spermatozoa	Ciereszko and Dabroski (1996); Lahnsteiner et al. (1997); Cabrita et al. (2001); Tekin et al. (2003); Bozkurt et al. (2005)
Bluefin tuna	*Thunnus thynnus*	Spermatozoa	Doi et al. (1982)
Giant freshwater prawn	*Macrobrachium rosenbergii*	Spermatophore	Chow (1982)
Atlantic salmon	*Salmo salar*	Spermatozoa	Stoss and Refstie (1983); Jodun et al. (2006)
Common carp	*Cyprinus carpio*	Spermatozoa	Kurokura et al. (1984); Magyary et al. (1996); Lahnsteiner et al. (2000); Betsy and Kumar (2015a, b, c, 2016); Betsy et al. (2016, 2017, 2019)
Red Tilapia	*Oreochromis* sp.	Spermatozoa	Chao et al. (1987)
Catla	*Catla catla*	Spermatozoa	Kumar (1989)
Mrigal	*Cirrhinus cirrhosus*	Spermatozoa	Kumar (1989); Sarder et al. (2009)
Mrigal	*C. mrigala*	Spermatozoa	Betsy and Kumar (2015c, d)
Rohu	*Labeo rohita*	Spermatozoa	Kumar (1989); Sarder et al. (2011)
Gilthead seabream	*Sparus aurata*	Spermatozoa	Fabbrocini et al. (2000); Cabrita et al. (2005a, b)
Yellowfin bream	*Acanthopagrus australis*	Spermatozoa	Thorogood and Blackshaw (1992)
Atlantic halibut	*Hippoglossus hippoglossus*	Spermatozoa	Billard et al. (1993)
Black grouper	*Epinephelus malabaricus*	Spermatozoa	Gwo (1993)
European catfish	*Silurus glanis*	Spermatozoa	Linhart et al. (1993)
Sea bass	*Lates calcarifer*	Spermatozoa	Palmer et al. (1993)
Cobia	*Rachycentron canadum*	Spermatozoa	Caylor et al. (1994)
White fish	*Coregonus* sp.	Spermatozoa	Piironen (1994)
Channel catfish	*Ictalurus punctatus*	Spermatozoa	Tiersch et al. (1994); Christensen and Tiersch (1997, 2005)
Northern pike	*Esox lucius*	Spermatozoa	Babiak et al. (1995); Lahnsteiner et al. (1998); Babiak et al. (1999)
Muskellunge	*Esoxmas quinongy*	Spermatozoa	Glogowski et al. (1999)
Mud crab	*Scylla serrata*	Spermatophore	Bhavanishankar and Subramoniam (1997)
Asp	*Aspius aspius*	Spermatozoa	Babiak et al. (1998)

(continued)

Table 6.1 (continued)

Species	Scientific name	Stage	References
Gold fish	*Carassius auratus*	Spermatozoa	Bercsényi et al. (1998)
Turbot	*Psetta maxima*	Spermatozoa	Suquet et al. (1998); Dreanno et al. (1997)
Turbot	*Scophthalmus maximus*	Spermatozoa	Chereguini et al. (2003)
Siberian sturgeon	*Acipenser baeri*	Spermatozoa	Glogowski et al. (2002)
Japanese eel	*Anguilla japonica*	Spermatozoa	Tanaka et al. (2002)
Mullet	*Liza parsia*	Spermatozoa	Bhavani et al. (2003)
European eel	*Anguilla anguilla*	Spermatozoa	Müller et al. (2004)
Matrinxã	*Brycon orthotaenia*	Spermatozoa	Melo and Godinho (2006)
Red seabream	*Pagrus major*	Spermatozoa	Liu et al. (2006)
Black tiger shrimp	*Penaeus monodon*	Spermatophore	Vuthiphandchai et al. (2007)
Brazilian flounder	*Paralichthys orbignyanus*	Spermatozoa	Lanes et al. (2008)
Pacific white shrimp	*Litopenaeus vannamei*	Spermatophore	Chao et al. (2009)
Grass carp	*Ctenopharyngodon idella*	Spermatozoa	Bozkurt et al. (2009)
Banana shrimp	*Penaeus merguiensis*	Spermatophore	Memon et al. (2011)
African catfish	*Clarias gariepinus*	Spermatozoa	Kamaruding et al. (2012)
Loach	*Misgurnus anguillicaudatus*	Spermatozoa	Yasui et al. (2009)

Table 6.2 Successful cryopreservation of embryo in fish species

Species	Scientific name	Stage	References
Common carp	*Cyprinus carpio*	Embryo	Zhang et al. (1989)
Catla	*Catla catla*	Embryo	Ahammad et al. (1998)
Mrigal	*Cirrhinus mrigala*	Embryo	Ahammad et al. (1998)
Rohu	*Labeo rohita*	Embryo	Ahammad et al. (1998)
Pacific oyster	*Crassostrea gigas*	Embryo	Rana et al. (1992); Gwo (1995)
Zebra fish	*Brachydanio rerio*	Embryo	Zhang et al. (1993); Adam (1995)
Rosy barb	*Puntius conchonius*	Embryo	Adam (1995)
Rainbow trout	*Oncorhynchus mykiss*	Blastomere	Calvi and Maisse (1998)
Sea perch	*Lateolabrax japonicus*	Embryo	Tian et al. (2003)
Japanese flounder	*Paralichthys olivaceus*	Embryo	Chen and Tian (2005)

6.5 Conclusion

In 1866, Montegazza suggested banks for frozen human sperm. He was of the opinion that a man dying on a battlefield may beget a legal heir with his semen frozen and stored at home. But his vision did not get a shape until 1963, wherein the 11th International Congress of Genetics in 1963, interest on the possibility of sperm banks was triggered and around a decade later, in the year 1977, the first commercial sperm bank was opened in California. It can be seen that it took around 150 years after Montegazza, during the Gulf war crisis in 1992; his vision became a reality when some service men opted to freeze and store specimens of their sperm before leaving for battle. The concept of human sperm bank during 1866 might have been a great mockery. But it has become a reality now and people recognize its importance. Similarly the use of cryopreserved spermatozoa in fish breeding might be a matter beyond one's imagination now. But the day is not very far off to make it reality, considering the usefulness of application of cryopreserved spermatozoa in aquaculture.

References

Adam MM (1995) Pre-freezing and cryo storage problems of rosy barbs and zebra fish embryos. PhD Thesis, Institute of Aquaculture, University of Stirling, pp 93

Agarwal NK, Saini V, Raghuvanshi SK (2013) Characterization and short-term storage of semen of a coldwater Himalayan fish species. Biojournal 8(1):1–8

Ahammad MM, Bhattacharyya D, Jana BB (1998) Effect of different concentrations of cryoprotectant and extender on the hatching of Indian major carp embryos (*Labeo rohita, Catla catla and Cirrhinus mrigala*) stored at low temperature. Cryobiology 37(4):318–324

Akash N (2001) Preservation of gametes and fertilization methodology in fringe-lipped carp, *Labeo fimbriatus* (Bloch). MFSc Thesis, University of Agricultural Sciences, Bangalore

Babiak I, Glogowski J, Luezynski MJ, Kucharezyk D, Luezynski M (1995) Cryopreservation of the milt of the northern pike. J Fish Biol 46(5):819–828

Babiak I, Glogowski J, Kujawa R, Kucharezyk D, Mamearz L (1998) Cryopreservation of sperm from Asp *Aspius aspius*. Prog Fish Cult 60:146–148

Babiak I, Fraser L, Dobosz S, Goryczko K, Kuzminski H, Strzezek J (1999) Computer-controlled freezing of rainbow trout *Onchorhynchus mykiss* (Walbaum) spermatozoa for routine programmes. Aquacult Res 30:707–710

Barrett I (1951) Fertility of salmonid eggs and sperm after storage. J Fish Res Board Canada 8 (3):125–133

Basavaraja N, Hegde SN (2005) Some characteristics and short-term preservation of spermatozoa of Deccan mahseer, *Tor khudree* (Sykes). Aquacult Res 36:422–430

Bennington NI (1936) Germ cell origin and spermatogenesis in the Siamese fighting fish, *Betta splendens*. J Morphol 60:103–125

Bercsényi M, Magyary I, Urbányi B et al (1998) Hatching out goldfish from common carp eggs: interspecific androgenesis between two cyprinid species. Genome 41:573–579

Best BP (2008) Scientific justification of cryonics practice. Rejuvenation Res 11(2):493–503. https://doi.org/10.1089/rej.2008.0661

Betsy CJ, Kumar JSS (2015a) Influence of egg yolk on the quality of cryopreserved spermatozoa of common carp, *Cyprinus carpio*. J Appl Aquaculture 27:40–49

Betsy CJ, Kumar JSS (2015b) Enhancing the life and potency of *Cyprinus carpio* spermatozoa during cryopreservation by glucose supplementation. J Aqua Trop 30(1–2):71–79

Betsy CJ, Kumar JSS (2015c) Evaluation of supplementation of glucose and egg yolk on the motility of *Cyprinus carpio* and *Cirrhinus mrigala* spermatozoa under short term storage. J Aqua Trop 30(3–4):135–147

Betsy CJ, Kumar JSS (2015d) Role of glucose in enhancing life and potency of *Cirrhinus mrigala* spermatozoa during cryopreservation. J Appl Nat Sci 7(1):159–164

Betsy CJ, Kumar JSS (2016) Effect of three different extenders on the spermatological parameters of *Cyprinus carpio*. J Aqua Trop 31(1–2):25–30

Betsy CJ, Kumar JSS, Jawahar KTP et al (2016) Captive broodstock development of common carp: influence of age on spermatogenesis and fertilization parameters. Indian J Fish 63(4):75–81

Betsy CJ, Kumar JSS, Jawahar KTP (2017) Effect of age on the spermatological and fertilization parameters of common carp brooders cryopreserved at three dilution ratios. Asian Fish Sci 30:195–205

Betsy CJ, Kumar JSS, Jawahar KTP (2019) Effect of different extenders and dilution ratios on the spermatological and fertilization parameters of Cyprinus carpio during cryopreservation. Biopreserv Biobank 17(2):157–162. https://doi.org/10.1089/bio.2018.0110

Bhavani CN, Noble D, Gopalakrishnan A, Sanil NK (2003) Cryopreservation of spermatozoa in the mullet, *Liza parsia:* effect of cryoprotectants on morphology and motility. J Mar Biol Assoc India 45(1):38–46

Bhavanishankar S, Subramoniam T (1997) Cryopreservation of spermatozoa of the edible mud crab *Scylla serrata* (Forskal). J Exp Zool 277:326–336

Billard R, Cosson J, Crim LW (1993) Motility of fresh and aged halibut sperm. Aquat Living Resour 6:67–75

Blaxter PC (1953) Sperm storage and cross fertilization of spring and autumn spawning herring. Nature 172:1189–1190

Bozkurt Y, Akcay E, Tekin N, Secer S (2005) Effect of freezing techniques, extenders and Cryoprotectants on the fertilization rate of frozen rainbow trout *(Oncorhynchus mykiss)* sperm. Isr J Aquacult Bamidgeh 57(2):125–130

Bozkurt Y, Ogretmen F, Secer FS (2009) Effect of different extenders and storage periods on motility and fertilization success of grass carp (*Ctenopharyngoden idella*) sperm during spawning season. Tarim Bilimleri Dergisi 15(3):277–284

Brofeldt P (1923) Ueber Transport von Fischrogen und Milchohne Wasser in Glastoepfen. Allgemeine Fischerei-Zeitung:166–169

Buyukhatipoglu S, Holtz W (1978) Preservation of trout sperm in liquid/frozen state. Aquaculture 14:49–56

Cabrita E, Robles V, Alvarez R, Herraez MP (2001) Cryopreservation of rainbow trout sperm in large volume straws: application to large scale fertilization. Aquaculture 201:301–314

Cabrita E, Robles V, Rebordinos L, Sarasquete C, Herráez MP (2005a) Evaluation of DNA damage in rainbow trout (*Oncorhynchus mykiss*) and gilthead sea bream (*Sparus aurata*) cryopreserved sperm. Cryobiology 50:144–153

Cabrita E, Robles V, Cuñado S et al (2005b) Evaluation of gilthead sea bream, *Sparus aurata*, sperm quality after cryopreservation in 5 ml macrotubes. Cryobiology 50:273–284

Calvi SL, Maisse G (1998) Cryopreservation of rainbow trout (*Oncorhynchus mykiss*) blastomeres: influence of embryo stage on post thaw survival rate. Cryobiology 36:255–262

Caylor RE, Biesiot PM, Franks JS (1994) Culture of cobia (*Rachycentron canadum*): cryopreservation of sperm and induced spawning. Aquaculture 125:81–92

Chao NH, Chen HP, Liao JC (1975) Study on cryogenic preservation of grey mullet sperm. Aquaculture 5:389–406

Chao NH, Chao WC, Liu KC, Liao IC (1987) The properties of tilapia sperm and its cryopreservation. J Fish Biol 30:107–118

Chao NH, Chen YN, Shieh WY, Lin KJ (2009) Cryopreservation of Spermatophores in the white shrimp (*Litopenaeus vannamei*). J Taiwan Fish Res 17(2):67–76

Chen SL, Tian YS (2005) Cryopreservation of flounder (*Paralichthys olivaceus*) embryos by vitrification. Theriogenology 63:1207–1219

Chereguini O, Garcia de la Banda I, Herrera M et al (2003) Cryopreservation of turbot *Scophthalmus maximus* (L.) sperm: fertilization and hatching rates. Aquac Res 34:739–747
Chow S (1982) Artificial insemination using preserved spermatophores in the palaemonid shrimp *Macrobrachium rosenbergii*. Bull Jpn Soc Sci Fish 48:1693–1695
Christensen JM, Tiersch TR (1997) Cryopreservation of channel catfish spermatozoa: effect of cryoprotectant, straw size, and formulation of extender. Theriogenology 47:639–645
Christensen MJ, Tiersch RT (2005) Cryopreservation of channel catfish sperm: effects of cryoprotectant exposure time, cooling rate, thawing conditions and male-to-male variation. Theriogenology 63:2103–2112
Ciereszko A, Dabroski K (1996) Effect of a sucrose-DMSO extender supplemented with pentoxifylline or blood plasma on fertilizing ability of cryopreserved rainbow trout sperm. Prog Fish Cult 58:143–145
de Quatrefages A (1853) Recherché sur la vitalete des spermatozoids de quelquesd'eaudouce. Annals des Sciences Naturelles 19:341–369
DeGraaf JD, Berlinsky DL (2004) Cryogenic and refrigerated storage of Atlantic cod (*Gadus morhua*) and haddock (*Melanogrammus aeglefinus*) spermatozoa. Aquaculture 234:527–540
DiLauro MN, Krise WF, Hendrix MA, Baker SE (1994) Short-term cold storage of Atlantic sturgeon sperm. Prog Fish Cult 56:143–144
Doi M, Hoshino T, Taki Y, Ogasawara Y (1982) Activity of the sperm of the Bluefin tuna *Thunnus thynnus* under fresh and preserved conditions. Bull Jap Soc Sci Fish 48(4):495–498
Dreanno C, Suquet M, Quemener L et al (1997) Cryopreservation of turbot (*Scophthalmus maximus*) spermatozoa. Theriogenology 48:589–603
Dushkina LA (1975) Viability of herring (*Clupea*) eggs and fertilizing capacity of herring sperm stored under various conditions. J Ichthyol 15:423–429
Fabbrocini A, Lavadera L, Rispoli S, Sansone G (2000) Cryopreservation of sea bream (*Sparus aurata*) spermatozoa. Cryobiology 40:46–53
Foote RH (1998) Artificial insemination to cloning: tracing 50 years of research. Ithaca
Gibson MB (1952) Storage of lake trout eggs and sperm. MS Report to Director, Ontario Fisheries Research Laboratory, Department of Zoology, University of Toronto
Glogowski J, Ciereszko R, Dabrowski K (1999) Cryopreservation of muskellunge and yellow perch semen. N Am J Aquac 61:258–262
Glogowski J, Kolman R, Szczepkowski M et al (2002) Fertilization rate of Siberian sturgeon (*Acipenser baeri*, Brandt) milt cryopreserved with methanol. Aquaculture 211:367–373
Gwo JC (1993) Cryopreservation of black grouper (*Epinephelus malabaricus*) spermatozoa. Theriogenology 39:1331–1342
Gwo JC (1995) Cryopreservation of oyster (*Crassostrea gigas*) embryos. Theriogenology 43:1163–1174
Hannah D (2016) The cryonics dilemma: will deep-frozen bodies be fit for new life?. The Guardian. https://www.theguardian.com/science/2016/nov/18/the-cryonics-dilemma-will-deep-frozen-bodies-be-fit-for-new-life
Hara S, Canto JT Jr, Almendras JME (1982) A comparative study of various extenders for milk fish *Chanoschanos* (Forsskal), sperm preservation. Aquaculture 28:339–346
Harvey B, Kelley RN (1984a) Control of spermatozoan motility in an euryhaline teleost acclimated to different salinities. Can J Zool 62(12):2674–2677
Harvey B, Kelley RN (1984b) Chilled storage of *Sarotherodon mossambicus* milt. Aquaculture 36:85–95
Henderson NE, Dewar JE (1959) Short-term storage of brook trout milt. Prog Fish Cult 21 (4):169–171
Iserson KV (2001) Chapter 7: Souls on ice. In: Death to dust: what happens to dead bodies? 2nd edn. Galen Press Ltd., Mountain View, p 821
Jodun WA, King K, Farrell P (2006) Methanol and egg yolk as cryoprotectants for Atlantic Salmon spermatozoa. N Am J Aquac 69:36–40

Kamaruding NA, Embong WKW, Abdullan RB (2012) Frozen-thawed sperm motility characteristics of African catfish (*Clarias gariepinus*) by using glycerol or DMSO based extender. Int J Environ Sci Dev 3(1):49–55

Kumar K (1989) Studies on cryogenic preservation of carp spermatozoa. In: Das P, Jhingran AG (eds) Fish genetics in India. Today and Tomorrow's Printers and Publishers, New Delhi, pp 199–206

Kurokura H, Hirano R, Tomita M, Iwahashi M (1984) Cryopreservation of carp sperm. Aquaculture 37:267–273

Lahnsteiner F, Weismann T, Patzner RA (1997) Methanol as cryoprotectant and the suitability of 1.2 mL and 5 mL straws for cryopreservation of semen from salmonid fishes. Aquaculture 145:281–294

Lahnsteiner F, Berger B, Weismann T, Patzner RA (1998) Physiology and biochemical determination of rainbow trout semen quality for cryopreservation. J Appl Aquac 6:47

Lahnsteiner F, Berger B, Horvath A et al (2000) Cryopreservation of spermatozoa in cyprinid fishes. Theriogenology 54:1477–1498

Lanes CFC, Okamoto M, Cavalcanti PV et al (2008) Cryopreservation of Brazilian flounder (*Paralichthys orbignyanus*) sperm. Aquaculture 275:361–365

Linhart O, Billard R, Proteau JP (1993) Cryopreservation of European catfish (*Silurus glanis* L.) sperm. Aquaculture 115:347–359

Liu Q, Li J, Zhang S et al (2006) An efficient methodology for cryopreservation of spermatozoa of red seabream, *Pagrus major*, with 2-mL Cryovials. J World Aquacult Soc 37(3):289–297

Luyet BJ, Gehenio MP (1940) Life and death at low temperatures. Biodynamica, Normandy

Magyary I, Urbanyi B, Horvath L (1996) Cryopreservation of common carp (*Cyprinus carpio* L.) sperm: II optimal conditions for fertilization. J Appl Ichthyol 12:117

Melo FCSA, Godinho HP (2006) A protocol for cryopreservation of spermatozoa of the fish *Brycon orthotaenia*. Anim Reprod 3(3):380–385

Memon AJ, Ikhwanuddin M, Talpur AD et al (2011) Assessment of sperm viability, determination fertilization capacity and hatching rate by artificial insemination of Banana shrimp, *Penaeus merguiensis* (De Man, 1888). Res J Appl Sci 6(3):174–178

Moen OM (2015) The case for cryonics. J Med Ethics 41(18):493–503. https://doi.org/10.1136/medethics-2015-102715

Moore AA (1987) Short-term storage and cryopreservation of walleye semen. Prog Fish Cult 49:40–43

Mounib MS, Hwang PC, Idler DR (1968) Cryogenic preservation of Atlantic cod (*Gadus morhua*) sperm. J Fish Res Board Canada 25:2623–2632

Müller T, Urbányi B, Váradi B et al (2004) Cryopreservation of sperm of farmed European eel *Anguilla anguilla*. J World Aquacult Soc 35:240–246

Palmer PJ, Blackshaw AW, Garrett RN (1993) Successful fertility experiments with cryopreserved spermatozoa of barramundi, *Lates calcarifer* (Bloch), using dimethylsulfoxide and glycerol as cryoprotectants. Rep Fertil Dev 5:285–293

Piironen J (1994) Composition and cryopreservation of sperm from some Finnish freshwater teleost fish. Finnish Fish Res 15:65–86

Polge C, Smith AU, Parkes AS (1949) Revival of spermatozoa after vitrification and dehydration at low temperatures. Nature 164:666

Rana KJ, McAndrew BJ, Musa MA (1992) Cryopreservation of oyster (*Crassostrea gigas*) eggs and embryos. In: Abstracts of the Workshop on gamete and embryo storage and cryopreservation in aquatic organisms, Paris

Rideout RM, Trippel EA, Litvak MK (2003) The development of haddock and Atlantic cod sperm cryopreservation techniques and the effect of sperm age on cryopreservation success. J Fish Biol 65:299–311

Robert H, David C (2014) "Brian failure and brain death". ACS surgery: principles and practice critical care. Decker Intellectual Properties Inc, pp 1–10

Sarder MRI, Rafiquzzaman SM, Sultana R, Faridul IM (2009) Cryopreservation of spermatozoa of mrigal, *Cirrhinus cirrhosus* with a view to minimize inbreeding and hybridization. J Bangl Agric Univ 7(1):211–215

Sarder MRI, Rahman AKMA, Samad MS et al (2011) Cryopreservation of sperm of *Labeo rohita* (Hamilton, 1822) and its use in fertilization and hatching of eggs. Prog Agri 22(1 & 2):123–137

Scheuring L (1924) Biologische and physiologischeuntersunchungen an Furellensperma. Arch Hydrobiol Suppl 4:181–318

Sherman JK (1964) Low temperature research on spermatozoa and eggs. Cryobiology 1 (2):103–129

Smith RT, Quistorff E (1943) Experiments with the spermatozoa of the steel head trout, *Salmo gairdneri* and the Chinook salmon, *Oncorhynchus tshawytscha*. Copeia 3:164–167

Spallanzani L (1776) Opuscoli di FisicaAnimale e vegetabile [...] aggiuntevialcuneletterelativeadessiopuscoli dal celebre Signor Bonnet di Ginevra e da altriscritteall'Autore. Modena

Stoss J, Refstie T (1983) Short-term storage and cryopreservation of milt from Atlantic salmon and sea trout. Aquaculture 30:229–236

Stoss J, Geries L, Holtz W (1987) The role of sample depth in storing chilled rainbow trout (*Salmo gairdneri*) semen under oxygen. Aquaculture 61:275–279

Suquet M, Dreanno C, Petton B et al (1998) Long-term effects of the cryopreservation of turbot (*Psetta maxima*) spermatozoa. Aquat Living Resour 11:45–48

Tanaka S, Zhang H, Horie N et al (2002) Long-term cryopreservation of sperm of Japanese eel. J Fish Biol 60:139–146

Tekin N, Seçer S, Akçay E, Bozkurt Y (2003) Cryopreservation of rainbow trout (*Oncorhynchus mykiss*) semen. Isr J Aquacult Bamidgeh 55(3):208–212

Thorogood J, Blackshaw AW (1992) Factors affecting the activation, motility and cryopreservation of the spermatozoa of the yellowfin bream, *Acanthopagrusaustralis* (Günther). Aquac Res 23:337–344

Tian YS, Chen SL, Yan AS et al (2003) Cryopreservation of the sea perch (*Lateolabrax japonicus*) embryos by vitrification. Acta Zool Sinica 49:843–850

Tiersch TR, Gouodie CA, Carmichael GJ (1994) Cryopreservation of channel catfish sperm: storage in cryoprotectants, fertilization trials and growth of channel catfish produced with cryopreserved sperm. Trans Am Fish Soc 123:580–586

Truscott B, Idler DR, Hoyle RJ, Freeman HC (1968) Sub-zero preservation of Atlantic salmon sperm. J Fish Res Board Canada 25:363–372

Vuthiphandchai V, Nimrat S, Kotcharat S, Bart AN (2007) Development of cryopreservation protocol for long term storage of black tiger shrimp (*Penaeus monodon*) spermatophores. Theriogenology 68:1192–1199

Wayman WR, Tiersch TR, Thomas RG (1998) Refrigerated storage and cryopreservation of sperm of red drum, *Sciaenops ocellatus* L. Aquac Res 29:267–273

Yasui GS, Lenin AR, Takafumi F, Arai K (2009) A sperm cryopreservation protocol for the loach *Misgurnus anguillicaudatus* and its applicability for other related species. Anim Reprod Sci 116 (3–4):335–345

Zhang XS, Zhao L, Hua TC, Zhu HY (1989) Study on the cryopreservation of common carp (*Cyprinus carpio*) embryos. Cryo Letters 10:271–278

Zhang T, Rawson D, Morris GJ (1993) Cryopreservation of pre-hatch embryos of Zebra fish, (*Brachydanio rerio*) embryos. Aquat Living Resour 6:143–153

7 Cryopreservation of Fish Gametes: An Overview

K. Thilak Pon Jawahar and Judith Betsy

Abstract

Cryopreservation of gametes is done in all sectors including livestock and humans with minor to major variations in the protocol. However, in fisheries, the protocol has been adapted from the field of livestock and is presently being used for various fish species with little modification in the technology. Hence in this chapter, the protocol used for cryopreservation of fish gametes and the mechanisms and problems involved in it are discussed. Ice formation during freezing is the major factor which decides the cryosurvival of spermatozoa. Hence, special emphasis has been made on the physical and chemical process of ice formation during freezing.

Keywords

Fish · Spermatozoa · Cryopreservation · Protocol · Cryoprotectant · Freezing

7.1 Introduction

Lazzaro Spallanzani the father of reproductive physiology found that freezing stallion semen did not kill sperm but held them in a dormant state until exposed to heat. Since 1780, investigators have been attempting to preserve semen and other cells by cooling. During the next 150 years, developments in the physics of cold temperature particularly preparation of liquid gases at temperatures far below those

K. Thilak Pon Jawahar (✉)
Frozen Semen Bank, Department of Animal Genetics and Breeding, Madras Veterinary College, Tamil Nadu Veterinary and Animal Sciences University, Chennai, India

J. Betsy
Department of Aquaculture, Fisheries College and Research Institute, Tamil Nadu Dr. J. Jayalalithaa Fisheries University, Tuticorin, India

J. Betsy, S. Kumar (eds.), *Cryopreservation of Fish Gametes*,
https://doi.org/10.1007/978-981-15-4025-7_7

ever recorded in nature, permitted a great extension of cryobiology and number of observations were made on the cells and tissues of plants and animals. The big break came in 1948 with the serendipitous discovery of cryoprotective action of glycerol on fowl spermatozoa by Polge and others. It was the spectacular discovery of Polge in 1949 that showed that the death of spermatozoa on freezing could be avoided if the cells were suspended in a medium containing cryoprotectant. This spectacular discovery removed at one stroke the main barrier opposing the preservation of living cells in frozen state. Since then several improved techniques have been developed in the cryopreservation of spermatozoa, yet only about 50% of spermatozoa survive deep freezing. With all improved techniques at hand today the problems of successfully freezing spermatozoa of all species have not been alleviated. The purpose of this chapter is to understand the principles of cryopreservation and behaviour of fish spermatozoa during freezing and thawing.

Artificial breeding in fish is a potential tool for the fish breeders and farmers to promote genetic merit among aquatic species. Accordingly, the artificial control techniques on gamete preservation are required for the successful year-round fish seedling production. A major technological breakthrough in this field is the advent of frozen milt technology permitting preservation of milt to several years and greater spread of superior germplasm. The artificial breeding of fish has been tested through the production of progeny using cryopreserved sperm. Successful cryopreservation of fish sperm have been achieved for more than 200 fish species including carp, salmonids, rainbow trout, catfish, cichlids, medakas, whitefish, pike, milkfish, grouper, cod and zebrafish. In spite of the commercial importance of the fishes in aquaculture, limited information is available concerning their sperm preservation. This is mainly because of specific specificity on the sperm properties, diluents and cryoprotectants, and lack of systematic studies from properties to preservation of sperm etc. Hence, the successes in fish milt cryopreservation are very much relied on factors such as having ample knowledge on the biology and reproductive biology of the particular species of interest, trained personnel in various aspects such as gamete cryopreservation, breeding methods, brood stock management and husbandry and larva rearing and nursing of the targeted species.

Cryopreservation has revolutionized artificial breeding and it is a valuable tool in assisted reproductive technology. Though this technique is widely adopted, the success rate on insemination/artificial breeding is low. Understanding the basic principles of cryobiology for each cell and suitably modifying the processing protocol will enhance the fertility rate. Fish sperm were among the first cell types successfully maintained through cryopreservation—the technique of freezing cells at extremely low temperatures so they remain genetically stable and metabolically inert. Thus, cryopreservation technology offers the best means for long-term storage of fish milt. Even though in general many successes have been achieved in fish semen cryopreservation, the technique remains as a method that is difficult to be standardized and used in all types of fishes. This is due to the fact that cryopreservation of sperms from different fish species required different conditions, where the protocol needs to be established individually. Even the "general protocol" of cryopreservation of fish sperm has been very well demonstrated by various researchers encompassed many variations observed when different species of fish are involved for the cryopreservation.

7.2 Protocol for Cryopreservation of Fish Milt

7.2.1 Selection of Donors for Milt Collection

Mature and healthy live ripe brooders should be selected by taking utmost care to avoid any injury preferably during the middle phase of breeding/spawning season. The brooders should be fed a diet adequate with reference to energy, protein, minerals and vitamins. The milt collector should have a fairly good knowledge on collection of good quality and quantity of milt without causing injury to the brooder. The milt collection area should be located nearer to the processing laboratory if possible and free from contamination and external disturbances. The fish from which milt is to be collected must be clean, and free from dirt to avoid possible contamination of milt during collection.

7.2.2 Collection of Milt

The fish from which milt is to be collected should be cleaned, wiped with soft cotton cloth or tissue paper to free from dirt to avoid contamination of milt from water, mucus, urine, blood, faecal matter etc. The milt can be stripped by giving gentle abdominal massage/pressure and collected directly in a clean, dry and sterilized cryovials (Fig. 7.1) or ampoules as quickly as possible without causing any stress to the donor and kept in a cool water bath/ice at 0–4 °C until evaluation and used for further processing. It is always better to discard the first one or two drops of milt to

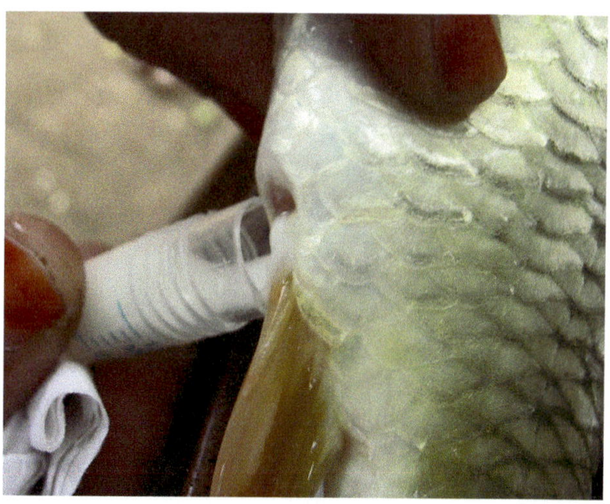

Fig. 7.1 Collection of milt in cryovials by gentle abdominal pressure (Photo courtesy: J. Stephen Sampath Kumar)

avoid contamination with blood, water, urine or the faeces as they significantly affect the quality of the milt and render the samples unfit for further processing. The frequency of collection must be arranged according to the capacity of the individual fish. The frequency of collection should be arranged according to the capacity of individual fish. The collections can take twice a week (one ejaculate each time) or once a week (two ejaculates per collection). Based on the volume and sperm density of the collected milt, the frequency may be increased. Fish donating milt of poor quality consistently should never be used for artificial breeding as the cause might be of hereditary in nature.

7.2.3 Evaluation of Fish Milt

Immediately after harvesting, each ejaculate should be examined routinely by the best methods available. In the laboratory, a number of sperm characteristics can be evaluated or measured. Unfortunately, till date no single test offers a high probability of predicting the freezability and fertility of spermatozoa. However, years of research and current field trials have shown that there is a high positive correlation between breeding efficiency and the available laboratory tests. The milt samples showing more than 70% sperm motility alone are used for cryopreservation. The most important milt characteristics evaluated routinely are described below.

7.2.3.1 Macroscopic Evaluation

Colour

The colour of fish milt normally ranges from milky white to creamy white. Immediately on collection, carefully examine the colour of the milt by visual appraisal. If the colour is abnormal, then the sample should be discarded. Pink or red colour is an indicative of presence of fresh blood due to mechanical or physical injuries at the time of harvesting and sometimes it may not be mixed up evenly in the milt. The brown colour indicates decomposed blood originating from diseased condition in the deeper parts of the genital tract and it will often be evenly spread in the milt. Yellowish or greyish or greenish colour of the milt is due to the presence of urine, faecal matter, pus which is an indication of inflammatory process in the genital tract. The contaminated semen is of poor quality, slowly loose viability and of no use for further processing.

Volume

The volume of milt can be measured in millilitres (ml) by collecting it directly into graduated cryovial/ampoule. The volume of the ejaculate varies from species to species, individual to individual within the same species and also between collections from the same individual depending more on the state of ripeness than the size. In general, the volume increases with age, body size, changes in reproductive health and frequency of collection. Mostly fish produce milt on reaching maturity. Any sudden deviation in volume in a routine schedule indicate either a

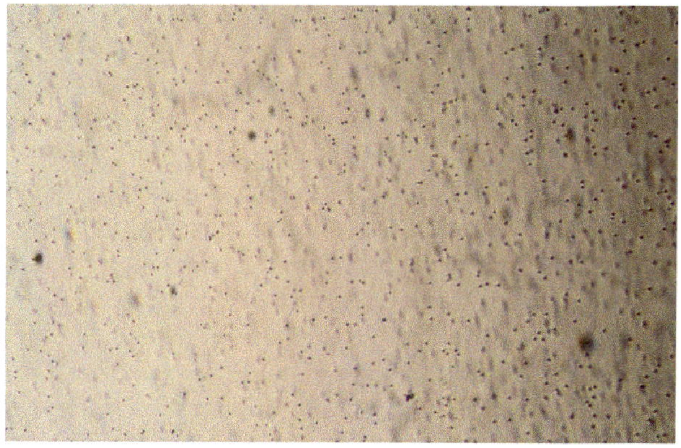

Fig. 7.2 Spermatozoa under phase contrast microscope (Photo courtesy: J. Stephen Sampath Kumar)

disease condition or improper handling and collection procedures adopted at the time of harvesting.

Consistency

The number of sperm cells in a given ejaculate affects the appearance of the milt. A sample that is clear and translucent is low in sperm number, whereas one that is thick and opaque is relatively high in sperm concentration. The viscosity and density of milt increase with increasing sperm concentration. By slightly tilting the collection ampoule and observing the density of milt, the consistency is graded as thick, medium and thin. The consistency of the milt harvested is variable factor both within and between individuals. The consistency must be homogenous throughout the ejaculate, even the smallest visible difference indicates the admixture of foreign materials like pus, faecal matter etc.

Presence of Foreign Matter

The milt harvested should be free from any foreign matter because its presence indicates the unsuitability of milt for further processing. The commonly seen foreign matters are blood, pus, urine, faecal matter, dust, scales etc.

7.2.3.2 Microscopic Evaluation

A phase contrast microscope is necessary for accurate evaluation of fish milt (Fig. 7.2).

Individual Progressive Motility

The assessment of motility of fish sperm essentially relied on subjective estimates of motility characteristics by the percentage of motile sperm cells, by the total duration of movement, or by a combination of both parameters. Individual sperm motility in the fish milt is evaluated by placing 1 μL of diluted milt sample and 1 μL of tap water on glass slide and observed under a microscope with 40× magnification.

Spermatozoa of fish spawning in brackish water and marine waters have more long-lasting motility duration. Otherwise, the motility of fish spermatozoa ranges from 30 to 300 s. This makes the study much complicated and almost all the species have immotile spermatozoa in their testes that are activated when they come in contact with the medium of lower or higher osmolality. Therefore, the motility is a factor that can be observed only for a limited period and all the observations are to be completed within that period.

The percentage of motile spermatozoa and the swimming vigour has usually been given a motility score corresponding to an arbitrary scale of criteria from 0 (immotile) to 5 (all spermatozoa vigorously motile). The motility rating has been defined in terms of percent moving spermatozoa in the field of view. This method is also used to evaluate the percentage of moving cells according to the percentage of rapid, vigorous and forward-moving motile spermatozoa.

The standard motility scores proposed for assessment are:

Progressive movement—Spermatozoa exhibiting various strong flagella movements with unidirectional cell movement; unable to follow the direction and pattern

Forward movement—Spermatozoa exhibiting various flagella movements with unidirectional cell movement that can be tracked under phase contrast microscope

Strong in loco vibration—Spermatozoa vibrating in a fixed place; with five to seven movements per second in a stationary position

Weak in loco vibration—Spermatozoa vibrating in a fixed place; with one to three movements per second without any movement from the place

Strong oscillation—Spermatozoa oscillating with two to three swings per second

Mild oscillation—Spermatozoa oscillating with one swing per second

Immotile—No movement and carried by the medium added

Normally the spermatozoa are not motile in the testes or seminal fluid. It becomes activated in fertilizing media and performs movement. In external fertilization, activating media is generally the river/pond water. Sperm motility is the most commonly used parameter to evaluate sperm quality, as sperm must be motile to achieve fertilization. The sperm showing progressively forward movement are only capable of fertilizing the ova. The motility is measured on a five-point scale from 0 to +4 as described by Agarwal (2011) and is given in Table 7.1.

Only those samples that initially show no motility but after activation with water/activating media (under microscope) show motility rating 3 and above are considered good for cryopreservation. After activation, sperm of freshwater species remain motile only for a brief period of seconds or minutes. The duration of motility refers to the total duration of forward motion and survival time of the spermatozoa at the time of breeding. The longer sperm motility duration is advantageous for maximum

Table 7.1 Motility measurement based on five-point scale

Sl. no.	Motility percentage	Rating
1	0 to <1%	0
2	1 to <25%	1
3	25 to <50%	2
4	50 to <75%	3
5	75 to 100%	4

Fig. 7.3 Analysing sperm concentration using haemocytometer (Photo courtesy: J. Stephen Sampath Kumar)

Table 7.2 Sperm concentration in different fish species

Sl. no.	Species	Sperm concentration per ml of milt
1	Indian major carps	2×10^7 to 3.5×10^7
2	Snow trout	2.94×10^8 to 9.76×10^8
3	*Schizothoraichthys richardsonii*	3.96×10^8
4	*S. progastus*	8.62×10^8
5	Mahseer (*Tor putitora*)	1.70×10^9

chances of fertilization. If the sperm motility duration is longer, cryopreservation is also generally easier and is a species specific feature.

Sperm Concentration/Sperm Density/Sperm Cell Count

The number of spermatozoa per unit volume of fish milt varies from zero in complete azoospermia to billion cells per millilitre in occasionally very thick samples. In general, concentration of spermatozoa varies with sexual development and maturity of the fish, nutrition status, reproductive health and size of the testes. In addition, there are real differences in spermatozoan concentration observed among individuals within the species, between species, among different age groups, different seasons of the year, different geographic locations and different phases of breeding.

The concentration of spermatozoa is estimated using haemocytometer (NAUBAEUR, Germany) (Fig. 7.3). The accuracy of the method varies with technical competence of the technician. Photoelectric colorimeter method can also be employed which is a fast and more reliable way of estimation of sperm concentration. Normally, the concentration of spermatozoa is high during the start of breeding season.

The number of motile viable spermatozoa has an important bearing on the fertilization success. It is an important parameter of quality assessment for cryopreservation of fish milt. The sperm concentration is a species-specific feature and the normal sperm concentration in some of the fish recorded is furnished in Table 7.2.

The samples having very less sperm concentration have reduced capacity to fertilize ovum and hence discarded. They are not used for cryopreservation or further processing in artificial breeding.

The sperm concentration of a milt sample can also be estimated by spermatocrit value which is an easy and quick method of assessing sperm concentration. For spermatocrit value estimation of fish semen, microhaematocrit capillary tubes are used. These capillaries are filled with raw milt by sucking and their both ends are sealed with haemoseal wax. These milt-filled capillaries are measured by a graduated scale in millimetres. Then these capillaries are placed in a haematocrit rotor and centrifuged. After centrifugation, the volume of packed cells is again measured by a graduated scale in millimetres. Thus, the spermatocrit value of semen is expressed as packed cell volume percentage. The centrifugation speed and time need to be standardized for each species separately.

Live and Dead Spermatozoa percent

Motility test indicate a rough idea of the liveability of the fish spermatozoa. A more accurate method of assessing the liveability is by estimating the percent live and dead spermatozoa using differential staining called Eosin Nigrosin technique. Eosin is the essential cellular stain which differentiates the dead and live spermatozoa. The live spermatozoa does not allow the eosin to penetrate its cell wall and hence remains unstained, while dead spermatozoa loose the permeability and allows stain to pass through and becomes pink in colour. Nigrosin provides a background so that the stained and unstained spermatozoa could be viewed clearly in the microscope. Depending on the degree of membrane disintegration, the spermatozoa will be coloured from pale pink to deep red. The percentage of dead spermatozoa should not exceed 20–25% in samples used for artificial breeding/cryopreservation.

Presence of Cells Other than Spermatozoa

The milt should be free from cells other than spermatozoa and cell debris. However, few epithelial cells and red blood cells are usually encountered in milt samples. The cells whose presence in the milt contradicted are spermatogonic cells, epithelial cells from the lining of urogenital tract, leucocytes (pus cells) and erythrocytes. The presence of spermatogonic cells is a sign of progressive testicular degeneration especially when accompanied by low sperm count. Trauma or infection causes an increase in the presence of epithelial cells in the semen. Pus indicates inflammation of testes, epididymes, seminal vesicles, cystitis, nephritis or an extra genital abscess.

7.2.3.3 Biochemical Evaluation

pH

The pH of normal milt ranges from 6.5 to 7.0. Good quality milt is more acidic than the poor sample with low sperm concentration. The pH of the milt decreases as the time between collection and measurement increases. On the other hand, bacterially contaminated milt, milt containing many dead spermatozoa, azoospermic condition, milt collected from infected genital organs has a higher pH. The milt may be alkaline when it carries pus.

The pH of the milt is best measured using a narrow range pH indicator strips or pH meter. The pH of milt varies with the species and is slightly alkaline in the freshwater fish. The pH is an important parameter for consideration during preparation of diluents/extenders for cryopreservation.

7.2.3.4 Properties of Fish Milt

The properties of milt and composition of its plasma are subject to change with season and frequency of milt collection, and with age, size and species of fish as well as the condition of objective fish. Therefore, it will be very difficult to compare certain results with other studies on the milt properties. Osmotic pressure of seminal plasma in freshwater fishes is generally lower than that of marine fishes. The osmotic pressure is direct factor influencing the initiation of sperm motility. Among the chemical composition of seminal plasma, Na^+ concentrations of freshwater fishes are lower than those of marine fishes, but on the contrary, K^+ concentrations in freshwater fishes are rather higher than those in marine fishes.

The progress in milt preservation and further significant developments can be achieved as a result of understanding the causes of cryoinjury. The process of cryopreservation represents an artificial interruption of the progress of the spermatozoa towards post-ejaculation maturation and fertilization. Thus, cryopreservation encompasses all stages from departure from body temperature until its restoration back to that temperature. The major problem is still that even with the best preservation techniques to date, post-thaw survival is restricted to about 50% of the sperm population.

7.2.4 Extenders for Freezing Fish Milt

The extender functions as a medium for cryoprotectant. It contains chemicals that minimize the cryoinjuries to cell due to formation of intracellular ice crystals during cryopreservation at the time of freezing and thawing of semen.

At least 50% of the sperm from an unselected population of fish species are rendered immotile during freezing and thawing. The damage is caused primarily by internal ice crystal formation that affects the structure of the spermatozoa and secondarily by the increase in solute concentration as water is withdrawn from the suspension medium or by interaction of these two physical factors. Hence, successful cryopreservation will depend upon prevention of the occurrence of the above three factors. This could be aimed by suspending the sperm in an ideal diluents/extender that could afford cryoprotection during freezing.

The basic components of diluents/extender for cryopreservation/freezing of fish spermatozoa are:

- Buffer
- Sugar

- Cryoprotectatnt
- Lipids
- Antibiotics

7.2.4.1 Buffer

For developing cryopreservation protocol for milt of any fish species, the prime requirement is to develop extender medium for dilution of milt, as undiluted one is not suitable for preservation. The extender inhibits the activation of spermatozoa and thereby preserves the vitality and viability of sperm by inhibiting the motility and conserving the energy.

A variety of inorganic and organic buffers are used for cryopreservation of fish milt. Reports on superiority of each buffer are numerous, the possibility of interactions with other factors in the milt makes it impossible to generalize these findings. A satisfactory buffer for one set of condition or a particular aquatic species may not be acceptable for others. However, an ideal buffer should prolong the life of milt, penetrate cells and act as a good intracellular buffer, isotonic, non-toxic during freezing and thawing, contains nutrients, stabilizing colloids and antioxidants, has antibacterial activity, provides better clarity under microscope for evaluation. Thus, each laboratory must specify which buffer under its conditions will provide maximum reproductive efficiency. Therefore, a good extender for preservation of milt of a species matches with the properties of the seminal fluid of that particular species. Further, the extender should be tested for its competency of inhibiting motility and extending life span of sperm. If motility hinders/inhibits, then that extender is not fit for preservation and should not be preferred.

With this concept, a long range of extenders/diluents/media have been used in different fish species such as Cortland's medium, NaCl solution, NaCl + urea solution, Fish Ringer's solution, KCl medium etc. Different salts, organic and inorganic compound are used for formulating the extender medium for a species. The functions of some of these components are:

Function of Different Salts used in Extenders (Agarwal 2011)

$NaHCO_3$, $KHCO_3$, TRIS—as buffer

NaCl—for tonicity

KCl—to prevent the sperm activation

Mg++, Ca++ ions—normal components of seminal fluid

Sugars—for source of energy

Glycine—for improved survival of sperm

Mannitol—to protect against toxic effect of DMSO

Lecithin, egg yolk, Promine D, bovine serum albumen (BSA)—to protect the membranes

The prepared extenders should be stored at 4–5 °C and used within 1 week.

7.2.4.2 Sugar

It provides glycolysable substrate to the sperm, prevents sperm agglutination, maintain required osmotic tension and electrolyte balance and gives added cryoprotection during freezing.

7.2.5 Cryoprotectant

Cryoprotectants are substances which greatly improve survival of cells after cryopreservation. Their mechanism of action is poorly understood and probably varies for different cryoprotectants. The possible modes of actions of cryoprotectants proposed are:

1. Change in size and shape of ice crystals formed
2. Salt buffering action that binds water and less ice crystals all formed
3. Gives more time for the cells to dehydrate
4. Interacts with cell membrane and makes them less brittle during freezing, thus preventing irreversible damage due to cracking
5. Lessen the effects of high concentration of molecules as cell dehydrates

The type of cryoprotectant and level of addition varies from cell to cell and species to species. Determining the cryoprotectant type and its right concentration in extender medium is an important step of developing cryopreservation protocol for a species. The cryoprotectant protects the spermatozoa during freezing by modifying the size and shape of the ice crystals formed, binds with the water and decreases the freezing point of the solution and hence less ice is formed, acts through salt buffering mechanism, reduces solute concentration and prevents denaturation of proteins and rupture of plasma membrane. Although a number of cryoprotectants are available, satisfactory results have been achieved with dimethyl sulphoxide (DMSO). The cryoprotectants are categorized into two viz., permeating and non-permeating cryoprotectants.

The permeating cryoprotectants are permeable to cell membrane and function by reducing the rate of diffusion of water from cell to extracellular ice crystal, reducing the cell volume change/salt concentration colligatively, lowering the homogenous nucleation temperature and reducing the rate of ice crystal growth. The common permeating cryoprotectants are DMSO, methanol, propylene glycol and glycerol.

DMSO and glycerol are most widely used permeating cryoprotectant to depress the freezing point of the extracellular medium, ameliorate the damaging effects of ice crystals and regulate the rate of cellular dehydration in the freshwater fishes. These cryoprotectants are also toxic to gametes when used in high concentration and prolonged exposure to cell during equilibration. Hence, it is always advisable to use cryoprotectants at optimum level at a concentration ranging between 5% and 10% for successful cryopreservation of fish sperm. However, the permeating cryoprotectants are often toxic to cells, and thus the choice of the types of cryoprotectant and their optimal concentration should be at a balance between protection and toxicity.

The non-permeating cryoprotectants are not permeable to cell membrane and function by depressing the freezing point and raising the glass transformation temperature of extracellular solution. Sucrose, glucose, dextran, egg yolk serum, skim milk and anti-freeze protein are of this category. Sometimes, a combination of different cryoprotectants in certain ratios could help improve the post-thaw motility.

7.2.6 Lipids

Addition of lipids minimize the effects of cold shock and thereby prevent the rupture of sperm plasma membrane. This reduces the leakage of intracellular enzymes, minerals, lipoproteins, ATO and other cellular material from spermatozoa. It also protects SH enzymes and anti-agglutinic factors present in the milt.

7.2.7 Antibiotics

Refrigeration greatly suppresses the multiplication of microorganism, but does not necessarily stop their growth. There are organisms that can survive an ultra-low temperature of −196 °C. Addition of antibiotics in the buffer serves the purpose to a certain extent. The antibiotics used should be relatively harmless to the sperm cells and broad spectrum too.

Introduction of frozen milt minimized the wastage of milt and have a detrimental effect on the breeding efficiency if inferior batch is distributed for artificial breeding. It is economically and biologically important that only milt with a high probability of successfully fertilizing fish eggs be processed and distributed. Hence, a very careful examination and selection of the milt sample before and after cryopreservation are the routine practice in the Cryobanks. Though variations in the milt quality are observed normally, it is influenced mostly by genetic, age, health and nutritional status, frequency of collection and method of milt collection.

7.2.8 Diluent Preparation

The mixture prepared by mixing of extender and cryoprotectant is called diluents. The diluent should be prepared fresh just prior to their use and checked with a semen sample to ensure that spermatozoa are not activated by the diluent. The concentration of cryoprotective agents (CPAs) in the diluent is subjective to the milt of the fish species for which diluent is required. The diluent is kept in refrigerator as it exhibits exothermic reaction.

7.2.9 Dilution of Fish Milt

The milt samples with good quality are subsequently diluted at an appropriate sperm to diluent ratio, with suitable extender solution and cryoprotectant. Sperm to diluent ratios ranged between 1:3 and 1:9 as reported in different literatures to produce best results in fish sperm cryopreservation studies. The type of diluents and dilution ratios workable to preserve sperm motility appear to vary among different fish species.

The diluents help to maintain sperm viability prior to and during the freezing process. This solution is a balanced salt buffer of specific pH and osmotic strength. Apart from salts, sometimes it is prepared with addition of organic compounds such as glucose. The nature of the effect of diluents solutions is based on the control of pH and salt concentration as well as the supply of energy, and can extend the functional life and fertilizing capability of the spermatozoa.

The evaluated milt that is fit for further processing is diluted with the different diluents in various semen-diluent ratios, to test the effective and suitable ratio for fertility trials. The dilution is important because it increases the volume of semen, so that it can be used for multiple inseminations. But this variable must be standardized for the efficacy of any protocol. Since sperm density shows high variability between species, within and between individuals of same species and during breeding seasons, the use of dilution ratio must be complimented with sperm concentration. The number of spermatozoa needed to fertilize one ovum is also species dependent. Therefore optimal sperm to egg ratio is determined first while calculating the semen-dilution ratio for cryopreserving the semen. As post-thaw viability and motility are usually lowered than the pre-freeze motility, this should also be compensated in calculating the viable sperm available to an ovum for fertilization from cryopreserved semen.

The diluent should be slowly added to the milt to avoid any shock (due to change in pH and Osmolality) to the sperm. After preparation, each sample is tested for its motility. If the sample shows reduced or no motility after mixing with diluents, then that diluent should be discarded and freshly prepared.

7.2.10 Equilibration of Fish Milt

Equilibration of fish milt is the pre-freeze storage period of fish spermatozoa at +5 °C done in a cold handling cabinet or refrigerator. It is the period/time given for the cryoprotectant to bring about its beneficial action on fish spermatozoa. During this period, the cryoprotectant binds with or permeates the fish spermatozoa and gives better cryoresistance to spermatozoa to withstand freezing stress. It protects the spermatozoa from damages caused during freezing and thawing. For effective protection during cooling, sufficient time should be given to facilitate the penetration of cryoprotectants into sperm cells. The equilibration time varies with the species and even for the same species it varies depending on the type and concentration of cryoprotectant/diluent used. The equilibration time should be kept to a minimum to avoid exhaustion. The exposure of fish spermatozoa for longer duration more than 45 min is not normally practised to avoid the toxic effects of cryoprotectants to the cells. Hence, the equilibration time is set at 15–30 min and not longer than 60 min at 5 °C. During this period itself, the filling and sealing of diluted milt in French mini straws/cryovials/ampoules should also be completed. At the end of the equilibration period, prefreeze motility is recorded and samples showing more than 70% pre freeze motility alone are taken for further processing in cryopreservation.

7.2.11 Different Packaging Systems

The ability of the spermatozoa to withstand freezing was discovered by British workers in 1949. However, much of the early technology with reference to storage and delivery systems was developed in the United States. Consequently, use of ampoules was developed in the United States. French, Danish and German were responsible for introduction of straw technique. Meanwhile, Japanese scientists developed a technique for freezing semen in pellets which has been adopted in some parts of the world for freezing semen of farm animals. Other packaging techniques reported in literatures have not become popular or economically feasible due to their poor freezability and fertility after artificial breeding.

7.2.11.1 Pellet Method

In this method, freezing is done by depositing 0.1–0.2 mL of diluted milt on the depressions created on dry ice i.e. solid carbon dioxide. The dry ice cooling method offers only one freezing rate of 30–35 °C per minute. This method is of limited use as milt is frozen as pellets that are prone to crumbling and cannot be hermetically sealed. The milt gets frozen and after 10 min the frozen pellets (like tablets) are collected in a goblet and placed in a canister and stored in liquid nitrogen in a cryocontainer at −196 °C. At the time of insemination, the frozen pellet is dissolved in a warm tube containing fresh diluents and used for artificial breeding. Pellet freezing is economical and occupies less storage space compared to other packaging units. However, identification of the frozen samples is difficult. Further, as the pellets stored are uncovered and directly exposed to liquid nitrogen, chances of contamination are more. Apart from these, when the pellets are handled by forceps while thawing, the pellets sometimes break, crumble and get attached to forceps resulting in loss of spermatozoa. This pellet method is extensively used for salmonids sperms.

7.2.11.2 Cryovial/Ampoule Method

This method is widely adopted. The cryovials or ampoules are sealed with screw caps, frozen and preserved in liquid nitrogen. For artificial breeding, the cryovial or ampoule is thawed, and the milt is used. Since milt is placed inside the cryovials or ampoules and sealed, contamination during storage is avoided. Identification of samples is also possible as the information required can be marked on the cryovials

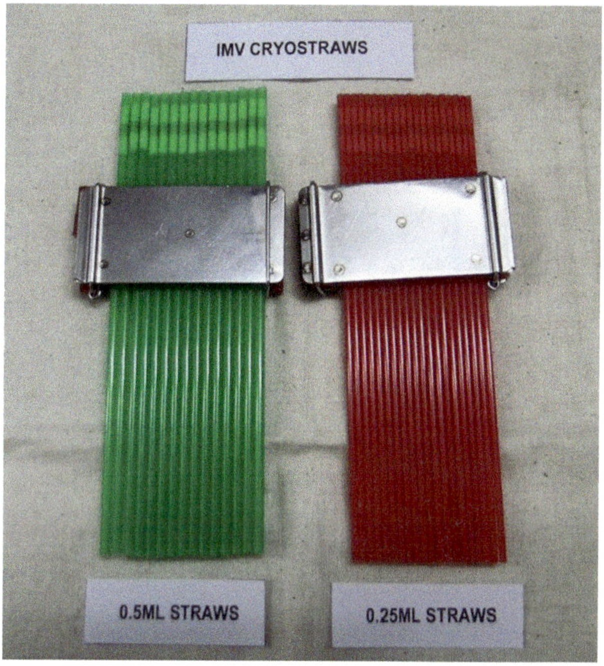

Fig. 7.4 IMV cryostraws of 025 mL and 0.5 mL volume (Photo courtesy: J. Stephen Sampath Kumar)

or ampoules. However, as the milt is frozen in a larger volume, the freezability and fertilizability may be less and moreover, they occupy more storage space in cryocontainers.

7.2.11.3 Straw Method

Plastic straws were first introduced in Denmark in 1940 for packaging semen of domestic mammals. The technique of freezing semen in straws using liquid nitrogen vapour was developed by Alder in 1960 and later modified and redefined by Cassou in 1965 made of polyvinylchloride called French straws. The French straws are available in different capacities viz. 0.25, 0.5 and 1.0 mL (Fig. 7.4). The commonly used one worldwide is 0.25 mL French mini straws. This straw method is more popular as it yields better post-thaw revival rate as the chances for recrystallization during thawing is less. Apart from these, German straws or Lanshut system developed in Germany and United States or Continental straws developed in the United States were also in use.

Straw method has got several advantages over other methods and more popular over other all over the world because milt processed is in thin film with greater surface area to volume ratio resulting in rapid heat exchange and better revival rate. Identification of sample by colour or straw, printing is possible and occupies less storage space in cryocontainers.

Fig. 7.5 Conventional freezing method using thermocole box (Photo courtesy: J. Stephen Sampath Kumar)

7.2.11.4 Printing, Filling and Sealing of Straws

The printing, filling and sealing of diluted fish milt into French mini straws can be carried out manually or automatically. When it is done automatically, machine will be doing all the operations within shorter duration and with least handling of milt.

7.2.12 Freezing of Straws/Cryovials

Freezing results as an outcome of cooling. The rate of cooling is important since it affects the rate of formation and size of ice crystals, as well as solution effects that occur during freezing. The slow cooling provide sufficient time for osmotic equilibrium to be maintained by cellular dehydration, but prolonged exposure of the unfrozen sperm to the hypertonic residual freezing solution may affect membrane integrity.

Liquid nitrogen vapour cooling method is widely used and easy for field application. Freezing of diluted milt is accomplished in straws in liquid nitrogen vapour conventionally in a thermocole box (Fig. 7.5) or LR 320 wide mouthed liquid nitrogen container or using programmable freezers. It is done at a rate of 10–45 °C per minute until the temperature reaches −70 °C. Control rate cooling (CRC) process is ideal as it provide uniform cooling rates and is particularly useful for establishing optimal cooling rates for semen of different fish species. But it requires sophisticated and costly system of computer assisted programmable cell freezers (Fig. 7.6).

In conventional freezing, cooling profiles are monitored by placing several sealed straws on a freezing rack at various heights above the liquid nitrogen to achieve desired cooling rates. After empirically establishing the optimal height (cooling rate) for cryopreservation of candidate fish species, milt filled straws are placed on freezing rack at a predetermined height above liquid nitrogen for freezing preferably at a height of 4 cm above the level of liquid nitrogen. The straws are allowed to cool for 10–15 min. During this period, the straws are super-cooled slowly from 0 to

Fig. 7.6 Programmable freezer (Photo courtesy: J. Stephen Sampath Kumar)

−120 °C. Then the straws are rapidly transferred in to a goblet/canister and are plunged directly into liquid nitrogen at −196 °C in a cryocan for long term storage.

While storing the straws, sample codes for fishes can also be given to the canisters for distinguishing the cryopreserved samples. Comprehensive records of cryopreserved materials should be maintained. It contains preservation details, location, species/strain, date and number of samples, origin, pre-freeze quality, colour coding of straws, identification/batch number, extender used, cryoprotectant used, canister position, sperm concentration, dilution rate, collection method, date of freezing, freezing method, cooling rate, post thaw quality etc.

7.2.13 Preliminary Storing

After primary freezing, the straws or ampoules or cryovials are plunged in to the plastic goblet or ampoule clips which in turn to be placed in a canister filled with liquid nitrogen. When the boil off ceases the goblet/canister is transferred into a liquid nitrogen container for temporary storage at −196 °C.

7.2.13.1 Storage and Handling of Frozen Milt

The cost of storing material at temperatures below freezing increases as temperature decreases. In general, the lower the temperature the longer the motility and fertility can be maintained. Several cryogenic agents have been used for freezing and preservation of fish milt viz. Solid carbon dioxide (−79 °C), liquid air (−183 °C), liquid nitrogen (−196 °C) etc.

Since 1964, liquid nitrogen has almost replaced all the other cryogenic agents in cryopreservation. Liquid nitrogen is the fourth coldest substance known which has a boiling point of −196 °C at atmospheric pressure. Being a liquid it comes into good contact with the surface of the packaging material used and hence a constant storage temperature is maintained throughout the packaging material. Liquid nitrogen is odourless, colourless and most inert substance; hence, it can be safely handled in preserving the biological materials. It is non-toxic and produces no toxic or irritating vapours. It has been reported that greater deterioration in motility and metabolic activity of spermatozoa occurs at −79 °C in solid carbon dioxide storage than at −196 °C in liquid nitrogen.

7.2.13.2 Precautions on Storage

The frozen genetic material should be kept completely submerged in liquid nitrogen. The cryocan in which the material kept should be periodically topped with liquid nitrogen. The cryovials/ampoules/straws should not be tightly packed and space for liquid nitrogen should be left. Always keep proper identification slips in the goblets to enable quick and easy identification.

The frozen germplasm should never be touched with fingers and must be handled by a stainless steel forceps or tweezers. To avoid any raise in temperature the forceps or tweezers should be precooled in liquid nitrogen to attain the temperature of liquid nitrogen before handling the cryovial or straw. The frozen milt is further exposed to elevated temperature when it is transferred or taken from storage container. The increase in temperature is determined by length of time exposed, ambient temperature, air circulation, solar radiation, level of liquid nitrogen in the container and the height to which canister is raised above the neck. The desired material should be picked out within 10 s. This will minimize the raise in temperature while handling. If the material is to be transferred from one container to another, then keep both the canisters of the containers submerged in liquid nitrogen kept in a thermocole box and complete the transfer as quickly as possible. The important factor enabling the successful storage of frozen material is maintaining them in low temperature.

7.2.14 Thawing of Milt

Thawing is the process of the revival of cryopreserved spermatozoa from dormant stage in ultra-low temperature (−196 °C) to fertilization temperature. During thawing, the same physiological processes take place in reverse order as in freezing. By this technique the fish spermatozoa are brought to the fertilizing temperature from the ultra-low temperature at −196 °C. Theoretically, the faster a sperm is frozen, the more rapidly it should be thawed for optimum survival. In rapid warming, less time is available for recrystalization to occur before complete thawing of frozen semen. Hence, to bypass crystallization zone, thawing must be rapid, progressive and uniform.

The fish spermatozoa do not withstand refreezing or repeated cooling and warming. There is no universally adopted system present with reference to the time and temperature allowed for thawing. The rate at which thawing occurs depends on the size and shape of packaging (vial/ampoule/straw), composition of the pack (Polypropylene/glass/plastic) and thawing medium (water/air). The most convenient method is to plunge the vial or straw in warm water (Fig. 7.7). Since thawing temperature has profound effects on sperm viability after cryopreservation, the

Fig. 7.7 Thawing of cryopreserved milt in water bath (Photo courtesy: J. Stephen Sampath Kumar)

most widely and commonly practised method is thawing at 37–40 °C for 30 s, which is suitable to get optimum revival and survival of fish spermatozoa. As soon as the straws are thawed, they should be removed from water bath and to minimize the risk of contamination wipe the external surface of straws by an alcohol-soaked gauze or clean dry tissue paper prior to cut open.

Type of packaging	Volume of semen	Duration of thawing
French mini straw	0.25 mL	30 s
Cryovial	1–3 mL	5 min

7.2.15 Post-thaw Evaluation

Post-thaw evaluation of frozen milt samples is assessed after 24 h of storage. After thawing, straws are cut open and percentage motility of thawed semen samples is estimated without delay because just after thawing, sperm become activated and remain motile only for a brief period. The percentage of post thaw motility usually decreased after cryopreservation and not affected with the time period at cryogenic temperature. Samples showing less than the desired satisfactory post thaw motility (40%) under a phase contrast microscope are discarded and good samples having a post thaw motility more than 40% progressive forward motility alone are transferred to permanent storage containers for long-term preservation.

Fertilization success of gametes and subsequent development of early life stages are the reliable measures of cryopreservation success. Therefore fertility/viability test of cryopreserved semen should be performed by artificial fertilization experiments and incubation and hatching of fertilized eggs in hatchery system to evaluate the success of cryopreservation technique. The fertilization rate and hatching percentage should be calculated and compared with the control treatments for evaluation of any cryopreservation protocol.

Remember, the most poor fertility results with frozen sperm insemination are due to improper handling or release of thawed spermatozoa not at the correct place and at right time.

7.3 What Happens when a Sperm Is Frozen and Thawed?

Simply freezing and thawing sperm or for that matter any mammalian cell in tissue fluid or compatible salt solutions, results in cell death. The lethality of freeze-thawing is apparently associated with the formation of ice rather than sub-zero temperatures since super cooling (cooling below the freezing point without ice formation) to temperatures as low as −15 °C permits survival, but freezing at this temperature and below does not. The damage to tissues during freezing and thawing is caused by:

1. Internal ice crystal formation that affects the structure of cells
2. The solute concentration that results as pure water is withdrawn from suspension media both inside and outside the cells (solution effect) and
3. Interactions of the above two physical factors

7.3.1 Physical and Chemical Basis of Ice Formation

Water is the primary constituent of biological fluids responsible for the internal transport of essential chemicals. Pure water freezes and forms ice crystals at 0 °C, whereas water containing ions and other substances in solution freezes at lower temperature depending on the concentration of these substances. As water in a solution is frozen, pure water crystals form, leaving behind greater liquid concentration of those substances in solution. This increases the osmotic pressure of the remaining solute which can damage the cells.

The major physical and chemical consequences of cryopreservation are the removal of pure water from solution to form ice and the resultant increased concentration of solutes in the residual fluid. These events and their effects on the cells are influenced by species, diluents/extenders and their osmolality and pH, level and type of cryoprotective agents, packaging unit, freezing rate, thawing rate etc.

The freezing point of a solution is based on the concentration of particles it contains as solutes. A given molecular weight of an electrolyte that dissociates into ions on entering into solution will lower the freezing point more than the same proportionate weight of a molecule that does not dissociate. As a solution cools to, through, and past the point at which ice forms, a continuing series of events takes place. The rate of cooling depends on the difference in temperature between the solution and its coolant and also on their relative mass.

Depending on the concentration of solutes and the rate of cooling, the temperature of the solution falls below the true freezing point of water until an ice crystal forms around a "seed". At that point the heat of fusion required to form ice must be removed from the solution, so its temperature shifts upward as the process takes place. As each molecule of pure water freezes out of the solution, the concentration of particles in that cell remains to increase, resulting in a continuous drop in the freezing point of that solution. These interrelated changes continue until the entire solvent and its contained solutes and other contents have completed the process through the eutectic point. Due to the longer exposure of the cells, the damaging effects of high solute concentrations should be most critical with slow rates of freezing. Similarly because cells may be exposed to damaging solute concentrations during rewarming, fast thawing should be beneficial.

7.3.2 Cryoinjury to Cells

Although the formation of extracellular ice crystals was once considered an important disruptive force, it does not seem to be a primary cause of cell damage during freezing. Extracellular ice is not in fact, particularly harmful. With few exceptions, however intracellular ice is lethal. In addition, recrystallization (the growth of large crystals at the expense of smaller ones) can occur at temperatures above −130 °C. Intracellular ice is considered to be a lethal event for most cell types including sperms and must therefore be avoided completely.

7.3.2.1 Influence of Cooling Rates

Maximum survival of cells is generally achieved at cooling rates between the extremes viz. at rates which are fast enough to minimize "solution effects" and yet not so fast as to cause intracellular ice. To interpret the importance of cooling rate let us consider the "environment" experienced by the cells during freezing.

With the onset of freezing the injurious aspects of "solution effects" begin. Solution effects arise from the concentration of solutes as water is frozen out of the solution in the form of pure ice. Thus, the water components of the extender which forms ice leaves the sperm bathed in increasing concentrations of the other extender components, and the collective effect of these concentrated components on cells represents the "solution effects".

When freezing is slow, ice formed is exclusively extracellular (outside the cell) and water is drawn from the cell in response to the increasing solute concentration outside the cell. This water contributes to the increasing proportion of extracellular ice. Therefore, during freezing the cell is actually being dehydrated as water is removed to form extracellular ice.

The ability of the cell membrane to permit this water to leave the cell is very important for survival during freezing. If the rate of cooling is too fast, water may not be able to leave the cell at a sufficient rate to keep up with the demand of the rapidly increasing concentration of extracellular solute. This results in super cooling of the less concentrated intracellular fluid with subsequent formation of intracellular ice.

Contrary to the usual impression, the challenge to cells during freezing is not their ability to endure the very low temperatures required for long storage (−196 °C) it is the lethality of an intermediate zone of temperature (−10 °C to −50 °C) that a cell must traverse twice, once during cooling and once during warming. Injury due to solution effects is encountered in this zone and is not only dependent upon solute concentrations but also time spent traversing the zone. Thus increasing the rate of cooling and rate of warming to minimize cell exposure to the intermediate temperature zone is a generally accepted principle in successful freezing of spermatozoa and many types of cells.

7.3.2.2 Mode of Solution Effect on Cryoinjury

The exact nature of injury inflicted upon cells by solution effect is not clear; however, the following hypotheses are proposed:

1. Cells are irreversibly injured when a lethal concentration of salts occurred in the unfrozen portion. The concentrated salt solution denatured components of the cell membrane.
2. One of the solution effects accompanying freezing is the removal of intracellular water with dehydration and shrinkage of the cell. In this minimum "cell volume theory" when the cell shrinks below a minimum volume, the structural integrity is jeopardized.
3. The fraction of the extracellular solution remaining unfrozen is most important. In this proposal freezing injury occurs when the unfrozen portion of the sample drops below 8–12%.

While the above theories may differ in how the solution effects or intracellular ice injures the cell, they do not argue the premise that the solution effect and/or intracellular ice are the hostile events leading to sperm injury and/or death during freezing and thawing.

7.3.2.3 Role of Cell Membrane Permeability

The permeability of the cell membrane to water and the solutes varies among cell types and may be dependent upon the composition of the cellular environment. In addition, cells vary considerably in their surface-to-volume ration. If formation of intracellular ice is lethal, then vulnerability of the cell to this event would depend upon how easily water could leave the cell to form extracellular ice during cooling. If cell membrane permeability of water was low or the cell is too large, the movement of water from inside the cell to the outside would be impaired when contrasted with a small cell having a high surface to volume ratio or one having high membrane permeability.

Sperm which has a very high surface to volume ratio contrasts with ovum or embryo where this ratio is very low. It is hypothesized that if the permeability of a cell to water, the temperature coefficient of that permeability and the size of the cell are known or can easily be estimated, it is possible to calculate the maximum cooling rate that will still provide enough time for a cell to dehydrate during cooling and thus avoid intracellular freezing.

7.3.2.4 Role of Freezing Rate on Cryosurvival of Sperm

Contrary to popular belief the challenge to cells during cryopreservation is not their ability to endure storage at low temperature, rather it is the lethality of an intermediate zone of temperature (15 to −60 °C) that the cells must traverse twice, once during freezing and once during thawing. As the cells are cooled to about −5 °C, both cells and surrounding medium remain unfrozen and gets super cooled. Between −5 and −10 °C ice forms in the external medium. The subsequent physical events in the cells depend on the cooling rate adopted in this zone. If the cells are cooled rapidly, intracellular ice crystals are formed and the cells are killed. However, if the cells are cooled too slowly, they will experience severe volume shrinkage and long-time exposure to high solute concentration causes cell injury. Therefore, either too high or too low a cooling rate can kill cells although the mechanisms underlying cell damage are different. Based upon this, an optimal cooling rate for cell cryosurvival should exist between the rapid and slow rates. The other biological parameter that plays vital role is water permeability and surface to volume ratio.

Normally the spermatozoa of all species responds well to freezing in static vapour freezers (LR 320, OM 320 etc.) since the sperm have wide tolerance to survive even at sub optimal freezing conditions around the ideal freezing curve. Static vapour freezing technique relies on the temperature difference between the liquid nitrogen and filled straws to vapourize the nitrogen. This produces a sufficient supply of vapour to cause a rapid fall of temperature within the straw. However, this method has a disadvantage in that apart from setting the initial conditions of vapour temperature and liquid nitrogen level in the freezing chamber, the operator has no further control over the freezing process. While poor freezers have sperm that require freezing conditions which approximate more closely to the ideal curve (perhaps due to possessing cell membranes less able to survive damage). In such cases, the computer-assisted programmable vapour biological freezer could be used for optimizing freezing conditions. Freezing conditions could be improved by optimizing cooling rates within the chamber especially between −2 and −20 °C. A slow cooling rate is needed between temperatures of 5 and −15 °C and more rapid cooling is needed between −15 and −25 °C. Loss of sperm below −30 °C was found to be small and cooling rates were not critical.

7.3.2.5 Optimizing Freezing Rate

If the cells are cooled too rapidly, water is not lost fast enough to maintain equilibrium; the cells become increasingly supercooled eventually attaining equilibrium by freezing intracellularly. In most cases, cells undergoing intracellular ice formation are killed. If the cells are cooled very slowly, cell loses water, gets dehydrated, shrinks and exposed to high solute concentration which causes cell injury and death. Therefore either too high or too low a cooling rate can kill cells although the mechanism of action is different. Based upon this, an optimal cooling rate for cell cryosurvival should exist between the "high" and "low" rates. The other biological parameter that plays vital role is water permeability and surface to volume ratio of the cell. Spermatozoa with a higher ratio can tolerate wide range of freezing rate (up to 100 °C per min).

7.4 Effects of Cryopreservation on Sperm

The spermatozoa have a diversity of functional attributes to be maintained and are accordingly regionally highly differentiated. The sperm lose their ability to function normally during freezing and thawing. The successful procedure should aim at preserving in the largest possible proportion of cells, the integrity of different structures with different cryobiological requirements. The salient changes that occur as a result of cryopreservation are:

1. Membrane damage—plasma membrane, outer acrosomal membrane, mitochondrial membrane
2. Structural alteration—cytoskeleton, axonemal element, mitochondria, nucleus
3. Functional changes—altered motility, reduced post thaw survival time, partially capacitated, acrosome reacted, homogenous population capable of achieving fertilization only in narrowly defined temporal and chemical circumstances

7.4.1 Merits of Frozen Milt in Aquaculture

- Permits maximum utilization of milt of a sire of outstanding high genetic merit thus providing sperm on demand and simplifying the timing of induced spawning
- Minimizes the number of males to be maintained and also eliminates the need to maintain live males
- No milt could be discarded because of age, as with milt stored in room/refrigerated temperature
- Removes barrier on time and distance in artificial breeding. International exchange of milt has come into existence now
- Selective mating/breeding is possible
- Insurance against loss of germ plasm by death
- Minimizes transportation cost and provides round the clock breeding facility to the breeders/farmers
- Entails the collection of sperm and eggs so that one male can fertilize eggs from several females and one female's eggs can be fertilized by sperm from several males thus can lead to a matrix where a group of select males can be mated with a group of select females to develop populations with distinctive traits.

7.4.2 Demerits of Frozen Milt in Aquaculture

- All the milt collected from individuals do not withstand rigors of freezing
- High initial investment cost
- Limits number of sires/males used and if proper care has not been taken up it may lead to inbreeding
- Requires better training of personnel both in the laboratory and field insemination
- Reduced or poor fertilization rate compared to other artificial breeding methods

7.5 Conclusion

Cryopreservation is a viable technology and plays major role in the genetic improvement of livestock. However, it is still in its infancy in fisheries sector. If understood and used properly, definitely it will be a successful technology that provide great unexploited potential for fish breeding. Hence, development of novel approaches and reliable methods to protect the spermatozoa of a very wide range of fish species during cryopreservation and thawing is the need of the hour. When this is achieved, cryopreserved fish spermatozoa could be used like fresh spermatozoa without loss of fertility and resolve major problems in aquaculture. Progress in these areas would undoubtedly provide a guaranty of the success and safety of the use of cryopreserved spermatozoa in aquaculture science.

Acknowledgement The author is immensely thankful to Dr. M. John Edwin, Retired Professor and Head, Department of Animal Genetics and Breeding, Madras Veterinary College, Chennai and acknowledges him for utilizing the manuscripts.

Reference

Agarwal NK (2011) Cryopreservation of fish semen. In: Himalayan aquatic biodiversity conservation & new tools in biotechnology. Transmedia Publication, pp 104–127

Cryopreservation and Short Duration Storage of Germ Cells and Male Gametes of Freshwater Fish

8

P. Routray

Abstract

Storage of fish germ cells and male gametes and utilization is a new frontier for aquaculture research and development. Cryopreservation and short-term preservation of fish milt and germ cells at low temperature may play a great role in brood stock upgradation by utilizing it at distantly located hatcheries. Fish seed production industry recorded a remarkable growth during the last several decades. The quality of seed is an important consideration for commercial aquaculturists. In hatchery seed production sector, storage of milt can facilitate selective breeding, hybridization and commercial seed production. It provides an instant milt component to a stripped egg mass under control condition. A success of stored milt depends upon the quality of milt and standard process of preservation. To keep the selected sample alive for certain duration needs preservation at an appropriate temperature with suitable extenders since only raw milt is unsuitable for preservation. Preservation may be non-cryogenic (short term) or cryogenic (long term). Gametes of improved stock are cryopreserved and utilized for quality seed production as well as upgrading the brood stock in carps. Gamete cryopreservation protocol for fish with special emphasis on freshwater carps is focused in this review.

Keywords

Cryopreservation · Extenders · Milt · Male gamete · Germ cell · Spermatozoa

P. Routray (✉)
Cryobiology Laboratory, ICAR-Central Institute of Freshwater Aquaculture, Bhubaneswar, India

J. Betsy, S. Kumar (eds.), *Cryopreservation of Fish Gametes*,
https://doi.org/10.1007/978-981-15-4025-7_8

8.1 Introduction

Cryopreservation and short duration storage (for few hours to few days) of germplasm that includes male gametes and germ cells (GCs) offers possibilities of preserving genome of endangered species besides its application in artificial propagation. Cryopreservation is a long-term storage method and as the name implies it uses ultra-low temperatures to preserve the structurally intact viable cells and tissues for prolonged period of time at a relatively low cost (Alikunhi et al. 1960). Cells of a wide variety of organisms have been successfully cryopreserved and their banking has been a regular practice in animal husbandry, medical practice and conservation science (Alikunhi et al. 1963; Arii et al. 1987). This technique is widely used to preserve gametes, embryos and embryonic cells of many animal species. Polge et al. (1949) reported the first successful cryopreservation of mammalian sperm cells. Sperm cell preservation of mammals and other aquatic animal species has become a regular practice around the world. Protocols for milt cryopreservation have been available for many fish species (Joseph and Jayaprakas 2014; Cabrita et al. 2010), however several variations do exist in the protocols in terms of the use of extenders, cryoprotective agents (CPAs), equilibration time, thawing and fertilization process. This is obvious since about 24,600 living species of fishes in 482 families and 57 orders have valid scientific descriptions (Nelson 1994). It has been estimated that there are approximately 23,000 bony fishes, 10,100 are entirely freshwater and 2500 moves between sea and freshwater during their life cycles (Helfman et al. 1997). Worldwide, 11% birds, 25% mammals and 34% of fish species are threatened (Lovell-Badge 2001; Vemuganti and Balasubramanian 2002) and aquatic ecosystem is becoming more vulnerable to disasters due to run off from agriculture and industrial sources, oil spills or sudden environmental changes leading to total elimination of stocks from different ecosystems. In this scenario, cryopreservation of germ cells (GC) offers hope for conservation of germplasm for future biotechnological application in aquaculture.

Cryopreservation of germ cells and male gametes is proven and repeatable protocols are being used worldwide. Cryopreservation of male gametes or spermatozoa is well established in a number fish species (Chao and Liao 2001; Cabrita et al. 2003; Suquet et al. 2000; Babiak et al. 1997; Horváth et al. 2003; Lahnsteiner 2000; Cloud 2000; Gwo 2000; Routray et al. 2003, 2008). The utilization of cryopreserved male gametes of fish is finding its use for various aquaculture purposes such as artificial insemination and stock upgradation of species also.

On the other hand, cryopreservation of germ cells (GC) also offers hope for conservation of germplasm for future biotechnological application in aquaculture. Fish like many other metazoans, have two distinct cell lineages, namely the immortal germ cells and mortal somatic cell. Primordial germ cells (PGCs) are germ line stem cells that give rise to gametes in vertebrates. They originate outside the embryo itself very early in development and migrate by a definite route into the genital ridges (Chiquoine 1954). The haploid gametes (oogonia and spermatogonia) are produced from germ cells through meiosis. Germ cells are only responsible for the transfer of genetic information to the next generation. So germ cells play a unique role in gamete production, heredity and evolution. Until recently, primordial germ cells (PGCs) have been studied in a number of organisms such as *Drosophila*, *anuran amphibian* and *nematode*. Among teleosts, zebra fish has been widely studied for specification, differentiation and migration of PGCs. In fish, as many other metazoans, germ cells are originated from primordial germ cells (PGCs). PGCs came from presumptive PGCs (pPGCs) (Kohji et al. 1984).

Here an attempt has been made to describe the status of short duration storage and cryopreservation of germ cells and male gametes of fish.

8.2 Short Duration Storage of Germ Cells and Male Gametes (Non-cryogenic)

Short-term preservation of carp milt at low temperature may play a great role in brood stock upgradation by utilizing it at distantly located hatcheries. One of the proven methods for best utilization of male gametes of fish during artificial insemination and controlled breeding programs is to use short duration preservation of milt for 1–2 days. The milt sample is diluted with extender and stored in a thermocol chamber or cool box with ice or in refrigerator. This is helpful for designing different sib breeding out of different male and female traits. Modified extender-C is found more easy and effective for such preservation. Milt samples are brought to room temperature before inseminating the egg sample for better result. The short-term preservation method is very handy where availability of liquid nitrogen is a constraint. This method has been employed at ICAR-CIFA to transport male gametes of catla from distant locations such as Andaman Nicobar Islands to Odisha by maintaining a temperature of 4 °C. The feasibility of transporting milt from one part of India to the other was carried out by transporting the back crossed variety milt from Andaman Islands to Bhubaneswar (ICAR-CIFA campus). It was interesting to know that the viability of carp (Back cross variety; Catla X Rohu) milt could be successfully transported in cold chain (4 °C) over ice. Upon fertilization with eggs collected from a catla female, 70% fertilization and 40% hatching could be obtained. It can be useful in artificial propagation and complement cryopreservation. Study conducted at CIFA revealed that Indian major carp semen can be successfully preserved for 18 h at 4 °C prior to artificial insemination. By this method, milt from Andaman Nicobar Islands could be brought to ICAR-CIFA and successful fertilization and seed production in catla was achieved. Similar short-term preservation and utilization of carp gametes were observed and found to be successful in different hatcheries including other countries too (Table 8.1). Similarly, germ cells isolated from fish could be stored for more than 24 h under refrigerated conditions.

Table 8.1 Utilization of short duration preserved milt (male gametes) of fish used for stock upgradation at different locations stored at 4 °C

Name of the farm/ place	Name of species	Average male weight (kg)	Average female weight (kg)	Duration after milt collection (h)	Milt: Extender-C dilution	Fertilization (%)
Tarahara, Nepal	Rohu	2.6	2.2	4	1:2	95
Tarahara, Nepal	Rohu	2.5	2.8	6	1:2	65
Tarahara, Nepal	Rohu	2.3	2.2	8	1:2	65
Janakpur, Nepal	Rohu	2.0	2.5	4	1:2	55
Janakpur, Nepal	Rohu	2.3	2.1	8	1:2	40
Tarahara, Nepal	Rohu	2.7	2.9	8	1:2	60
Dambulla, Sri Lanka	Catla	3.9	4.5	12	1:2	50
Dambulla, Sri Lanka	Catla	4.2	5.0	14	1:2	65
Dambulla, Sri Lanka	Catla	4.6	4.2	14	1:2	70
ICAR-CIFA, India	Rohu	2.9	2.5	18	1:2	72
ICAR-CIFA, India	Catla	3.2	3.5	13	1:2	70
ICAR-CIFA, India	Mrigal	2.3	2.8	20	1:2	80
Kailash Hatchery, Mayurbhanj, Odisha, India	Catla and Rohu	2.5	2.1	19	1:2	85
Mr. Batakrushna Sahoo, Sarakana, Odisha, India	Catla and Rohu	2.5	2.1	19	1:2	85
Yash Hatchery, Bihar, India	Rohu and Catla	2.3	2.8	20	1:2	82

8.3 Cryopreservation of Germ Cells and Male Gametes

Protocols for milt cryopreservation have been available for many fish species; however, fish eggs and embryos have not been successfully cryopreserved till now. There are reports about the cryopreservation of GCs in different fish species viz. rainbow trout (Okutsu et al. 2006), Siberian sturgeon (Pšenička et al. 2012), *Tinca tinca* (Linhartová et al. 2014), catla and rohu (Patra et al. 2016) and zebrafish (Higaki et al. 2009). GCs are believed to be well-suited to cryopreservation due to small size and a high level of sexual plasticity, which allows them to differentiate into fully functional gonads of both sexes (Okutsu et al. 2006). Spermatozoa can be stored for years together in liquid nitrogen (−196 °C) where all the biochemical activities of a living cell ceases. The milt is diluted in extender and mixed with cryoprotectant.

Cryoprotectant is a chemical that protects the cell from chilling injury during the course of freezing. Carps are the mainstay of freshwater aquaculture in the Indian sub-continent. Long-term preservation of teleost gametes and that of carps in particular could be highly beneficial for fish farming and wildlife preservation (Rall 1993; Routray 2003). Successful freezing of sperm has been achieved in several species of carps which belong to the cyprinid family. Cryopreservation of common carp sperm has been extensively studied. In early trials fertility of thawed sperm was either not tested (Sneed and Clemens 1956) or was low (Kossman 1973; Mockzarski 1977; Stein and Bayrl 1978). The first practical results were reported by Kurokura et al. (1984). Koi carp sperm were frozen in a methanol-dry ice bath in 0.5 mL straws. Freezing with 15% dimethyl sulfoxide (DMSO) resulted in 69% fertilization against a fresh sperm control of 83%. Cryopreservation of carp milt has also been reported by Kumar (1988). Long-term preservation of Indian major carp milt and their utilization in seed production have also been reported (Gupta et al. 1995; Routray 2003).

The general protocol of cryopreservation of small cells and spermatozoa is as follows: (1) loading, which is the equilibrium of the specimen in cryoprotective agents (CPA); (2) dehydration in a solution (extender + CPA); (3) plunging the specimen in liquid nitrogen; (4) warming the specimen and (5) unloading of CPAs, by progressive dilution from the specimens. Wilmut (1972) and Whittingham et al. (1972) reported firstly that mouse embryos could be successfully frozen at −196 °C. Although the general protocols were similar with that of spermatozoa, the slow cooling and thawing rates were preferred for better survival in embryos.

8.4 Conventional Methods

Conventional methods of cryopreservation, such as the use of a Styrofoam box filled with liquid nitrogen or a programmable freezer, rely on comparatively slow, controlled cooling during early ice formation. Such a method is being used at CIFA, Bhubaneswar and also being advocated for hatchery managers due to its convenience and cost-effectiveness in cryopreserving carp spermatozoa. This method has been proven suitable for carp sperm cryopreservation in bulk.

For Indian major carps, milt samples were equilibrated for 30–40 min at 4 °C including filling and sealing of French straws/visotubes. These straws were kept over liquid nitrogen for 5 min over tripod stand in a manual cryofreezer (CIFACRYO) developed at CIFA. The liquid nitrogen vapour in the chamber provide a good enough freezing rate for freezing milt samples filled in 0.25–0.5 mL straws. Thus, cryopreserved spermatozoa have been utilized in hatcheries across India (Gupta et al. 1995; Routray et al. 2007, 2008).

8.5 Vitrification

By following the glass transition principle, liquid systems are rapidly cooled by avoiding ice crystal formation which is called as vitrification. In this process, the solution forms an amorphous glass as a result of rapid cooling by direct immersion of the visotubes/straws containing the milt samples into liquid nitrogen. The glass retains the normal molecular/ionic distributions of a liquid but remains in an extremely viscous form. The glass is devoid of all ice crystals, and the spermatozoa are not subjected to the physical damage due to the ice crystal formation during freezing which is peculiar in conventional cryopreservation method. Vitrification is a simple and less expensive alternative to conventional freezing. Vitrification solutions are generally cryoprotectants of high concentrations which effectively dehydrate the cells prior to the initiation of the cooling process. Vitrification solutions can be of permeating or semi-permeating type or a combination of both. Vitrification solutions share three common properties. First, these vitrification solutions contain a combination of low and high molecular weight cryoprotectants. Low molecular weight cryoprotectants penetrate cell membranes and protect the cytoplasm from damage during freezing. Higher molecular weight cryoprotectants do not pass across the cell membranes; however, they are effective extracellular dehydration agents. Secondly, the final overall concentration of the cryoprotective agents in the mixture is high, to enhance vitrification and thus, avoiding lethal ice crystal formation. Finally, the standard vitrification solution contains an isotonic level of saline.

8.6 Protocol for Cryopreservation

8.6.1 Collection of Brood Fish

Matured males showing free oozing of semen upon gentle abdominal pressure were selected for semen collection. Each time, four to six males were injected intraperitoneally with Ovaprim (Salmon GnRH+ Domperidone) at a rate of 0.2 mL/kg. Semen collection was carried out 5 h after the hormone injections. Milt samples were collected in ice cooled and sterilized test tubes. During the semen collection, attention was paid to prevent contaminations by faecal matter, urine, blood or scales and to maintain the temperature of the collected semen at 4 °C. The collected semen was evaluated for sperm yield/kg body weight, motility, pH, spermatocrit percentage and sperm count and biochemical composition of seminal plasma.

8.6.2 Equilibration with Extender

For successful cryopreservation, it is essential to prevent any activation of spermatozoa during preservation. Undiluted sperm is unsuitable for storage at cryogenic temperatures, so it should be diluted with an appropriate medium (Scott

Table 8.2 Extenders used in male gamete cryopreservation

Name	Composition	References
Extender-C	NaCl–750 mg, KCl–20 mg, $CaCl_2$–20 mg, $NaHCO_3$–20 mg. Added to 100 mL distilled water	Gupta et al. (1995)
Modified extender-C	NaCl–750 mg, KCl–70 mg, $CaCl_2$–20 mg, $NaHCO_3$–20 mg, D-glucose 10 mg. Added to 100 mL distilled water	Routray et al. (2006)
Yamamoto ringer	NaCl–750 mg, KCl–20 mg, $CaCl_2$–20 mg, $NaHCO_3$ added to adjust pH to 7.3. Added to 100 mL distilled water	Yamamoto (1975)
Freshwater fish ringer	NaCl–128 mM, KCl–2.7 mM, $CaCl_2.2H_2O$–1.5 mM, $NaHCO_3$–2.4 mM	Kurokura et al. (1984)
Mounib	$KHCO_3$–100 mM, Sucrose–125 mM, Glutathione–6.5 mM	Mounib (1978)
Frog ringer	NaCl–111.2 mM, KCl–1.9 mM, $CaCl_2$–0.8 mM, $NaHCO_3$–2.4 mM, NaH_2PO_4–0.1 mM, Glucose–11.1 mM	Sneed and Clemens (1956)
Stein and Bayrle	NaCl–128 mM, KCl–5.1 mM, $NaHCO_3$–23.8 mM, Glucose–5.6 mM, Egg yolk–20%	Stein and Bayrle (1978)

and Baynes 1980; Stoss 1983). Because motility of fish spermatozoa is mostly a one-time event, this medium should not induce motility and at the same time must not interfere with the ability of the spermatozoa to be activated subsequently during utilization. Media that satisfies these conditions is called "Extender". The chemical components of extender are NaCl–750 mg, KCl–20 mg, $CaCl_2$–20 mg, $NaHCO_3$–20 mg added to 100 mL distilled water and adjust the osmolality in the range of 280–300 mOsm/kg. The extender may be prepared in advance and, immediately before use (Routray et al. 2002). In carps, extender-C and modified extender-C with minor modifications have been in use in India (Gupta et al. 1995; Routray 2003). A list of extenders generally used for fish semen cryopreservation is presented in Table 8.2.

8.6.3 Cryoprotectants Used for Cryopreservation

To protect the cells from cryoinjuries, it is essential to protect them with the help of chemicals known as cryoprotective agents (CPAs) or cryoprotectants. The CPA used in this protocol is DMSO (dimethyl sulfoxide), which gives consistent result. The CPA should not be mixed directly in the milt sample. It is mixed in the ratio (DMSO 15 mL + Extender 65 mL) and kept in refrigerator due to its exothermal reaction. The CPA should be added prior to mixing with semen samples and the percentage of DMSO should not be more than 20% in any case.

8.6.4 Thawing/Warming and Fertilization

Thawing is an important step in cryoprotocol; if optimum conditions are not followed, ice formation takes place due to recrystallization. In the case of carp spermatozoa, fast thawing is preferable, because slow thawing can recrystallize small intercellular crystals that may damage the cells. In the present protocol, the cryomilt samples are thawed in warm water at 38 ± 1 °C. Visotubes and French straws take approximately 65–70 and 8–10 s, respectively, until slush formation. After slush formation, it is important to keep the milt decanted from the visotubes on to a fertilization plate for nearly 5 s before mixing with the unfertilized eggs. This is done to bring both gametes to isothermal conditions at room temperature. Eggs fertilized with the cryomilt are washed thoroughly 5–10 times in freshwater and then placed in flow-through incubation chambers to wash out excess CPA and for hatching.

References

Alikunhi KH, Vijayalakshmanan MA, Ibrahim KH (1960) Preliminary observations on the spawning of Indian carps, induced by injection of pituitary hormones. Indian J Fish 7:1–19

Alikunhi KH, Sukumaran KK, Parameswaram S (1963) Induced spawning of the Chinese grass carp, *Ctenopharyngodon idellus* (C. and V.) and silver carp (*Hypophthalmichthys molitrix*) (C. and V.) in ponds at Cuttack, India. Curr Sci 32:103–106

Arii N, Namai K, Fujiya G, Nakazawa T (1987) Cryoprotection of medaka embryos during development. Zool Sci 4:813–818

Babiak I, Glogowski J, Brzuska E, Szumiec J, Adamek J (1997) Cryopreservation of sperm of common carp, *Cyprinus carpio* L. Aquac Res 28:567–571

Cabrita E, Robles V, Cherguinni O, Wallace JC, Herráez MP (2003) Effect of different cryoprotectants and vitrificant solutions on the hatching rate of turbot embryos (*Scophthalmus maximus*). Cryobiology 47:204–213

Cabrita E, Sarasquete C, Martínez-Páramo S, Robles V, Beirão J, Pérez-Cerezales S, Herráez MP (2010) Cryopreservation of fish sperm: applications and perspectives. J Appl Ichthyol 26:621–830

Chao N-H, Liao IC (2001) Cryopreservation of finfish and shellfish gametes and embryos. Aquaculture 197:161–189

Chiquoine AD (1954) The identification, origin and migration of the primordial germ cells in the mouse embryo. Anat Rec 118:135–146

Cloud JG (2000) Cryopreservation of sperm of steelhead rainbow trout after refrigerated storage. In: Tiersch TR, Mazik PM (eds) Cryopreservation in aquatic species. World Aquaculture Society, Baton Rouge, pp 101–103

Gupta SD, Rath SC, Dasgupta S (1995) Fertilization efficiency of cryopreserved carp spermatozoa over four years at different time intervals after thawing. Geobios 22:208–211

Gwo J-C (2000) Cryopreservation of sperm of some marine fishes. In: Tiersch TR, Mazik PM (eds) Cryopreservation in aquatic species. World Aquaculture Society, Baton Rouge, pp 138–160

Helfman GS, Collette BB, Facey DE (1997) The diversity of fishes. 528 pp. Blackwell Science: Oxford. ISBN 0 86542 256 7 (hard cover). Anim Conserv 1(2):151–151. https://doi.org/10.1017/S1367943098000092

Higaki S, Mochizuki K, Baba H, Akashi Y, Yamaha E, Katagiri S, Takahashi Y (2009) Feasibility of cryopreservation of zebrafish, *Danio rerio* primordial germ cells by whole embryo freezing. Jpn J Vet Res 57:119–128

Horváth Á, Miskolczi E, Urbányi B (2003) Cryopreservation of common carp sperm. Aquat Living Resour 16:457–460

Joseph A, Jayaprakas V (2014) Factors affecting the motility and short term preservation of spermatozoa of two species of indigenous ornamental fishes, *Rasbora daniconius* and *Puntius filamentosus*. J Aquat Biol Fish 2:52–61

Kohji I, Takashi O, Sakiko N (1984) Differentiation of presumptive primordial germ cell (pPGC)-like cells in explants into PGCs in experimental tadpoles. Dev Biol 103:258–262

Kossman H (1973) Versuch zur Konservierung des karpfenspermas (*Cyprinus carpio*). Arch Fisch 23:125–128

Kumar KA (1988) Comparative study of various extenders for cryopreservation of carp spermatozoa. J Anim Sci 58:1355–1360

Kurokura H, Hirano R, Tomita M, Iwahashi M (1984) Cryopreservation of carp sperm. Aquaculture 37(3):267–273

Lahnsteiner F (2000) Cryopreservation protocols for sperm of salmonid fishes. In: Tiersch TR, Mazik PM (eds) Cryopreservation in aquatic species. World Aquaculture Society, Baton Rouge, pp 91–100

Linhartová Z, Rodina M, Guralp H, Gazo I, Saito T, Pšenička M (2014) Isolation and cryopreservation of early stages of germ cells of tench, *Tinca tinca*. Czeh J Anim Sci 59:381–390

Lovell-Badge R (2001) The future of stem cell research. Nature 414:88

Mockzarski M (1977) Deep freezing of carp (*Cyprinus carpio* L.) sperm. Bull Acad Pol Ser Biol 24:187–190

Mounib MS (1978) Cryogenic preservation of fish and mammalian spermatozoa. Reproduction 53 (1):13–18

Nelson JS (1994) Fishes of the world, 3rd edn. Wiley, New York, p 620

Okutsu T, Yano A, Nagasawa K, Shikina S, Kobayashi T, Takeuchi Y, Yoshizaki G (2006) Manipulation of fish germ cell: visualization, cryopreservation and transplantation. J Reprod Dev 52:685–693

Patra S, Mishra G, Dash SK, Verma D, Nandi S, Jayasankar P, Routray P (2016) Transplantation worthiness of cryopreserved germ cells of Indian major carp Rohu, *Labeo rohita*. Curr Sci 111:739–746. https://doi.org/10.18520/cs/v111/i4/739-746

Polge C, Smith AU, Parks AS (1949) Revival of spermatozoa after vitrification and dehydration at low temperatures. Nature 164:666

Pšenička M, Saito T, Rodina M, Linhartovà Z, Flajshans M (2012) Isolation and cryopreservation of spermatogonia and oogonia of sturgeons for the conservation of genetic resources. In: Methodology, 29th edn. Faculty of Fisheries and Protection of Waters, University of South Bohemia in ČeskéBudějovice, Vodňany

Rall WF (1993) Genetic change in hatchery populations. In: Cloud JG, Thorgaard GH (eds) Genetic conservation of salmonid fishes. Plenum Press, New York, p 137

Routray P (2003) Evaluation of cryopreserved milt of a selected carp, *Labeo calbasu* for fertilization and seed production. Ph.D. Thesis, Sambalpur University, Sambalpur

Routray P, Suzuki T, Strüssmann CA, Takai R (2002) Factors affecting the uptake of DMSO by the eggs and embryos of medaka, Oryzias latipes. Theriogenology 58:1483–1496

Routray PS, Gupta D, Behera MK (2003) Cryogenics in fish hatchery technology. Fishing Chimes 23:7–9

Routray P, Choudhary AK, Dash SN, Verma DK, Dash C, Swain P, Jena JK, Gupta SD, Sarangi N (2006) Cryopreservation of dead fish spermatozoa several hours after death of Indian major carp, *Labeo rohita* and its successful utilization in fish production. Aquaculture 261:1204–1211

Routray P, Verma DK, Sarkar SK, Sarangi N (2007) Recent advances in carp seed production and milt cryopreservation. Fish Physiol Biochem 33(4):413–427

Routray P, Dash SN, Dash C, Swain P, Sarkar SK, Sarangi N (2008) Cryopreservation of silver barb *Puntius gonionotus* (Bleeker) spermatozoa: effect of extender composition, cryoprotective agents and freezing rate on their post thawing fertilization ability. Aquac Res 39:1597–1605

Scott AP, Baynes SM (1980) A review of the biology, handling and storage of salmonid spermatozoa. J Fish Biol 17:707–739

Sneed KE, Clemens HP (1956) Survival of fish sperm after freezing and storage at low temperatures. Progress Fish Cult 18(3):99–103. https://doi.org/10.1577/1548-8659(1956)18[99:SOFSAF]2.0.CO;2

Stein H, Bayrle H (1978) Cryopreservation of the sperm of some freshwater teleosts. Annales de Biologie Animale Biochimie Biophysique 18(4):1073–1076

Stoss J (1983) Fish gamete preservation and spermatozoan physiology. In: Hooar WS, Randall DJ (eds) Fish physiology, vol 9B. Academic, New York, pp 305–350

Suquet M, Dreanno C, Fauvel C, Cosson J, Billard R (2000) Cryopreservation of sperm in marine fish. Aquac Res 31(3):231–243

Vemuganti GS, Balasubramanian D (2002) Heralding the dawn of cultured adult stem cell transplantation. Indian J Biotechnol 1:39–49

Whittingham DG, Leibo SP, Mazur P (1972) Survival of mouse embryos frozen to −196 °C to −296 °C. Science 178:411–414

Wilmut I (1972) The effects of cooling rate, warming rate, cryoprotective agent and stage of development on survival of mouse embryos during freezing and thawing. Life Sci 11:1071–1079

Yamamoto T (1975) Fertilization in Oryzias egg. In: Yamamoto T (ed) Medaka (killifish) biology and strains. Keigaku, Tokyo, pp 81–95

Cryopreservation of Marine Fish Sperm

9

Tian Yongsheng, Zhang Jingjing, Li Zhentong, Li Ziqi,
and Wang Linna

Abstract

This paper reviews the progresses made in sperm cryopreservation of marine fish, including 40 species belonging to the genera Gadiformes, Salmoniformes, Perciformes, Pleuronectiformes, and *Epinephelus*. The factors that affect the efficiency of sperm cryopreservation mainly include formula of sperm diluent, type of antifreeze, and cryopreservation method. There are differences in the optimal formula of sperm diluent among species, various fish sperm diluents are prepared by the researchers using physiological salt, glucose, sucrose, and fetal bovine serum. The commonly used sperm diluents include Hanks, Ringer's, MPRS, Ts-2, ELS-3, and MFs-3, and the data from past studies suggest that the sperm diluent plays a primary role in the survival of frozen sperm. Cryoprotectants can protect sperm from the freezing damage during the freezing process; dimethyl sulfoxide (DMSO) and 1,2-propylene glycol (PG) are the most commonly used cryoprotectants, they show the best effect at 8–15% concentration to cryopreserve sperm. The sperm cryopreservation methods include programmed cooling and the vitrification method. The research on sperm cryopreservation technology has effectively solved the industrial problems of reproductive isolation caused by the unsynchronized maturation of male and female gonads and geographical distribution in fish. Many countries, such as the United States, Canada, and China, have established various fish sperm cryobank. The

T. Yongsheng (✉) · W. Linna
Yellow Sea Fisheries Research Institute, Chinese Academy of Fishery Sciences, Qingdao, China

Laboratory for Marine Fisheries Science and Food Production Processes, Qingdao National Laboratory for Marine Science and Technology, Qingdao, China
e-mail: tianys@ysfri.ac.cn

Z. Jingjing · L. Zhentong · L. Ziqi
Yellow Sea Fisheries Research Institute, Chinese Academy of Fishery Sciences, Qingdao, China

College of Fisheries and Life Science, Shanghai Ocean University, Shanghai, China

J. Betsy, S. Kumar (eds.), *Cryopreservation of Fish Gametes*,
https://doi.org/10.1007/978-981-15-4025-7_9

cryopreserved sperm have been applied in gynogenesis induction in flatfish, hybrids breeding in grouper, and mass breeding of fish fry.

Keywords

Fish · Marine · Spermatozoa · DMSO · Cryobank

9.1 Introduction

There are more than 20,000 species of marine fish in the world, and over 2100 species are endemic to China. The total capture and farming volumes of marine fish for human consumption across the world in 2017 were 66.3 million and 2.25 million tons, respectively, of which 7.65 million and 1.42 million tons were harvested in China. With changes in the marine environment, increase in seawater temperature, exploiting of marine resource, excessive capturing of marine fish, and industrial pollution discharge, the number of fish has decreased or has become miniaturized. In more extreme cases, species are facing extinction through these environmental changes. According to the report of the IUCN Red List of Threatened Species (www.iucnredlist.org/), there are 35 threatened fish species across the world, including Nassar Grouper (*Epinephelus striatus*), Atlantic Goliath Grouper (*Epinephelus itajara*), Angelshark (*Squatina squatina*), and Hectors Dolphin (*Cephalorhynchus hectori*). Therefore, the cryopreservation of germplasm resources of marine fish is a major mission to preserve fish species.

Cryopreservation of fish sperm is an important method to achieve long-term preservation of germplasm. By establishing cryopreserved sperm stock of various fish species, they can be protected from the effects of overfishing, environmental pollution, and ecological deterioration. In addition, cryopreservation of genetic material can prevent germplasm degradation and genetic homogenization that results from long-term breeding and inbreeding, and relieve reproductive isolation caused by geographical distribution, reproductive temperature, and time. It can also be beneficial for species with unsynchronized maturation of the sexes, self-crossing of sex-transformed individuals, and can facilitate hybridization of different breeds. For farming purposes, cryopreservation can provide abundant sperm for fish sex regulation, selection and crossbreeding, inducing the development of gynogenesis and androgenesis, artificial reproduction, collecting genetic material for biotechnology research, and breeding of high-quality stocks. For these reasons, the development and application of cryopreservation of fish sperm are integral to aquaculture and scientific research development.

9.2 Overview of Cryopreservation of Marine Fish Sperm

Cryopreservation of fish sperm began in the 1950s, where sperm nests of Atlantic herring were preserved with dry ice by the British scholar Blaxter (1953). Afterward, scholars from various countries conducted sperm cryopreservation studies on fish such as Gadiformes, Salmoniformes, Perciformes, Pleuronectiformes, and Epinephelus (Table 9.1).

9.2.1 Cryopreservation of Gadiformes Fish Sperm

Mounib (1978) cryopreserved cod semen at −196 °C, achieving a fertility rate of 80–89%. Sperm from haddock (*Melanogrammus aeglefinus*) and Atlantic cod (*Gadus morhua*) were cryopreserved in MME + 10% DMSO by DeGraaf and Berlinsky (2004), and the survival rates were 63.0% and 66.0%, respectively, indicating that cod sperm can be cryopreserved for a long time and still fertilize eggs after thawing. Rideout (2004) cryopreserved haddock and Atlantic cod sperm using propylene glycol (PG), dimethyl sulfoxide (DMSO), and glycerol, and obtained a survival rate of 62.5 and 56.3% using PG, suggesting that PG is superior to the other antifreeze reagents that were tested. The physiological indexes of sperm coherence (SEM), seminal plasma osmotic pressure, pH, protein concentration, antitrypsin activity, and total antioxidant capacity of Atlantic cod sperm were measured to evaluate the quality and performance of sperm that were cryopreserved. At the same time, Atlantic cod sperm was cryopreserved with different antifreeze agents, further indicating that sperm can have higher survival rate (66%) when using propylene glycol (Butts et al. 2010, 2011).

9.2.2 Cryopreservation of Salmoniformes Fish Sperm

The sperm of Coho salmon (*Oncorhynchus kisutch*), Chinook salmon (*O. tshawytscha*) and rainbow trout (*O. mykiss*) cryopreserved by Ott (1975), and the cryopreserved sperm had a fertilization rate of 29–83%. The sperm of *O. gorbuscha* and *Salmo trutta* were cryopreserved by Philpott (1993), and the fertilization rate was 44–85%. Erdahl and Graham (1987) froze squid sperm and obtained a fertilization rate greater than 90%. Mounib (1978) cryopreserved Atlantic salmon sperm, and after 1 year, the fertilization rate was 80%. Yang (2018) used Ringer's and 10% methanol to develop a method for cryopreservation and transportation of Atlantic Salmon (*Salmo salar*) sperm and obtained a sperm survival rate of 36%. Cryopreservation significantly affected the Atlantic salmon sperm DNA fragmentation, sperm cell membrane, and mitochondrial membrane potential, and the correlation between mitochondrial membrane potential and sperm motility to fertilization rate was 0.75 and 0.59, respectively (Figueroa et al. 2016). High concentration of seminal plasma in sperm cryopreservation solution for vitrification and cryopreservation can improve the quality of sperm, with DNA fragmentation,

Table 9.1 Studies on sperm cryopreservation in marine fish species

Common name	Species	Extender composition	Cryoprotectant	Motility after thawing (%)	Authors
Atlantic herring	*Clupea pallasi*	Sea water	12.5% glycerol	80–85	Blaxter (1953)
Atlantic cod	*Gadus morhua*	125 mM sucrose, 100 mM $KHCO_3$, 6.5 mM reduced glutathione	10% PG	56.3 ± 4.1	Rideout et al. (2004)
Ocean pout	*Macrozoarces americanus*	1.45 mM $CaCl_2$, 0.84 mM $MgSO_4$, 10.25 mM $KHCO_3$, 183 mM NaCl, 0.15 mM glucose	20% DMSO	20–25	Yao et al. (2000)
Haddock	*Melanogrammus aeglefinus*	125 mM sucrose, 100 mM $KHCO_3$, 6.5 mM reduced glutathione	10% PG	62.5 ± 3.2	Rideout et al. (2004)
Southern hake	*Merluccius australis*	Stopmilt®	1.2 M DMSO + 0.3 M sucrose +2% BSA	–	E. Figueroa et al. unpublished data
Atlantic salmon	*Salmo salar*	Cortland	1.3 M DMSO + 0.3 M glucose + 2% bovine serum albumin	58.5 ± 5.3	Figueroa et al. (2016)
Amago salmon	*Oncorhynchus masou ishikawae*	300 mM glucose	10% DMSO	18.5 ± 2	Ohta et al. (1995)
Brook trout	*Salvelinus fontinalis*	5.85 g L^{-1} NaCl, 0.255 g L^{-1} KCl, 0.33 g L^{-1} $NaHCO_3$, 0.25 g L^{-1} Na_2HPO_4, 0.145 g L^{-1} $CaCl_2$ H_2O, 0.2 g L^{-1} $MgCl_2$ $6H_2O$, 0.1 g L^{-1} citric acid, 10 g L^{-1} glucose, 10 mL KOH (1.271/100 mL), 10 mL bicine (5.31/100 mL)	10–20% DMSO, DMA, glyerol	84.5	Glogowski et al. (1996)

Spotted seatrout	*Cynoscion nebulosus*	5.26 g L^{-1} NaCl, 0.26 g L^{-1} KCl, 0.33 g L^{-1} $NaHCO_3$, 0.04 g L^{-1} Na_2HPO_4, 0.04 g L^{-1} KH_2PO_4, 0.13 g L^{-1} $MgSO_4$ $7H_20$, 0.66 g L^{-1} glucose; 200 mOsMol kg^{-1}	15% DMSO + 15% EG + 10% Gly + 1% X-1000tm + 1% Z-1000tm	73 ± 21	Cuevas-Uribe et al. (2013)
Large yellow croaker	*Pseudosciaena crocea*	Cortland	10% DMSO or EG	87.00 ± 2.45 87.50 ± 2.52	Jiang et al. (2011)
European sea bass	*Dicentrarchus labrax*	59.83 mM NaCl, 1.47 mM KCl, 12.91 mM $MgCl_2$, 3.51 mM $CaCl_2$, 20 mM $NaHCO_3$, 0.44 mM glucose, 1% BSA, 1 mL mM hypotaurine; pH 7.7 taurine or 1 mL mM hypotaurine; pH 7.7 hypotaurine; pH 7.7	10% DMSO	30.1 ± 3.2	Martınez-Paramo et al. (2013)
Red drum	*Sciaenops ocellatus*	5.26 g L^{-1} NaCl, 0.26 g L^{-1} KCl, 0.33 g L^{-1} $NaHCO_3$, 0.04 g L^{-1} Na_2HPO_4, 0.04 g L^{-1} KH_2PO_4, 0.13 g L^{-1} $MgSO_4$ $7H_20$, 0.66 g L^{-1} glucose; 200 mOsMol kg^{-1}	15% DMSO + 15% EG + 10% Gly + 1% X-1000tm + 1% Z-1000tm	30	Cuevas-Uribe et al. (2013)
Golden kingklip	*Genypterus blacodes*	Stopmilt®	1.2 M DMSO + 0.3 M sucrose + 1% BSA	–	E. Figueroa et al. unpublished data
Patagonian blenny	*Eleginops maclovinus*	Stopmilt®	1.5 DMSO + 0.4 M glucose + 2% BSA	–	E. Figueroa et al. unpublished data
Sea perch	*Lateolabrax japonicus*	60.35 mM NaCl, 1.80 mM NaH_2PO_4, 3 mM $NaHCO_3$, 5.23 mM KCl, 1.3 mM $CaCl_2$ $2H_2O$,	10% DMSO	68.3 ± 4.4	Ji et al. (2004)

(continued)

Table 9.1 (continued)

Common name	Species	Extender composition	Cryoprotectant	Motility after thawing (%)	Authors
		1.13 mM $MgCl_2$ $6H_2O$, 55.55 mM glucose			
Pacific bluefin tuna	*Thunnus orientalis*	171.12 mM NaCl	20% DMSO or 10–20% Gly or 10% methanol	84.3 ± 5.7–93.3 ± 8.2	Gwo et al. (2005)
Red sea bream	*Pagrus major*	8.00 g L^{-1} NaCl, 0.40 g L^{-1} KCl, 0.14 g L^{-1} $CaCl_2$, 0.10 g L^{-1} $MgSO_4$ $7H_2O$, 0.10 g L^{-1} $MgCl_2$ $6H_2O$, 0.06 g L^{-1} Na_2HPO_4 $12H_2O$, 0.35 g L^{-1} $NaHCO_3$, 1.00 g L^{-1} glucose	15% DMSO	87.67 ± 2.52	Chen et al. (2010)
Gilthead sea bream	*Sparus aurata*	10.01 mg/mL $KHCO_3$, 1.99 mg/mL reduced gluthatione, 42.78 mg/mL sucrose, 10 mg/mL 101 BSA, 10% Me_2SO, pH 7.8	10% DMSO + 10% glycerol	60 ± 3	Zilli et al. (2018)
Lane snapper	*Lutjanus synagris*	7.89 g L^{-1} NaCl, 1.19 g L^{-1} KCl, 0.22 g L^{-1} $CaCl_2$, 0.73 g L^{-1} $MgCl_2$, 0.08 g L^{-1} NaH_2PO_4, 0.84 g L^{-1} $NaHCO_3$; pH 8.2, 172 mOsMol kg^{-1}	10% DMSO	98 ± 3	Sanches et al. (2015)
Red snapper	*Lutjanus campechanus*	5.26 g L^{-1} NaCl, 0.26 g L^{-1} KCl, 0.33 g L^{-1} $NaHCO_3$, 0.04 g L^{-1} Na_2HPO_4, 0.04 g L^{-1} KH_2PO_4, 0.13 g L^{-1} $MgSO_4$ $7H_2O$, 0.66 g L^{-1} glucose; 200 mOsMol kg^{-1}	15% DMSO + 15% EG + 10% Gly + 1% X-1000tm + 1% Z-1000tm	38	Cuevas-Uribe et al. (2013)
Mangrove red snapper	*Lutjanus argentimaculatus*	7.50 g L^{-1} NaCl, 0.20 g L^{-1} $NaHCO_3$, 0.20 g L^{-1} KCL, 0.20 g L^{-1} $CaCl_2$ $2H_2O$, 5.00 g L^{-1} glucose; pH 7.9, 315 mOsMol kg^{-1}	10% DMSO	91.1 ± 2.2	Vuthiphandchai et al. (2009)

Mutton snapper	*Lutjanus analis*	7.89 g L^{-1} NaCl, 1.19 g L^{-1} KCl, 0.22 g L^{-1} $CaCl_2$, 0.73 g L^{-1} $MgCl_2$, 0.08 g L^{-1} NaH_2PO_4, 0.84 g L^{-1} $NaHCO_3$; pH 8.2, 172 mOsMol kg^{-1}	10% DMSO	90.1	Sanches et al. (2013)
Flounder	*Paralichthys olivaceus*	24.72 g L^{-1} NaCl, 0.67 g L^{-1} KCl, 1.36 g L^{-1} $CaCl_2$ $2H_2O$, 4.66 g L^{-1} $MgCl_2$ $6H_2O$, 6.29 g L^{-1} $MgSO_4$ $7H_2O$, 0.18 g L^{-1} $NaHCO_3$; pH 8.2, 205 mOsmol kg^{-1}	12% Gly	76.20 ± 10.0	Zhang et al. (2003)
Brazilian flounder	*Paralichthys orbignyanus*	110 mM sucrose, 100 mM $KHCO_3$, 10 mM Tris-Cl; PH 8.2335 mOsMol kg^{-1}	10% DMSO	50 ± 6	Lanes et al. (2008)
Summer flounder	*Paralichthys dentatus*	7.25 g L^{-1} NaCl, 0.38 g L^{-1} KCl, 0.18 g L^{-1} $CaCl_2$, 1.00 g L^{-1} $NaHCO_3$, 0.23 g L^{-1} $MgSO_4$ $7H_2O$, 0.41 g L^{-1} NaH_2PO_4 H_2O, 1.00 g L^{-1} glucose	15% DMSO or 15% PG	78.00 ± 4.70, 60 ± 7.90	Liu et al. (2015)
Spotted halibut	*Verasper variegatus*	110 mM sucrose, 100 mM $KHCO_3$ and 10 mM Tris-Cl; pH 8.2, 335 mOsMol kg^{-1}	13.3% DMSO or 13.3% PG	75.83 ± 4.91	Tian et al. (2008)
Winter flounder	*Pseudopleuronectes americanus*	137 mM NaCl, 11 mM KCl, 4 mM Na_2HPO_4 $7H_2O$; pH 7.7	PG	80–100	Rideout et al. (2003)
Atlantic halibut	*Hippoglossus hippoglossus*	100 mM $KHCO_3$, 125 mM sucrose; 315 mOsMol kg^{-1} or HBSS, 273 mOsm	10–15% DMSO	58.4 ± 23.6, 52.2 ± 27.2, 63.3 ± 30.3	Ding et al. (2011)
Turbot	*Scophthalmus maximus*	110 mM sucrose, 100 mM $KHCO_3$ and 10 mM Tris-Cl; pH 8.2, 335 mOsMol kg^{-1}	10% DMSO	70.1 ± 8.9	Chen et al. (2004)

(continued)

Table 9.1 (continued)

Common name	Species	Extender composition	Cryoprotectant	Motility after thawing (%)	Authors
Japanese bitterling	*Macroramphosus scolopax*	FBS	10% methanol + 95 or 90% fetal bovine serum	19.3 ± 2.5	Ohta et al. (2001)
Dusky grouper	*Epinephelus marginatus*	1% NaCl, 10 mg/mL BSA	10% DMSO	36.8 ± 10.2	Cabrita et al. (2008)
Seven grouper	*E. septemfasciatus*	135 mM NaCl, 2 mM KCl, 2.3 mM $MgCl_2$, 1.3 mM $CaCl_2$, 20 mM $NaHCO_3$, 20 mM TAPS–NaOH, 5.6% glucose, 13% Trehalose, 95% FBS	5% DMSO	77.6 ± 8.5	Koh et al. (2010)
Red-spotted grouper	*E. akaara*	26.32 g/L NaCl, 1.32 g/L KCl, 0.65 g/L $MgSO_4{\cdot}7H_2O$, 0.18 g/L $Na_2HPO_4{\cdot}7H_2O$, 0.18 g/L KH_2PO_4, 1.15 g $NaHCO_3$, 3.30 g/L glucose	10% DMSO cholesterol	77.1 ± 5.0	Liu et al. (2011)
Seven grouper	*E. septemfasciatus*	60 g/L glucose, 10 g/L NaCl, 0.5 g/L $NaHCO_3$	10% DMSO or 10% PG	76.7 ± 0.0 or 75.0 ± 5.0	Tian et al. (2013)
Longtooth grouper	*E. bruneus*	10 g/L NaCl, 0.22 g/L KCl, 0.25 g/L $CaCl_2$, 0.74 g/L $MgSO_4.7H_2O$, 1.19 g/L HEPES ($C_8H_{18}N_2O_4S$), 0.9 g/L glucose, 0.1 g/L streptomycin, 100,000 unit penicillin G	5.0% DMSO	99.5 ± 0.8	Oh et al. (2013)
Kelp grouper	*E. moara*	9 g/L NaCl, 10 g/L $KHCO_3$; 10% FBS	10% PG	57.2 ± 3.7	Qi et al. (2014)
Giant grouper	*E. lanceolatus*	111.23 mM NaCl, 2.38 mM $KHCO_3$, 1.88 mM KCl, 2.38 mM $NaHCO_3$, $MgCl_2{\cdot}6H_2O$, 0.82 mM $CaCl_2{\cdot}2H_2O$, 0.082 mM NaH_2PO_4, 277.37 mM glucose, 10% FBS	15% DMSO	63.9 ± 4.2–74.8 ± 12.7	Tian et al. (2015)

Giant grouper	*E. lanceolatus*	7.25 g/L NaCl, 0.38 g/L KCl, 1.00 g/L $NaHCO_3$, 0.23 g/L $MgSO_4 \cdot 7H_2O$, 0.41 g/L $NaH_2PO_4 \cdot H_2O$, 0.1 g glucose	PG (8% or 12%) DMSO (10% or 12%)	70.8 ± 5.9 or 72.3 ± 5.0 73.4 ± 5.1 or 70.7 ± 5.0	Liu et al. (2015)
Orange-spotted grouper	*E. coioides*	Same as above	DMSO (8% or 10%)	81.7 ± 4.7 or 80.8 ± 5.9	Liu et al. (2015)
Seven grouper	*E. septemfasciatus*	Same as above	10% PG	72.3 ± 4.3	Liu et al. (2015)
Kelp grouper	*E. moara*	Same as above	10% PG	71.7 ± 5.1	Liu et al. (2015)
Red-spotted grouper	*E. akaara*	0.1 g/L KCl, 9.0 g/L NaCl, 0.06 g $CaCl_2$, 0.08 g/L $MgCl_2$, 0.5 g/L $NaHCO_2$, 0.02 g/L albumin	10% DMSO	85.0 ± 2.9	Ahn et al. (2018)
Brown-marbled grouper	*E. fuscoguttatus*	135 mM NaCl, 2 mM KCl_2, 2.3 mM $MgCl_2$, 1.3 mM $CaCl_2$, 20 mM $NaHCO_3$, 20 mM HEPES–NaOH, 85% FBS	15% PG	76.7 ± 8.8	Yusoff et al. (2018)

plasma membrane integrity, mitochondrial membrane integrity, motility, and fertility rates at 9.2%, 98.6%, 47.2%, 44.1%, and 46.2%, respectively (Figueroa et al. 2015). Seungki and Goro (2016) cryopreserved the endangered Manchurian trout (*Brachymystax lenok*) spermatogonia, and transplanted the thawed testicular cells into allogeneic triploid hatchlings. The transplanted spermatogonia migrated and were incorporated into the recipient gonads, where they underwent gametogenesis.

9.2.3 Cryopreservation of Perciformes Fish Sperm

Pullin (1972) used liquid nitrogen to preserve the sputum of black sea bream (*Acanthopagrus schlegelii*) for 315 days, and obtained a fertilization rate of 20–39%. Liu et al. (2007a) used cryopreserved red seabream (*Pagrus major*) sperm in 15% DMSO. After thawing, sperm motility, fertilization rate, and hatching rate were 81.0%, 92.8%, and 91.8%, respectively. Authors found that the sperm motility decreased gradually with storage time from 87.67 to 50.67% after 1–73 months. Sperm fertilization rate did not decrease significantly during 26 months (60.33%), but decreased significantly after 48 months of storage. The sperm fertilization rate was 39.56% in sperm that were stored for 73 months (Chen et al. 2010). Sansone (2002) studied the equilibrium temperature and freezing rate of cryopreserved sea bass (*Dicentrarchus labrax*) sperm, which was equilibrated for 6 h at 0–2 °C and cooled at a rate of 15 °C/min, and found that these two parameters are important for the vitality of frozen sperm. Eszter Kása (2017) used Cryotop and Straw vectors to cryopreserve Eurasian perch sperm, and obtained a survival rate of 26.4%, and fertilized egg development rate of 2.45%. Ji et al. (2004) used MPRS sperm dilution to cryopreserve sperm from sea perch (*Lateolabrax japonicas*). There was no significant difference between frozen sperm activity and fresh sperm, and the fertilization rate and hatching rate reached 84.8% and 70.1%, respectively. Martínez-Páramo et al. (2013) studied the effects of two sulfur-containing amino acids, taurine and hypotaurine, on the cryopreservation of European perch sperm. It was found that adding 1 mM taurine and 1 mM hypotaurine to the cryopreservation solution improved sperm quality and sperm motility and reduced DNA damage in cryopreserved sperm. The cryopreservation technology of Perciformes sperm is relatively well established, and the long-term preservation of sperm can be achieved by various methods.

9.2.4 Cryopreservation of *Epinephelus* Fish Sperm

At present, sperm from more than ten species of grouper have been cryopreserved, including Seven grouper (*Epinephelus septemfasciatus*) (Koh et al. 2010; Tian et al. 2013), Giant grouper (*Epinephelus lanceolatus*) (Tian et al. 2015), Red-spotted grouper (*Epinephelus akaara*) (Liu et al. 2011; Ahn et al. 2018), Dusky grouper (*Epinephelus marginatus*) (Cabrita et al. 2008), Longtooth Grouper (*Epinephelus bruneus*) (Imaizumi et al. 2005), Black grouper (*Epinephelus malabaricus*) (Gwo

1993), Greasy grouper (*Epinephelus tauvina*) (Withler and Lim 1982), Orange-spotted grouper (*Epinephelus coioides*) (Kiriyakit et al. 2011), Kelp grouper (*Epinephelus moara*) (Qi et al. 2014), and Brown-marbled grouper (*Epinephelus fuscoguttatus*) (Yusoff et al. 2018). Cryopreservation solution used for these species is listed in Table 9.1, and often use 5.0–15% DMSO or 8–15% 1–2 propylene glycol as antifreeze. Except for *E. marginatus* (36.8%), sperm activity after cryopreservation is high in these species, ranging from 57.24% (*E. moara*) to 99.5% (*E. bruneus*), and cryopreservation has been applied for crossbreeding *E. lanceolatus* with a variety of grouper species and large-scale cultivation of its fry (Tian et al. 2017a, b).

9.2.5 Flounder Sperm Cryopreservation

Zhang et al. (2003) used DMSO, Gly, and methanol antifreeze to cryopreserve sperm of *Paralichthys olivaceus*, and found that DMSO had the highest motility (60.5%). Chen et al. (2004) used the sperm diluent TS-2 + 10% DMSO to cryopreserve turbot (*Scophthalmus maximus*) sperm, and obtained the fertilization and hatching rates of 70.1% and 46.8%, respectively. Ji et al. (2005) used MPRS to cryopreserve sperm of stone flounder and Japanese flounder, and the survival rate after freezing was above 70%. Tian et al. (2008) used the sperm diluent TS-2 and MPRS to freeze *Verasper variegatus* sperm. The fertilization and hatching rates of the frozen sperm were 34.52% and 23.53%, respectively. Tian et al. (2009a) cryopreserved the sperm of *Cynoglossus semilaevis*, and the thawed sperm had 53.5% motility, 55% fertilization rate, and 35% hatching rate. Jiang et al. (2014) used the diluent SFs-4 + 20% DMSO to freeze the sperm of *Platychthys stellatus*, and the sperm activity reached 66.67% after freezing. Song et al. (2016) used sperm cryopreservation solution to cryopreserve sperm of *Pseudopleuronectes yokohamae* with MFs-3 + 20% 1,2-propylene glycol, or 20% ethylene glycol, and the sperm motility after freezing was 95.26%, the fertilization rate and hatching rate were 80.08% and 77.44%, respectively. Cuevas-Uribe (2017) used a variety of vitrification fluids to cryopreserve Southern Flounder (*Paralichthys lethostigma*) sperm and obtained a survival rate of 28% and membrane integrity of 11.0%. Tian et al. (2006) used MPRS to cryopreserve the sperm of Summer flounder (*Paralichthys dentatus*) and was able to fertilize *P. olivaceus* eggs using thawed sperm, with 81.63% fertilization rate and 98% hatching rate. The sperm of Atlantic halibut (*Hippoglossus hippoglossus, L.*) was frozen through the large-volume freezing method using Hanks' balanced salt Solution (HBSS) with 10–15% DMSO, which enabled the commercialization of frozen sperm (Ding et al. 2011). The above method is effective in cryopreserving of flounder sperm with 10–15% DMSO. Ten percent of 1,2-propylene glycol is effective to cryopreserve some fish. The main component of diluted sperm is NaCl, KCl, and glucose, adjusting the solution to the osmotic pressure of different types of fish sperm.

9.3 Sperm Cryopreservation Dilution Screening

Fish sperm dilution is a key factor in successfully cryopreserving sperm. Due to the different physiology and fluid components of sperm from different species, the composition of seminal plasma is also different. In order to maintain the integrity of sperm structure and survival rate in vitro, the allocation of sperm diluent has strict requirements. The solution used in diluting sperm is generally composed of physiological salt, disaccharides, monosaccharaides, and calf serum. Its function is to maintain the osmotic pressure and pH of sperm, and to provide nutrients and energy for sperm. Physiological salts mainly include NaCl, KCl, Na_2CO_3, $MgSO_4$, $KHCO_3$, $CaCl_2$, $MgCl_2$, KH_2PO_4, K_2HPO_4, and Tris–Cl, and these compounds provide suitable osmotic pressure and ionic components to maintain the integrity of the sperm. Nutritional solution mainly include glucose, sucrose, trehalose, fructose, lactose, and serum. These components provide nutrition, but also help maintain the extracellular osmotic pressure. In addition, antibiotics may be added to prevent the spread of bacteria or viruses (Oh et al. 2013). The typical dilutions using to cryopreserve grouper sperm are ELRS-3, ELS-3, EMS (Tian et al. 2013, 2015), and typical sperm dilution of flatfish is TS-2, SFs-4 and MFs-3 (Ji et al. 2004; Jiang et al. 2014; Song et al. 2016). The main salt ions used in the dilution are Na^+, K^+, Ca^{2+}, and Mg^{2+}, and the salt composition can be adjusted according to the physiological characteristics of different fish species.

Na^+ generally accounts for a large proportion in sperm dilution, and is the main source of Na^+/K^+ metabolism inside and outside the cell. K^+ inhibits the movement of calcium and magnesium ions, thus inhibiting the movement of the sperm. It can also partially relieve the inhibition of potassium ions, which is necessary for sperm activation. Tris–Cl buffers the pH of the solution to prevent excessive lactic acid production during sperm metabolism.

In vitro fertilized fish sperm can metabolize tricarboxylic acid (Cardiner 1978), which can utilize exogenous carbohydrates, especially glucose, galactose, and fructose, through oxidation. Glucose has a significant effect on prolonging sperm life.

$KHCO_3$, KCl, $NaHCO_3$, $MgCl_2{\cdot}6H_2O$, NaCl, $CaCl_2{\cdot}2H_2O$, NaH_2PO4, glucose, TRIS alkali, and fetal bovine serum (FBS) were used in the screening of cryopreservation diluent for giant grouper, the sperm dilution EM1-2, TS-2, MPRS, ELS1, ELS2, ELS3, ELRS0, ELRS1, ELRS2, ELRS3, ELRS4, ELRS5, and ELRS6 were screened for their effect on sperm physiological parameters, such as fast and slow motion ratio, exercise time, vitality, and longevity analysis. ELS3 and ELRS3 yield the best results, where cryopreserved sperm viability reached 51.1–69.44% A total of 13 kinds of sperm dilution solutions EM1-2, TS-2, MPRS, ELS1, ELS2, ELS3, ELRS0, ELRS1, ELRS2, ELRS3, ELRS4, ELRS5, and ELRS6 were prepared by using $KHCO_3$, KCl, $NaHCO_3$, $MgCl_2{\cdot}6H_2O$, NaCl, $CaCl_2{\cdot}2H_2O$, NaH_2PO4, glucose, TRIS alkali, and fetal bovine serum (FBS), ELRS3, ELRS4, ELRS5, and ELRS6. Through the analysis of physiological parameters such as fast and slow movement ratio, movement time, vitality and longevity of sperm, two diluents,

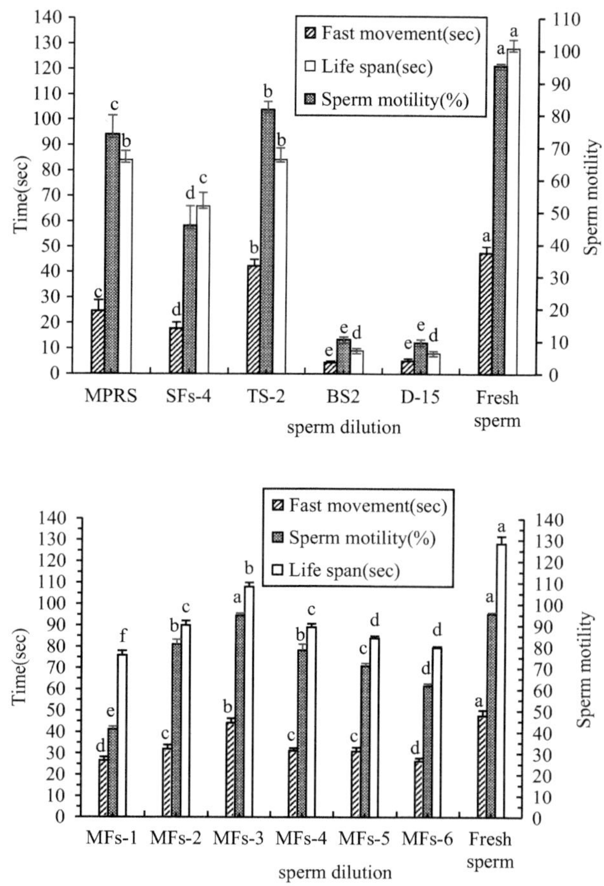

Fig. 9.1 The effects of dilution solutions MPRS, Ts-2, D-15, sfs-4, BS2, and MFs1~10 on the cryopreservation of *Pleuronectes yokohamae* sperm (Song et al. 2016)

ELS3 and ELRS3, were selected, and the sperm viability could reach 51.1–69.44% (Tian et al. 2015).

In the cryopreservation of Marbled Flounder (*Pleuronectes yokohamae*), 15 kinds of dilution solutions, such as MPRS, Ts-2, D-15, SFs-4, BS2, and MFs1~10 were prepared using various salts and sugar sources. Of the solutions that were tested, sperm cryopreserved in MFs-3 had 95.26% motility, 46 s fast movement time and 124.33 s life span, the results indicated the repeatability and availability of sperm diluent (Song et al. 2016) (Fig. 9.1).

9.4 Cryoprotectants of Fish Spermatozoa

Cryoprotectants protect cells from cryodamage by decreasing the ice point during the freezing process. In 1949, Polge found the protective ability of glycerol, where sperm frozen in solution containing glycerol had reduced damage after freezing and thawing (Lovelock and Polge 1954). Since then, a variety of chemical substances with properties of cryoprotection have been found. All cryoprotectants are either permeating or non-permeating cryoprotectants, based on whether the solute can penetrate the cell membrane or not. Almost all permeating cryoprotectants are micromolecules that can enter the cell interior with high penetration rate, such as dimethyl sulfoxide (DMSO), propylene glycol (PG), glycerol (Gly), methanol (MeOH), and ethylene glycol (EG). They can enable the cells to shrink rapidly from the osmotic loss of water, and bond the water molecules in solutions through hydration, which increases the viscosity of solutions. Through these changes, permeating cryoprotectants protect the cells by impeding the crystallization of small intracellular ice crystals that can form during cooling.

However, various permeating cryoprotectants have differences in the concentration, permeability of cells and influence on hydrone activity. Non-permeating cryoprotectants, can dissolve in water, but cannot enter the cell, and include glucan, sucrose, trehalose, fructose, polyvinyl pyrollidone (PVP), albumin, polyethylene glycol (PEG), and hydroxyethyl starch (HES). They increase the concentration of extracellular solution and rapidly remove water from cells, and reduce the concentration of electrolyte in the solution, reducing the excessive osmotic damage that can be caused by salt ions.

Ten to fifteen percent DMSO and PG are widely used to cryopreserve marine fish sperm. DMSO is one of the most common cryoprotectants, due to the low molecular weight and strong permeability. PG is less toxic than DMSO, and has obvious positive effect in the sperm cryopreservation of Marbled Flounder (*Pseudopleuronectes yokohamae*) (Song et al. 2016), Blue grouper (*Epinephelus tukula*), and Camouflage grouper (*Epinephelus polyphekadion*). PG increases the number of surviving cells during freezing, enhances the fluidity of cytomembrane, dehydrates in low temperature, and reduces the intracellular ice crystal formation by rearranging the membrane lipids and proteins (Holt 2000). FBS provides a variety of macromolecule proteins, hormones, and lipids, and protects the cellular membrane and maintain the internal structure of cells (Ahn et al. 2018).

The effects of cryoprotectants, including DMSO, DMAC, PG, and Gly, were compared in the cryopreservation of *E. septemfasciatus* sperm. DMSO and PG had similar effects on sperm motility, longevity, and fast motion time ($P < 0.05$, Fig. 9.2) (Tian et al. 2015).

The effects of dimethyl sulfoxide (DMSO), 1,2-propylene glycol (PG), ethylene glycol (EG), methanol (MeOH), glycerol (Gly), and dimethyl formamide (DMF) on sperm cryopreservation of marbled flounder were compared, and the results showed that PG and EG led to higher sperm motility and longevity (Fig. 9.3) (Song et al. 2016).

9.5 Sperm Cryopreservation Methods

Programmed and vitrification cryopreservation are the principal methods to cryopreserve fish sperm.

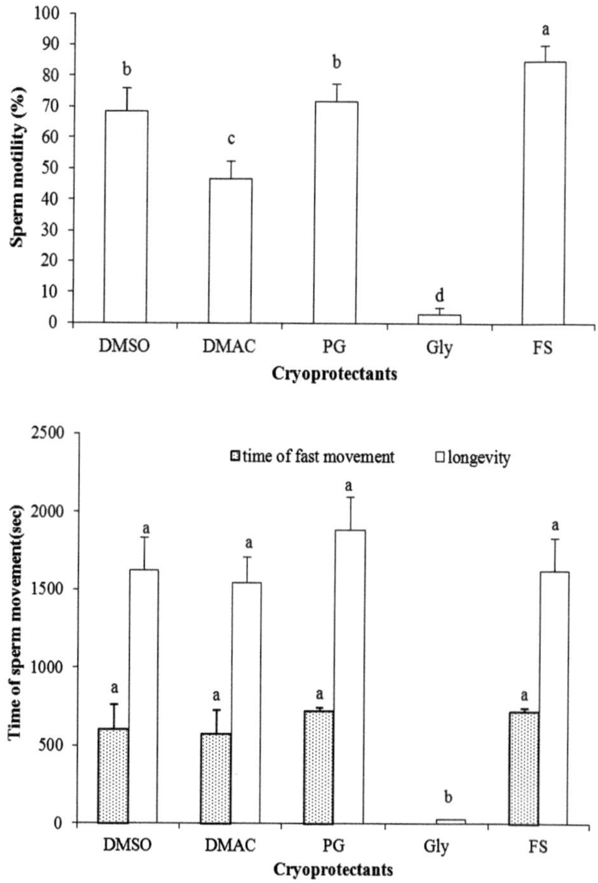

Fig. 9.2 Motilities of thawed sperm and fresh sperm (FS) of seven-band grouper *E. septemfasciatus* cryopreserved in different cryoprotectants (**a**) and fast-moving time and longevity (**b**). ($n = 3$, $P < 0.05$). The values were expressed as mean ± S.D (Tian et al. 2015)

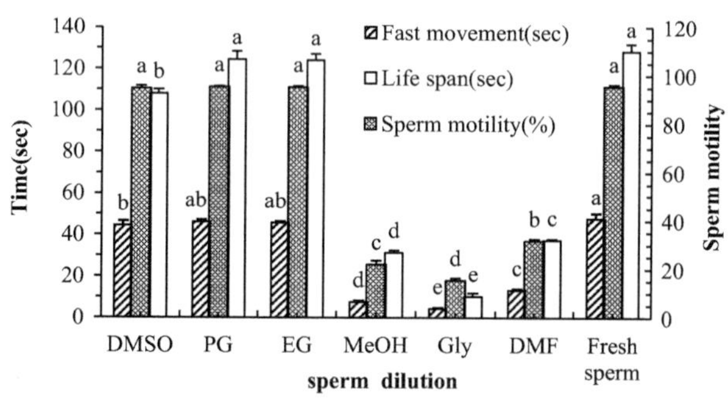

Fig. 9.3 The effects of cryoprotectants DMSO, PG, EG, MeOH, Gly, and DMF on the cryopreservation of *Pleuronectes yokohamae* ($n = 3$, $P < 0.05$) (Song et al. 2016)

9.5.1 Programmed Cryopreservation of Fish Sperm

Programmed cryopreservation is conducted through the use of a program-controlled refrigerating instrument. Different applications were used in the cryopreservation of sperm from different species. Cells can accumulate damages caused by ice crystals at 0 to −60 °C during the freezing process, which is the temperature range that causes the most damage during cryopreservation (Hua and Ren 1994). Therefore, it is crucial to identify methods that allow for cells to survive these crucial temperatures.

Sansone (2002) found that sea bass sperm diluted with 10% ethylene glycol (EG) for 6 h at 0–2 °C, and then frozen at 15 °C/min showed similar motility as fresh sperm. Koh (2010) screened the movement rate of seven-band grouper sperm and found that freezing sperm to −40 °C at 49.8 ± 1.7 °C min^{-1}, then placing the sperm in liquid nitrogen led to the highest movement rate after thawing. Liu (2016) cryopreserved grouper sperm by mixing sperm in a 2-mL cryotube, and samples were equilibrated for 5 min at 0 °C, frozen from 0 to −150 °C at a cooling rate of 20 °C/min with a Kryo-360 to 1.7 programmable freezer (Planer Plc., Middlesex, UK), and then placed in liquid nitrogen. The three steps method was applied to cryopreserve perch and turbot sperm as follows: (1) sperm was equilibrated in cryotubes (1.8 mL) for 30 min at 4 °C, and then placed into cryotubes in the fabric bags and hung in a liquid nitrogen (LN) tank vapor at a position of 6–10 cm above the LN surface (−180 °C) for 10 min; (2) the fabric bags were lowered to 5 cm above the LN surface for 5 min; (3) fabric bags were immersed into the LN (−196 °C) for long-term preservation (Ji et al. 2004; Chen et al. 2004). Tian (2013, 2015) cryopreserved grouper sperm by mixing sperm and cryopreservation solution at a 1:1 ratio, and equilibrated the samples for 5–10 min at room temperature, placed the mixture into 2-mL cryotubes, and placed 5–10 cryotubes in a fabric bag. Next, the fabric bags were hung in the LN vapor at a position 5–10 cm above the LN surface for 10 min, and plunged into LN for preservation. This method yielded a high number of surviving sperm after thawing. The method above is simple and convenient, and suitable for large-scale cryopreservation of fish sperm in aquafarms, especially without the use of a program-controlled machine.

9.5.2 Vitrification Cryopreservation of Fish Sperm

Vitrification cryopreservation uses a high concentration of vitrification solution to dilute sperm and uses straws of 250 μL as the main carrier, which are plunged into LN after equilibrating at 4 °C or room temperature. The principal of vitrification cryopreservation is moving the samples through the ice crystal formation zone (0–60 °C) at a rapid speed by accessing the LN temperature zone (−196 °C) directly. This allows the solution to enter a vitrification state that is between solid and liquid states, which can reduce damage on cells (Li et al. 1998). However, this method is not suitable for large-scale sperm cryopreservation due to the small volume of the straw, and the high toxicity of the vitrification solution.

In a study where marine European eel (*Anguilla anguilla*) sperm were cryopreserved using vitrification, the ASMA analysis found that the cryopreserved and fresh sperm had similar head area and perimeter (Kása et al. 2017). Longtooth grouper (*E. bruneus*) sperm was cryopreserved by vitrification utilizing the straws, which were suspended for 3 min at 3.5 cm above the liquid nitrogen surface

(−76 °C) before being immersed into liquid nitrogen, and the survival rate of the frozen sperm was 66.3% (Lim and Le 2013).

Rafael Cuevas-Uribe (2015) studied the survival rate and membrane integrity of marine fish sperm that were cryopreserved by vitrification and tested the properties of 29 vitrification solutions. Vitrification solution with 15% DMSO + 15% ethylene glycol + 10% glycerol + 1% X-1000™ + 1% Z-1000™ had an average post-thaw rate of 58% motility and 19% membrane integrity for spotted seatrout, 38% motility and 9% membrane integrity for red snapper (*Lutjanus campechanus*), and 30% motility and 19% membrane integrity for red drum (*Sciaenops ocellatus*). Vitrification offers an alternative to conventional cryopreservation.

Zilli (2018) screened three different vitrification devices (loops, drops, and cut straws), and found that dropping samples directly into liquid nitrogen in a 20 μL spermatozoa suspension (dropwise method) diluted with Mounib buffer containing 10% Me_2SO + 10% glycerol was most effective. Adding anti-freezing proteins AFPI and AFPIII to the vitrification solution improved the sperm motility.

9.6 Establishment and Application of the Fish Sperm Cryobank

Sperm cryobanks play an important role in the long-term preservation of fish germplasm genetic resources. Scientists in the United States have established a sperm cryobank for salmon and trout in Northwestern United States. They preserved the sperm of 500 salmon and 150 rainbow trout individuals from wild populations, as well as more than 10,000 sperm samples (Cloud et al. 2000). Since 1990, the World Fish Trust and the Government of Canada have started to build a sperm cryobank of Canadian trout, and they cryopreserved sperm of *Oncorhynchus nerka*, *O. tshawytscha*, *O. kisutch*, *O. mykiss*, *S. salar*, and other fish in more than 7000 tubes, establishing a complete management and evaluation system of the sperm cryobank (Harvey 2000). In addition, they also established a database management system of salmon and trout sperm cryobank (Kincaid 2000). Consejo Nacional de Ciencia Tecnologia (CONACYT) and Secretary of Environment and Natural Resources (SEMARNAT) are working to establish the California aquatic species germplasm cryobank (GBAS) (Eugenio-gonzaliz et al. 2009). In China, the Chinese Academy of Sciences Yellow Sea Fisheries Research Institute (Chen et al. 2007), the Yangtze River Fisheries Research Institute (Liu et al. 2007a, b), Institute of Oceanology (Liu et al. 2011), and other agencies have studied the sperm cryopreservation technology of freshwater and seawater fish, establishing a corresponding sperm cryobank.

In recent years, we cooperated with the domestic research institutes and enterprises to collect and cryopreserve sperm of flatfish, grouper, and cold water fish in Shandong, Jiangsu, Liaoning, Fujian, Guangdong, Hainan, Xinjiang, and other regions, and have created a fish sperm cryobank containing 30 species of marine and freshwater fishes in Yellow Sea Fisheries Research Institute. Up to 10,000 mL of sperm from 17 species of Perciformes (including 10 species of

Table 9.2 Sources and volume of the frozen sperm in sperm cryobank in Yellow Sea Fisheries Research Institute

Species	Volume (mL)	Sperm motility (%)	References
E. septemfasciatus	150	76.67	Tian et al. (2015)
E. moara	500	57.24±3.69	Qi et al. (2014)
E. akaara	100	75.0-80.0	
E. lanceolatus	3000	74.75±12.7	Tian et al. (2015)
E. fuscoguttatus	500	60.0-85.0	
E. bruneus	150	60.0-85.0	
E. awoar	150	60.0-85.0	
E. tukula	200	85.67±5.13	
E. polyphekadion	100	60.0-75.0	
E. coioides	200	70.0-80.0	
Cromileptes altivelis	*50*	70.0-80.0	
Plectropomus lepardus lacepede	*100*	60.0-70.0	
Pagrosomus major	*200*	75.0-85.0	
Sebastes schlegeli	*50*	50.0-60.0	
Lateolabrax japonicus	500	68.3±4.4	Ji et al. (2004)
Centropristis striata	50	60.0	
Siniperca chuatsi	700	70.0-80.0	
Scophthalmus maximus	100	78.3±7.6	Chen et al. (2004)
Paralichthys lethostigma	30	60.0-75.0	
P. dentatus	100	60.0-70.0	Tian et al. (2006)
P. olivaceus	500	70.0-85.0	
Verasper variegatus	100	70.3±5.6	Tian et al. (2008)
Platichthys stellatus	500	61.67±5.00	Jiang et al. (2014)
Pseudopleuronectes yokohamae	500	95.26±0.39	Song et al. (2016)
Cynoglossus semilaevis	100	55.0±5.0	Tian et al. (2009a, b)
Esox lucius	200	40.0-50.0	
Lota lota	1000	26.67±5.77	Tian et al. (2014)
Hucho taimen	1000	18.33±10.6	Tian et al. (2016b)
Thamnaconus septentrionalis	20	50.0-60.0	
Total	11,500		

Epinephelus), 8 species of Pleuronectiformes, and 3 species of cold water freshwater fish were collected and cryopreserved in China (Table 9.2).

Cryopreserved fish sperm play an important role in the induction of gynogenesis. For example, gynogenetic diploid larvae were induced by cryopreserved sperm of perch in *C. semilaevis, S. maximus, Verasper moseri,* and *P. stellatus*, resulting in a yield of 2.5%, 34.8%, 40.68%, and 0.01% respectively. The cryopreserved sperm provided rich genetic resources for fish sex control and breeding (Chen et al. 2009; Su et al. 2008; Yang et al. 2009; Duan et al. 2017).

Cryopreserved sperm can also be used in selective breeding, family establishment, and crossbreeding research. For example, a pure breeding line of flounder was established using frozen sperm of perch (Tian et al. 2017a, b), a large number of families of *P. olivaceus* and *P. stellatus* were established by using cryopreserved sperm (Tian et al. 2009b, 2016a), grouper breeding varieties with excellent growth characteristics were developed by hybridization of cryopreserved sperm of *E. lanceolatus* with eggs of various groupers (Tian et al. 2017a, b). This suggests that the cryopreserved sperm have played an important role in these areas.

The sperm cryopreservation technology plays an important role in the industrial aquaculture, especially for grouper. The establishment of the grouper sperm cryobank solved the problem of sexually asynchronous spawning seasons within and between grouper populations, and eliminated the bottleneck of insufficient quantity of grouper sperm. It also overcame the reproductive isolation between species caused by geographical distribution, temperature, and season-dependent reproductive barriers. Thus, this technology allowed for the hybridization of distant grouper species, providing a safeguard for cultivation of excellent new varieties of grouper that can grow rapidly under a wide range of suitable temperature and with good meat quality. This technology also provided rich sources for large-scale grouper breeding, and the frozen sperms were widely applied in crossbreeding and seedling cultivation. The extensively collected and cryopreserved sperm of *E. lanceolatus* was artificial hybridized with eggs of the *E. moara* breeding population, and a large number of families were established, and the inheritance, additive, and dominant effects of hybrids were analyzed (Tian et al. 2017a, b), a high-performing hybrid variety "yunlong grouper" (*E. moara* (♀) × *E. lanceolatus* (♂)) with fast growth, wide range of suitable temperature and good meat quality, was obtained through this cross. The growth rate of "yunlong grouper" is 3.08 times that of its female parent *E. moara*, and is 1.37 times that of the extensive breeding hybrid *E. fuscoguttatus* (♀) × *E. lanceolatus* (♂), which can grow to 700–1250 g during its first year (Wang et al. 2018; Tang et al. 2018; Li et al. 2019). At present, this new variety has been extensively cultured in Shandong, Hebei, Tianjin, Fujian, Guangdong, Hainan, Japan, and South Korea, producing significant economic benefits.

Acknowledgments Authors acknowledge the following projects: Key Research and Development Plan of Shandong (2019GHY112063), Superior Seed Project of Shandong (2019LZGC020), and Central Public-interest Scientific Institution Basal Research Fund, YSFRI, CAFS (NO. 20603022019002, 20603022017013).

References

Ahn JY, Park JY, Lim HK (2018) Effects of different diluents, cryoprotective agents, and freezing rates on sperm cryopreservation in *Epinephelus akaara*. Cryobiology 83:60–64. https://doi.org/10.1016/j.cryobiol.2018.06.003

Blaxter THS (1953) Sperm storage and cross-fertilization of spring and autumn spawning herring. Nature 172:1189–1190. https://doi.org/10.1038/1721189b0

Butts IAE, Babiak I, Ciereszko A et al (2011) Semen characteristics and their ability to predict sperm cryopreservation potential of Atlantic cod, *Gadus morhua* L. Theriogenology 75:1290–1300. https://doi.org/10.1016/j.theriogenology.2010.11.044

Butts IAE, Litvak MK, Kaspar V, Trippel EA (2010) Cryopreservation of Atlantic cod *Gadus morhua* L. spermatozoa: effects of extender composition and freezing rate on sperm motility, velocity and morphology. Cryobiology 61:174–181. https://doi.org/10.1016/j.cryobiol.2010.07.001

Cabrita E, Engrola S, Conceicao LEC et al (2008) Successful cryopreservation of sperm from a sex-reversed dusky grouper, *Epinephelus marginatus*. Aquaculture 287:152–157. https://doi.org/10.1016/j.aquaculture.2008.10.019

Cardiner RW (1978) Utilization of extracellular glucose by spermatozoa of two viviparous fishes. Comp Biochem Physiol 59A:165–168

Chen SL, Ji XS, Yu GC et al (2004) Cryopreservation of sperm from turbot (*Scophthalmus maximus*) and application to large-scale fertilization. Aquaculture 236:547–556. https://doi.org/10.1016/j.aquaculture.2003.10.027

Chen SL, Tian YS, Li J et al (2007) Theory and techniques of fish spermatozoa and embryos cryopreservation. China Agriculture Press, Beijing

Chen SL, Tian YS, Yang JF et al (2009) Artificial gynogenesis and sex determination in half-smooth tongue sole (*Cynoglossus semilaevis*). Mar Biotechnol 11:243–251. https://doi.org/10.1007/s10126-008-9139-0

Chen YK, Liu QH, Li J et al (2010) Effect of long-term cryopreservation on physiological characteristics, antioxidant activities and lipid peroxidation of red seabream (*Pagrus major*) sperm. Cryobiology 61(2):189–193. https://doi.org/10.1016/j.cryobiol.2010.07.003

Cuevas-Uribe R, Chesney EJ, Daly J et al (2013) Vitrification of sperm from marine fishes: effect on motility and membrane integrity. Aquac Res 46(7):1770–1784. https://doi.org/10.1111/are.12337

Cloud JG, Armstrong R, Wheeler P et al (2000) The northwest salmonid germplasm repository. In: Cryopreservation in aquatic species. World Aquaculture Society, Baton Rouge, pp 338–342

Cuevas-Uribe R, Chesney EJ, Daly J, Tiersch TR (2015) Vitrification of sperm from marine fishes: effect on motility and membrane integrity. Aquac Res 46(7):1770–1784. https://doi.org/10.1111/are.12337

Cuevas-Uribe R, Hu E, Daniels H et al (2017) Vitrification as an alternative approach for sperm cryopreservation in marine fishes. N Am J Aquac 79(2):187–196. https://doi.org/10.1080/15222055.2017.1281855

DeGraaf JD, Berlinsky DL (2004) Cryogenic and refrigerated storage of Atlantic cod (Gadus morhua) and haddock (*Melanogrammus aeglefinus*) spermatozoa. Aquaculture 234:527–540. https://doi.org/10.1016/j.aquaculture.2003.11.037

Ding F, Lall SP, Li J et al (2011) Cryopreservation of sperm from Atlantic halibut (*Hippoglossus hippoglossus, L.*) for commercial application. Cryobiology 63(1):56–60. https://doi.org/10.1016/j.cryobiol.2011.04.009

Duan HM, Tian YS, Li WL et al (2017) Comparison of gynogenetic diploid and haploid embryonic development and hybridization of starry flounder *Platichthys stellatusr*. J Fish Sci China 24 (3):477–487. https://doi.org/10.3724/SP.J.1118.2017.16299

Erdahl AW, Graham EF (1987) Fertility of teleost semen as affected by dilution and storage in a seminal plasma-mimicking medium. Aquaculture 60(3–4):311–321. https://doi.org/10.1016/0044-8486(87)90296-1

Eugenio-Gonzaliz MA, Padilla-Zarate G, Oca CMD, Paniagua-Chavez CG (2009) Information technologies supporting the operation of the germplasm bank of aquatic species of Baja California, Mexico. Rev Fish Sci 17(1):8–17. https://doi.org/10.1080/10641260802008031

Figueroa E, Merino O, Risopatrón J et al (2015) Effect of seminal plasma on Atlantic salmon (*Salmo salar*) sperm vitrification. Theriogenology 83(2):238–245. https://doi.org/10.1016/j.theriogenology.2014.09.015

Figueroa E, Valdebenito I, Merino O et al (2016) Cryopreservation of Atlantic salmon *Salmo salar* sperm: effects on sperm physiology. J Fish Biol 89(3):1537–1550. https://doi.org/10.1111/jfb.13052

Glogowski J, Babiak I, Goryczko K et al (1996) Activity of aspartate amino transferase and acid phosphatase in cryopreserved trout sperm. Reprod Nutr Dev 8:1179–1184. https://doi.org/10.1071/rd9961179

Gwo HH, Weng TS, Fan LS et al (2005) Development of cryopreservation procedures for semen of Pacific bluefin tuna *Thunnus orientalis*. Aquaculture 249:205–211. https://doi.org/10.1016/j.aquaculture.2005.03.034

Gwo JC (1993) Cryopreservation of black grouper *Epinephelus malabaricus* spermatozoa. Theriogenology 39:1331–1342. https://doi.org/10.1016/0093-691X(93)90235-W

Harvey B (2000) The application of cryopreservation on fish genetic conservation. In: Advances in world aquaculture. World Aquaculture Society, Baton Rouge, pp 332–337

Holt WV (2000) Basic aspect of frozen storage of semen. Anim Reprod Sci 62(1–3):3–22. https://doi.org/10.1016/S0378-4320(00)00152-4

Hua ZZ, Ren HS (1994) Cryogenic biomedical technology. Science Press, Beijing, pp 20–88

Imaizumi H, Hotta T, Ohta H (2005) Cryopreservation of kelp grouper *Epinephelus bruneus* sperm and comparison of fertility of fresh and cryopreserved sperm. Aquat Sci 53:405–411. https://doi.org/10.1111/j.1444-2906.2005.00985.x

Ji XS, Chena SL, Tiana YS et al (2004) Cryopreservation of sea perch (*Lateolabrax japonicus*) spermatozoa and feasibility for production-scale fertilization. Aquaculture 241:517–528. https://doi.org/10.1016/j.aquaculture.2004.07.012

Jiang J, Tian YS, Wang B et al (2014) Cryopreservation of spermatozoa in starry flounder (*Platichthys stellatus*) and its analysis of the physiological characteristics. J Agric Biotechnol 22(1):17–26. https://doi.org/10.3969/j.issn.1674-7968.2014.01.003

Ji XS, Chen SL, Zhao Y, Deng H (2005) Cryopreservation of stone flounder and Japanese flounder sperms and its application to the breeding. Mar Fish Res 26(1):13–17. https://doi.org/10.3969/j.issn.1000-7075.2005.01.003

Jiang JH, Yan JQ, Zhu JQ et al (2011) Sperm cryopreservation and the cytoarchitecture damage detection of *Pseudosciaena crocea*. J Agric Biotechnol 19(4):725–733. https://doi.org/10.3969/j.issn.1674-7968.2011.04.019

Kása E, Bernáth G, Kollár T et al (2017) Development of sperm vitrification protocols for freshwater fish (*Eurasian perch*, *Perca fluviatilis*) and marine fish (*European eel*, *Anguilla anguilla*). Gen Comp Endocrinol 245:102–107. https://doi.org/10.1016/j.ygcen.2016.05.010

Kincaid HL (2000) Development of databases for germplasm repositories. World Aquaculture Society, Baton Rouge, pp 323–331

Kiriyakit A, Gallardo WG, Bart AN (2011) Successful hybridization of groupers (*Epinephelus coioides* × *Epinephelus lanceolatus*) using cryopreserved sperm. Aquaculture 320:106–112. https://doi.org/10.1016/j.aquaculture.2011.05.012

Koh ICC, Yokoi KI, Tsuji M et al (2010) Cryopreservation of sperm from seven-band grouper, *Epinephelus septemfasciatus*. Cryobiology 61(3):0–267. https://doi.org/10.1016/j.cryobiol.2010.09.003

Lanes CFC, Okamoto M, Cavalcanti PV et al (2008) Cryopreservation of Brazilian flounder (*Paralichthys orbignyanus*) sperm. Aquaculture 275:361–365. https://doi.org/10.1016/j.aquaculture.2007.12.025

Li GW, Zhen CY, Tang B (1998) Cryobiology. Hunan Science and Technology Press, Changsha, pp 55–100

Li ZT, Tian YS, Tang J et al (2019) Growth characteristics and comparative analysis of Yunlong grouper with *Epinephelus moara* and Zhenzhulongdan grouper. J Fish China 43(4):1005–1017. https://doi.org/10.19663/j.issn2095-9869.20180730001

Lim HK, Le MH (2013) Evaluation of extenders and cryoprotectants on motility and morphology of longtooth grouper (*Epinephelus bruneus*) sperm. Theriogenology 79:867–871. https://doi.org/10.1016/j.theriogenology.2013.01.004

Liu L, Linhart O, Wei QW et al (2007b) Comparative study of activating mediums for the cryopreserved sperm of several sturgeons using CASA. J Fish China 31(6):711–720. https://doi.org/10.3321/j.issn:1000-0615.2007.06.002

Liu QH, Li J, Zhang SC et al (2007a) Flow cytometry and ultrastructure of cryopreserved red seabream (*Pagrus major*) sperm. Theriogenology 67(6):1168–1174. https://doi.org/10.1016/j.theriogenology.2006.12.013

Liu QH, Lu G, Che K et al (2011) Sperm cryopreservation of the endangered red spotted grouper, *Epinephelus akaara*, with a special emphasis on membrane lipids. Aquaculture 318:185–190. https://doi.org/10.1016/j.aquaculture.2011.05.025

Liu QH, Ma DY, Xu SH et al (2015) Summer flounder (*Paralichthys dentatus*) sperm cryopreservation and application in interspecific hybridization with olive flounder (P olivaceus). Theriogenology 83(4):703–710. https://doi.org/10.1016/j.theriogenology.2014.11.004

Liu QH, Xiao ZZ, Wang XY et al (2016) Sperm cryopreservation in different grouper subspecies and application in interspecific hybridization. Theriogenology 85:1399–1407. https://doi.org/10.1016/j.theriogenology.2015.12.023

Lovelock JE, Polge C (1954) The immobilization of spermatozoa by freezing and thawing and the protective action of glycerol. Biochem J 58(4):618–622. https://doi.org/10.1042/bj0580618

Martínez-Páramo S, Diogo P, Dinis MT et al (2013) Effect of two sulfur-containing amino acids, taurine and hypotaurine in European sea bass (Dicentrarchus labrax) sperm cryopreservation. Cryobiology 66(3):333–338. https://doi.org/10.1016/j.cryobiol.2013.04.001

Mounib MS (1978) Cryogenic preservation of fish and mammalian spermatozoa. J Reprod Fertil 53:13–18. https://doi.org/10.1530/jrf.0.0530013

Oh SR, Lee CH, Kang HC et al (2013) Evaluation of fertilizing ability using frozen thawed sperm in the Longtooth grouper, *Epinephelus bruneus*. Dev Reprod 17(4):345–351. https://doi.org/10.12717/DR.2013.17.4.345

Ohta H, Kawamura K, Unuma T et al (2001) Cryopreservation of the sperm of the Japanese bitterling. J Fish Biol 58:670–681. https://doi.org/10.1111/j.1095-8649.2001.tb00521.x

Ohta H, Shimma H, Hirose K (1995) Relationship between fertilityand motility of cryopreserved spermatozoa of the amago salmon Oncorhynchus masou ishikawae. Fish Sci 61:886–887. https://doi.org/10.2331/fishsci.61.886

Ott AG (1975) Cryopreservation of Pacific salmon and steelhead trout sperm. Oregon State University, Dissertation

Philpott M (1993) The dangers of disease transmission by artificial insemination and embryo transfer. Br Vet J 149:339–369. https://doi.org/10.1016/S0007-1935(05)80075-2

Pullin RSV (1972) The storage of plaice (Pleuronectes platessa) sperm at low temperatures. Aquaculture 1:279–283. https://doi.org/10.1016/0044-8486(72)90029-4

Qi WS, Jiang J, Tian YS et al (2014) Sperm cryopreservation of kelp grouper *Epinephelus moara*. Progress Fish Sci 35(1):26–33. https://doi.org/10.3969/j.issn.1000-7075.2014.01.004

Rideout RM, Litvak MK, Trippel EA (2003) The development of a sperm cryopreservation protocol for winter flounder *Pseudopleuronectes americanus* (Walbaum): evaluation of cryoprotectants and diluents. Aquac Res 34(8):653–659. https://doi.org/10.1046/j.1365-2109.2003.00879.x

Rideout RM, Trippel EA, Litvak MK (2004) The development of haddock and Atlantic cod sperm cryopreservation techniques and the effect of sperm age on cryopreservation success. J Fish Biol 65:299–311. https://doi.org/10.1111/j.0022-1112.2004.00449.x

Sanches E, Oliveira I, Serralheiro P et al (2013) Cryopreservation of mutton snapper (Lutjanus analis) sperm. Ann Braz Acad Sci 85(3):1083–1091. https://doi.org/10.1590/S0001-37652013005000047

Sanches E, Oliveira I, Serralheiro P et al (2015) Sperm cryopreservation of lane snapper Lutjanus synagris (Linnaeus, 1758). Braz J Biol 75(3):662–669. https://doi.org/10.1590/1519-6984.20613

Sansone G, Fabbrocini A, Ieropoli S et al (2002) Effects of extender composition, cooling rate, and freezing on the motility of sea bass (*Dicentrarchus labrax* L.) spermatozoa after thawing. Cryobiology 44:229–239. https://doi.org/10.1016/S0011-2240(02)00026-3

Seungki L, Goro Y (2016) Successful cryopreservation of spermatogonia in critically endangered Manchurian trout (Brachymystax lenok). Cryobiology 72:165–168. https://doi.org/10.1016/j.cryobiol.2016.01.004

Song LN, Tian YS, Li XK et al (2016) Cryopreservation of marbled flounder *(Pseudopleuronectes yokohamae*) sperm and analysis of its physiological characteristics. J Agric Biotechnol 24 (4):584–592. https://doi.org/10.3969/j.issn.1674-7968.2016.04.013

Su PZ, Chen SL, Yang JF et al (2008) Induction of gynogenesis in *Scophthalmus maximus* by heterologous sperms of *Lateolabrax japonicas*. J Fish Sci China 15(5):715–721. https://doi.org/10.3321/j.issn:1005-8737.2008.05.001

Tang J, Tian YS, Li ZT et al (2018) Analysis of genetic characters in *Epinephelus moara, E lanceolaus* and their hybrids. J Agric Biotechnol 26(5):819–829. https://doi.org/10.3969/j.issn.1674-7968.2018.05.010

Tian YS, Chen SL, Ji XS et al (2008) Cryopreservation of spotted halibut (*Verasper variegatus*) sperm. Aquaculture 284(1–4):268–271. https://doi.org/10.1016/j.aquaculture.2008.07.047

Tian YS, Chen SL, Liu BW, Wang B (2006) Embryonic and postembryonic development of hybrid produced with frozen sperms of *Paralichthys dentatus* and eggs of *Paralichthys olivace*us. J Fish China 30(4):433–443. https://doi.org/10.3321/j.issn:1000-0615.2006.04.001

Tian YS, Chen ZF, Duan HM et al (2017a) The family line establishment of the hybrid *Epinephelus moara*♀×*E. lanceolatus* by using cryopreserved sperm and the related genetic effect analysis. J Fish China 41(10):1–12. https://doi.org/10.11964/jfc.20161210630

Tian YS, Duan HM, Li XK et al (2017b) Growth and genetic analysis among three homologous inbred strains of *Paralichtys olivaceus*. J Fish Sci China 24(1):11–21. https://doi.org/10.3724/SP.J.1118.2017.16163

Tian YS, Chen SL, Ji XS et al (2009a) Cryopreservation of tongue sole *Cynoglossus semilaevis* sperm. Mar Fish Res 30(6):97–102. https://doi.org/10.3969/j.issn.1000-7075.2009.06.013

Tian YS, Jiang J, Ma Y et al (2014) Spermatozoa cryopreservation of the burbot (*Lota lota*) from Irtysh river. J Agric Biotechnol 22(9):1149–1156. https://doi.org/10.3969/j.issn.1674-7968.2014.09.012

Tian YS, Jiang J, Wang N et al (2015) Sperm of the giant grouper: cryopreservation, physiological and morphological analysis and application in hybridizations with red-spotted grouper. J Reprod Dev 61(4):333–339. https://doi.org/10.1262/jrd.2014-087

Tian YS, Li XK, Duan HM et al (2016a) The family lines establishment and genetic effects analysis of the starry flounder *Platichthys stellatus*. Acta Oceanol Sin 38(6):21–31. https://doi.org/10.3969/j.issn.0253-4193.2016.06.003

Tian YS, Ma Y, Xie ZN et al (2016b) Cryopreservation of *Hucho taimen* spermatozoa. J Agric Biotechnol 24(1):90–97. https://doi.org/10.3969/j.issn.1674-7968.2016.01.012

Tian Y, Qi W, Jiang J et al (2013) Sperm cryopreservation of sex-reversed seven-band grouper, *Epinephelus septemfasciatus*. Anim Reprod Sci 137(3–4):230–236. https://doi.org/10.1016/j.anireprosci.2013.01.014

Tian YS, Xu TJ, Chen SL et al (2009b) Parent effects and estimation of genetic parameters for three Japanese flounder breeding populations. Acta Oceanol Sinia 31(6):119–129. https://doi.org/10.3321/j.issn:0253-4193.2009.06.013

Vuthiphandchai V, Chomphuthawach S, Nimrat S (2009) Cryopreservation of red snapper (*Lutjanus argentimaculatus*) sperm: effect of cryoprotectants and cooling rates on sperm motility, sperm viability, and fertilization capacity. Theriogenology 72(1):129–138. https://doi.org/10.1016/j.theriogenology.2009.02.013

Wang LN, Tian YS, Tang J et al (2018) Analysis and quality evaluation of nutritional components in the muscle of *Epinehelus moara, E. lanceolatus* and hybrid "Yunlong grouper". J Fish China 42(7):1085–1093. https://doi.org/10.11964/jfc.20171011001

Withler FC, Lim LC (1982) Preliminary observations of chilled and deep-frozen storage of grouper *Epinephelus tauvina* sperm. Aquaculture 27:289–392. https://doi.org/10.1016/0044-8486(82)90125-9

Yusoff M, Hassan BN, Ikhwanuddin M et al (2018) Successful sperm cryopreservation of the brown-marbled grouper, *Epinephelus fuscoguttatus*, using propylene glycol as cryoprotectant. Cryobiology 81:168–173. https://doi.org/10.1016/j.cryobiol.2018.01.005

Yang H, Hu E, Buchanan JT, Tiersch TR (2018) A strategy for sperm cryopreservation of Atlantic salmon, *Salmo salar*, for remote commercial-scale high-throughput processing. J World Aquac Soc 49(1):96–112. https://doi.org/10.1111/jwas.12431

Yang JF, Chen SL, Su PZ et al (2009) Study on gynogenesis induced by heterogenous sperms in flounder *Verasper moseri*. J Fish Sci China 33(3):372–379. https://doi.org/10.3321/j.issn:1000-0615.2009.03.003

Yao Z, Crim LW, Richardson GF et al (2000) Motility, fertility and ultrastructural changes of ocean pout (*Macrozoarces americanus L.*) sperm after cryopreservation. Aquaculture 181:361–375. https://doi.org/10.1016/s0044-8486(99)00240-9

Zhang YZ, Zhang SC, Liu XZ et al (2003) Cryopreservation of flounder (*Paralichthys olivaceus*) sperm with a practical methodology. Theriogenology 60:989–996. https://doi.org/10.1016/s0093-691x(03)00097-9

Zilli L, Bianchi A, Sabbagh M et al (2018) Development of sea bream (*Sparus aurata*) semen vitrification protocols. Theriogenology 110:103–109. https://doi.org/10.1016/j.theriogenology.2017.12.039

10 Sperm Cryopreservation in Crustaceans

Karina Morales-Ueno and Carmen Guadalupe Paniagua-Chávez

Abstract

Sperm cryopreservation facilities and reproductive management adds flexibility to breeding program design; it also allows mobilization over long distances, extends sperm half-life for several years, enables the use of selected males and prevent the reduction of genetic variability caused by inbreeding or environmental disasters. Crustaceans hold over 30% value of the aquaculture market, despite its commercial importance, procedures granting its genetic diversity protection are not fully developed or widespread. To date, few papers have been reported on the cryopreservation of sperm from crustaceans, this is due to the diverse morphology between sperm and spermatophore among crustacean species, together with its different fertilization strategies. To assess the viability of sperm, it is common to use the microscopic examination of membrane integrity. However, the detailed reproductive characteristics of crustaceans have not been fully explained and not all interactions between their sperm and cryoprotectants are known. Therefore, the "gold test" to verify the viability of cryopreserved/thawed sperm is to test its fertilization rate.

Keywords

Crustacea · Sperm · Spermatophore · Cryopreservation · Reproduction

K. Morales-Ueno · C. G. Paniagua-Chávez (✉)
National Subsystem for Aquatic Genetic Resources of México (SUBNARGENA) at Centro de Investigación CientÚfica y de Educación Superior de Ensenada, B.C. (CICESE), Ensenada Baja California, México
e-mail: cpaniagu@cicese.mx

J. Betsy, S. Kumar (eds.), *Cryopreservation of Fish Gametes*,
https://doi.org/10.1007/978-981-15-4025-7_10

10.1 Introduction

Cryopreservation is one of the most important conservation techniques used to preserve viable cells for long term. This technique keeps cell physiognomy and metabolic activity in stasis, avoiding severe enzymatic degradation. The genetic material within sperm cells also remains unaffected. Therefore, cryopreservation is also used as a tool to preserve genetic variation in domestic and wild animal populations.

Successful sperm cryopreservation has been commercially developed for aquaculture species, especially for fish. For example, Atlantic salmon milt cryopreservation began in the 1990s, allowing gene banking for multiple valuable domestic and wild populations (O'Reilly and Doyle 2007). Despite the clear advantages of semen cryopreservation in aquatic species, its use has not yet been extended in all species. In crustaceans, there are few available studies on sperm cryopreservation, mostly centered on shrimp, specifically in *Penaeus vannamei*, perhaps due to its commercial importance, its relatively easy husbandry, and reproductive management (Chow et al. 1991; Leung-Trujillo and Lawrence 1991; Alfaro and Lozano 1993; Heitzmann et al. 1993; Wang et al. 1995; Ceballos-Vázquez et al. 2004; Alfaro et al. 2007; Alfaro-Montoya 2010; Ulate and Alfaro-Montoya 2010; Aungsuchawan et al. 2011; Garza-Torres et al. 2011).

10.2 Cryobiology

Cryopreservation is based on the physical principle established by Svante August Arrhenius, which indicates the speed of chemical reactions increases with temperature in the proportional ratio to the concentration of electrical molecules. Then, all cellular chemicals and biological reactions may be delayed by decreasing their reaction temperature. The lower the temperature at which a cell is stored, the lower the chemical reaction rate within the cell. Consequently, the translational movement of the cells (i.e., the molecular movements postulated for a diatomic molecule including translation, rotation, and vibration) when reaches −196 °C is so limited that cells can be stored for a long time (Day and Stacey 2007; Katkov et al. 2012).

In most animals, sperm is usually cryopreserved in aqueous cryoprotectant solutions. These cryoprotectants are substances that protect the gamete from mechanical or chemical damage caused by low temperatures exposure. When sperm and aqueous solutions freeze, some of their components acquire a solid phase composed of ice. In contrast, the rest of the components acquire an amorphous vitreous structure without the formation of ice crystals. This phenomena is known as vitrification. The final proportion of ice crystals and vitrified material that forms in the sample are determined by the concentration of cryoprotectants and the freezing rate (Mazur 2004; Tucker and Liebermann 2007).

Intracellular ice formation must be avoided during cooling and thawing, and can be prevented by adding non-permeable cryoprotectants to dehydrate the cell or permeable cryoprotectants, which in high concentration, increase intracellular

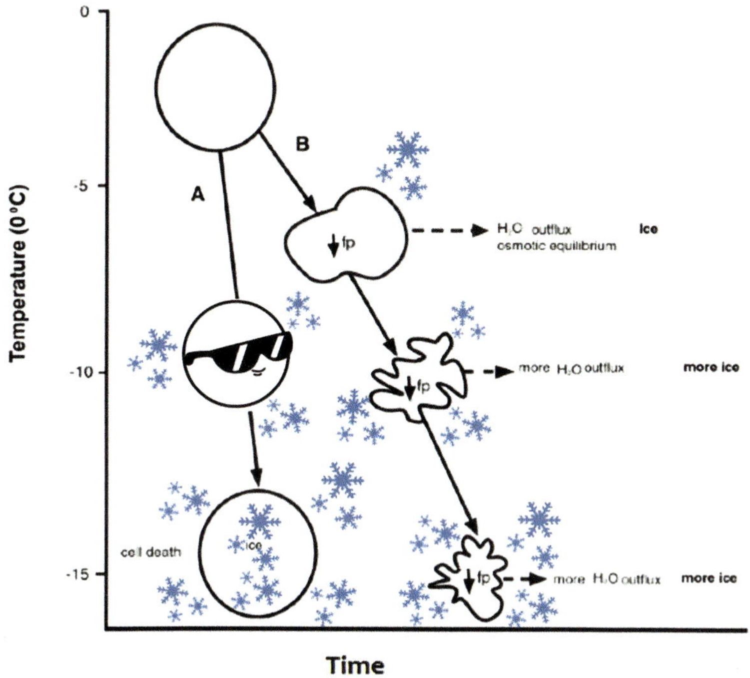

Fig. 10.1 Intracellular ice formation depends on solute concentration. (A) low solute concentration and fast freezing rate produce large extracellular ice and intracellular supercooling at −10 °C. As temperature drops, fragile balance is lost, and large ice crystals form inside and outside the cell space. (B) High solute concentration promotes cell dehydration, decrease its freezing point (fp), and form smaller ice crystals at the extracellular matrix producing vitrification of gamete's inner content

viscosity (Fuller et al. 2004; Day and Stacey 2007). Osmotic spermatozoa dehydration is explained in Fig. 10.1. Before cooling, solute concentration into the cell is low. When the temperature decreases relatively fast and reaches −10 °C, spermatozoa is super cold (e.g., few extracellular ice crystals and super-cold liquid in intracellular space are formed). As the temperature continues dropping, the balance is lost, and large lethal ice crystals form inside and outside the cell (scenario A). In scenario B, the medium around the cells contains non-permeating cryoprotectants. While temperature decreases at a slower rate, spermatozoa dehydrate, and smaller ice crystals around the gamete form. At the end of the cooling process, spermatozoa inner content is vitrified, and extracellular ice crystals remain small, resulting in a less damaged gamete.

The formation of intracellular ice should be avoided during cryopreservation to prevent any irreversible mechanical damage of membrane or other cellular structures. The formation of small ice crystals at extracellular space usually does not pose a risk and is not considered harmful. However, extracellular ice formation will not always be an innocuous event. Such a phenomenon can be equally lethal if the size of the ice crystals formed is enormous or if there is no enough space between

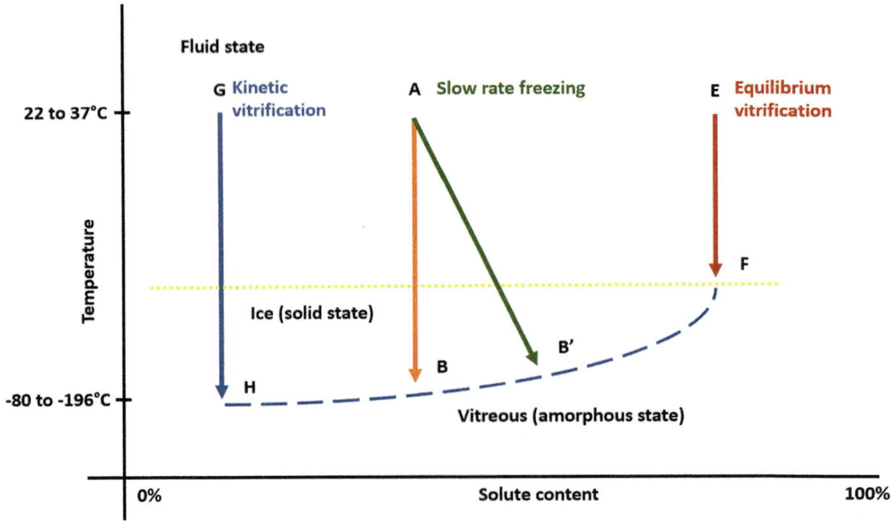

Fig. 10.2 Three main cryopreservation methods. Yellow-Dotted line separates liquid (fluid state) from ice (solid state). Vitrification zone (amorphous state) is located below the long-dashed, blue-dashed line. (1) green Line A to B′, slow freezing rate. Here, vitrification takes place at a higher temperature as the cryoprotectant concentration increases. If the concentration is low, the reaction shifts to the left (orange line A to B). (2) red Line E to F, equilibrium vitrification. High viscosity of cryoprotectant prevents intracellular and extracellular ice formation. Therefore, cells can be vitrified at a lower temperature. (3) blue Line G to H kinetic vitrification requires reaching a lower temperature than the first two methods. It can also be achieved with the use of very low and even without any cryoprotectant. However, it requires ultrafast freezing temperatures to prevent lethal ice crystals formation. Image modified (Katkov et al. 2012)

gametes to cause mechanical extracellular damaged by ice formation (Fuller et al. 2004; Day and Stacey 2007).

Sperm cryopreservation techniques can be divided into slow rate freezing, equilibrium vitrification, and kinetic vitrification (Fuller et al. 2004).

Slow rate freezing: This is also known as equilibrium freezing and is achieved by combining slow freezing rate (<2 °C/min) and moderate solute concentration to reach intracellular vitrification and to produce small ice crystals at the extracellular space (Fig. 10.2, green line, A to B′). Another alternative is to use less cryoprotectants concentration to achieve the same results. However, it requires faster freezing rates (orange line, A to B). This alternative is time-consuming and usually requires the use of specialized equipment (programmed controlled freezer chambers) and highly trained personnel (Foote 2002; Vajta and Nagy 2006).

Equilibrium vitrification: The technique requires the use of high solute content combined with a fast freezing rate ranging between 1000 and 30,000 °C/min (Fig. 10.2, red line, E to F). The use of high viscosity cryoprotectants prevents intracellular and extracellular ice formation. Consequently, the cell can reach vitrification state within a wide range of temperatures. Procedure is comparatively faster than slow rate freezing, and rapid freezing rates are easily achieved by plunging the sample directly into liquid nitrogen (Moawad et al. 2008; Mochida et al. 2013).

Kinetic vitrification. This technique requires reaching a much lower temperature and can be obtained by the use of low to moderate cryoprotectant concentration, and sometimes without any (Fig. 10.2, blue line, G to H). However, it demands the use of ultrafast freezing rates (>50,000 °C/min) to prevent lethal ice formation. To achieve ultrafast freezing rates into the cells, they must not only be rapidly introduced into liquid nitrogen, but they also require a volume as smallest as possible (~20 μL or less). The procedure is quick and straightforward, requiring minimum time to add cryoprotectants and freeze (Hamawaki et al. 1999; Isachenko et al. 2003; Endo et al. 2012).

10.3 Extender Solutions

Extenders are solutions formulated to suspend and keep the sperm viable. It can be used to suspend sperm and facilitate its counting or to create several doses before being cryopreserved. Extenders can mimic the composition of seminal plasma or a physiological solution. The formulation of these solutions could be complex to very simple, but the use of simple formulations is preferred. Most of the extenders used on crustaceans have virtually the same formulation and are relatively easy to make (Salazar et al. 2009; Cloud 2011; Wayman et al. 2011). Extenders enable short-term sperm transport and storage, prevent bacterial or fungal contamination (when antibiotics or antifungals are added), stabilize pH, provide nutrients, and facilitates the incorporation of other substances, such as cryoprotectants. The Cornell extender solution, developed in 1950 by Foote and Bratton, is the most popular animal extender solution. It was initially developed to dilute bull sperm but later was used as a standard extender in other terrestrial and aquatic animal species (Foote 2002). The antibiotic mixture developed by Foote and Bratton is trendy and regularly incorporated on extenders developed for aquatic animals. This mixture prevents bacterial/fungal growth and prevents transmission of infectious diseases (Tiersch and Green 2011). Later, more extender solutions were specifically designed for use in aquatic organisms (Kwantong and Bart 2003; Muchlisin 2005; Park and Chapman 2005; Mansour et al. 2006; Cloud 2011; Glenn III et al. 2011; Gwo 2011; Wayman et al. 2011; Gaitán-Espitia et al. 2013).

10.4 Cryoprotectants

Cryoprotectants are solutions developed to protect cells during cryopreservation process. As explained before, its function is to avoid lethal ice formation outside and inside the spermatozoa during freezing (Fuller et al. 2004; Cabrita et al. 2009). Cryoprotectants are divided into permeable and non permeable. Cryoprotectant permeability depends mainly on its molecular weight and spermatozoa's membrane arrangement. Therefore, cryoprotectants with low molecular weight can cross the cell membrane easily. Permeable cryoprotectants include dimethyl sulfoxide (DMSO), butanediol, propylene glycol, ethylene glycol, propanediol, methanol, formamide, acetamide, and glycerol. Most of these compounds are alcohols and promote spermatozoa dehydration while stabilizing their structures by replacing the water inside, which prevents the later formation of ice crystals in the intracellular space (Mazur 2004; Kopeika et al. 2007; Cloud 2011).

Non-permeable cryoprotectants have high molecular weight. These molecules stay in the extracellular space and are usually combined with one or more permeable cryoprotectants. Non-permeable cryoprotectants promote spermatozoa dehydration by increasing its osmotic gradient. They produce a synergistic effect when used in combination with permeable cryoprotectants (Fuller et al. 2004; Day and Stacey 2007).

Most used non-permeable cryoprotectants are sugars, such as glucose, sucrose, trehalose, sorbitol, and dextrose, but others such as skim milk, antifreeze proteins, and egg yolk are also used. Protective mechanisms of high molecular weight compounds such as disaccharides are not fully known. However, it has been proposed that they could coat the cell membrane, protecting it from the deleterious process by forming hydrogen bridges with water, reducing its biological reactivity. Relatively large molecules such as dextran act as thickening agents and facilitate vitrification since they can supercool the solutions. Consequently, the use of permeable cryoprotectants is needed in a smaller amount (Cabrita et al. 2003; Fuller et al. 2004; Day and Stacey 2007).

10.5 Freezing Containers for Sperm Cryopreservation

The most used freezing containers for long-term storing of sperm are French straws, which are made of inert material resistant to sudden changes in temperature (Pacey and Tomlinson 2009). These devices can be found in different colors, materials, and sizes, ranging from 0.25 to 5 mL (Fig. 10.3). After French straws, second most used freezing container is cryovials (Fig. 10.4), these are also made of biocompatible materials, having different closing systems: snap cap, external thread screw, or internal thread screw.

Some sperm cryopreservation processes, such as vitrification, require the use of tiny volume (less than 20 μL). Such a small sample volume requires the use of specialized devices, handling, processing, and storage. Before the development of cryogenic devices to handle very small samples, the devices used to handle small samples were merely adaptations of existing ones. One strategy was to use 0.25- or 0.5-mL French straws, with sample filling only a fraction of total volume. Later, straws or capillary glass tubes were heated and manually stretched to produce thin capillary tubes. The final product was later known as open stretched straw (OPS). Another alternative was to cut straws along its longitudinal axis. Later, more devices were adapted or developed for such small samples: miniature nylon handles, copper or gold grids (used in electron microscopy), plastic pipette tips, aluminum foil, surfaces made from different metals, crystal tubes, and vials, among others. Some

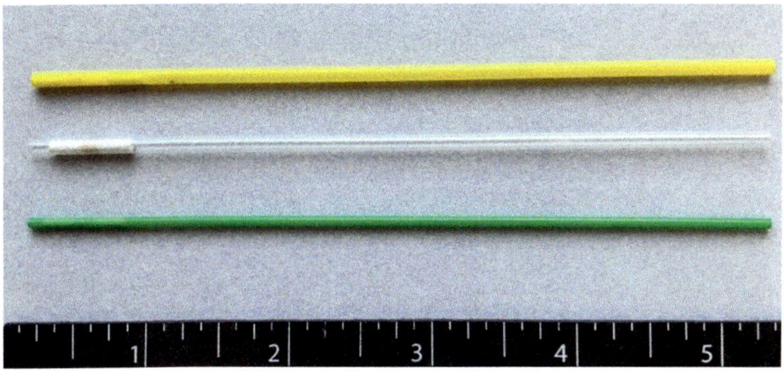

Fig. 10.3 Yellow (top) and transparent (middle) 0.5-mL French straws. Green (bottom) features same technical specifications as the 0.5-mL straws, except that it has a capacity of 0.25 mL and the external surface of the plug has a different color

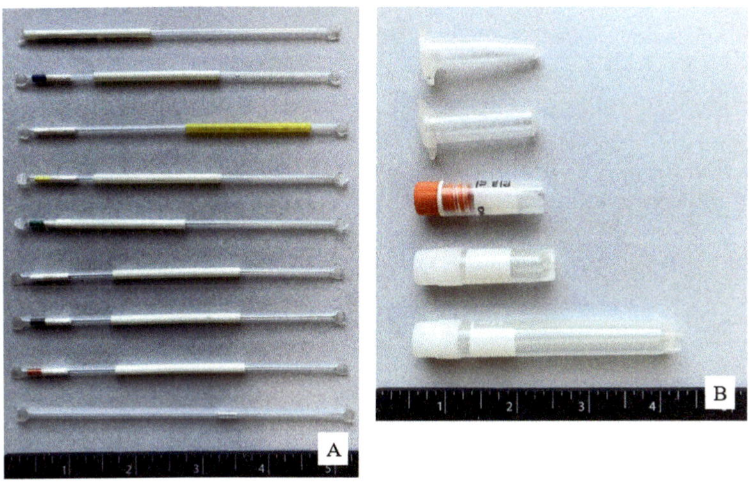

Fig. 10.4 (**a**) Different variations of same capacity High French straws. Some straws comprise two parts separated by a white plug. Heat sealed high security straws is leak proof in liquid nitrogen and mechanically resistant to high pressures. Colored inserts are used for simple identification in cryogenic environments. (**b**) Different type and capacity cryovials, from top to bottom: 1.5-mL snap cap, 2-mL snap cap, 1-mL internal thread screw, 1.8-mL external thread screw, and 4.5-mL external thread screw. All straws and vials are made of biocompatible materials

of the devices developed were patented and are now known by their trade names, such as CryoTip, Cryopette, Rapid-I, JY Straw, Cryotop, Cryoloop (Fig. 10.5), Cryoleaf, Fiber plug or Vitri-Inga (Tucker and Liebermann 2007; Saragusty and Arav 2011).

A combination of small-volume samples and suitable freezing containers are needed to reach ultrafast freezing speeds necessary for vitrification (Table 10.1, Fig. 10.6). These allow fast and homogeneous heat diffusion when immersed in

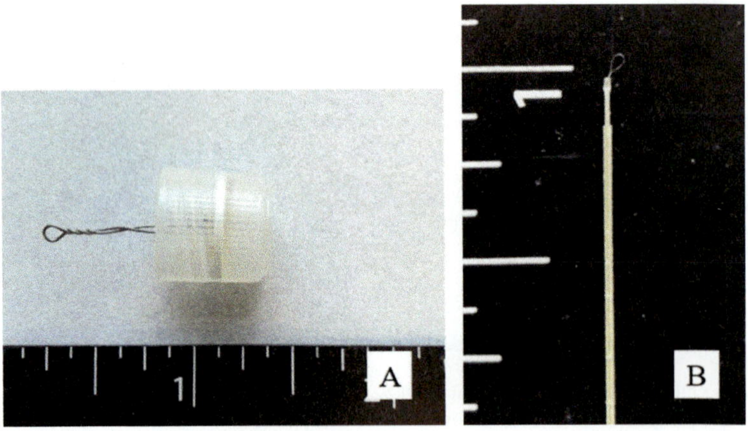

Fig. 10.5 (**a**) Handmade stainless steel CryoLoop. (**b**) Mounted 20-μm diameter nylon CryoLoop

Table 10.1 Comparative freezing rates between devices with different design and volume

Device/sample volume	Cooling speed (°C/min)	References
0.25-mL French straw/0.25-mL	~2500	Tucker and Liebermann (2007)
Cryotop/~10-μL	~12,000	Kuwayama et al. (2005a)
0.25-mL French straw/1 μL	~15,000	Tucker and Liebermann (2007)
Open pulled straw (OPS)/1 μL	~16,700	Kuwayama et al. (2005b)
Cryotip/~1-μL	~20,000	Vajta and Nagy (2006)
Cryoloop/1-μL	~20,000–700,000	AbdelHafez et al. (2011)

Initial cooling temperature, 25 °C; final temperature, −196 °C; initial thawing temperature, −196 °C; final temperature, 37 °C.

liquid nitrogen (Tucker and Liebermann 2007; Pegg 2007). In crustacean, sperm volume is of just few tens of microliters. In addition, sperm of most external fertilization crustaceans is highly viscous and adhesive, therefore development of specially designed containers is critical not only to handle small volumes, but also to reach ultrafast freezing speeds and facilitate the loading and unloading processes of the device.

10.6 Crustacean Reproductive Biology

10.6.1 Crustacean Reproductive System

Progress of visible sexual differentiation in crustaceans varies, but in most of the species begins at some point of its postlarval development and culminates with the complete growth of external sexually dimorphic structures (e.g., petasma, abdomen shape) (Giese et al. 1982; Cabrita et al. 2009). In most crustacean's species, spermatogenesis begins in the testis, where formation of syncytial clones of the

Fig. 10.6 Terminal ampule showing the characteristic milky white color. Arrow indicates gonadal maturity in male *P. vannamei*

germ cells takes place, followed by the differentiation of sperm in the collecting tubule and proximal vas deferens, culminating in the formation of mature sperm in the vas deferens middle, ascending and descending regions (Subramoniam 1993; Alfaro et al. 2007). In shrimps, secondary spermatophore layers are also added in the proximal vas deferens. The rest of the spermatophore components (rigid structures identified as tube and wings) develop within the terminal ampule (Campos-Ramos et al. 2006; Alfaro-Montoya 2010; Peralta Martínez et al. 2013). In crabs, vas deferens are followed by seminal vesicles, where spermatophores are formed, followed by the accessory glands and ejaculatory duct (Subramoniam 1993; He et al. 2015).

In most shrimps, male sexual maturity could be simply confirmed by examining its ventral region. The spermatophores are easily located in a milky white color structure known as the terminal ampule, at the end of the vas deferens (Fig. 10.6) (Martínez Córdova et al. 1999). In other crustaceans, such as crabs, sexual maturity could take years, and sperm maturity is usually seasonal, indicating that sperm production fluctuates throughout the year. For instance, in *Eriocheir sinensis*, there is no sperm production from December to April, and its peak sperm production is expected from August to October. Nonetheless, crab gender can be determined by examining its ventral area (Tiersch and Green 2011; Wang et al. 2015).

Crustaceans are, for the most part, dioecious and nor sperm neither spermatophores are ever released directly into the water. Instead, male gametes could be transferred to the female in two different ways: externally, placed near female's gonophore, or internally, introduced into simple or complex seminal receptacles (e.g., spermatheca) (Subramoniam 1993; McLay and López Greco 2011). Although few crustacean species have specialized gonopods to introduce sperm or spermatophores directly inside female's body during mating, most crustaceans rely on spermatophore transfer strategy (Chow et al. 1982; Giese et al. 1982; Bauer and Min 1993; López Greco et al. 2007; Sal Moyano et al. 2009; Simeo et al. 2009; Alfaro-Montoya 2010; Fransozo et al. 2016).

Briefly, spermatophores are described as carrier structures containing sperm. Spermatophore shape and structure differ broadly among crustaceans. Finding shapes from symmetrical to asymmetrical, spherical to elongated, and wrapped within single or successive keratinized and mucilaginous protective layers. Brachuryans' spermatophores are usually symmetrical and microscopic (Fig. 10.7). Spermatozoa are suspended in viscous seminal plasma, everything covered by an outer thick and thin inner layer. Some brachyuran crabs need to discharge abundant quantities of fluid to bring out several spermatophores at the time during mating. Anomurans' spermatophores evolved to make epizoic fertilization more efficient. One of its distinctive structures consists of a stalk with a sperm ampule at one end and a glutinous pedestal at the other (Fig. 10.7). The later designed to be firmly attached outside the female's sternal region (Malek and Bawab 1974; Bauer 1976; Sandifer et al. 1984; Dougherty et al. 1986; El-Sherief

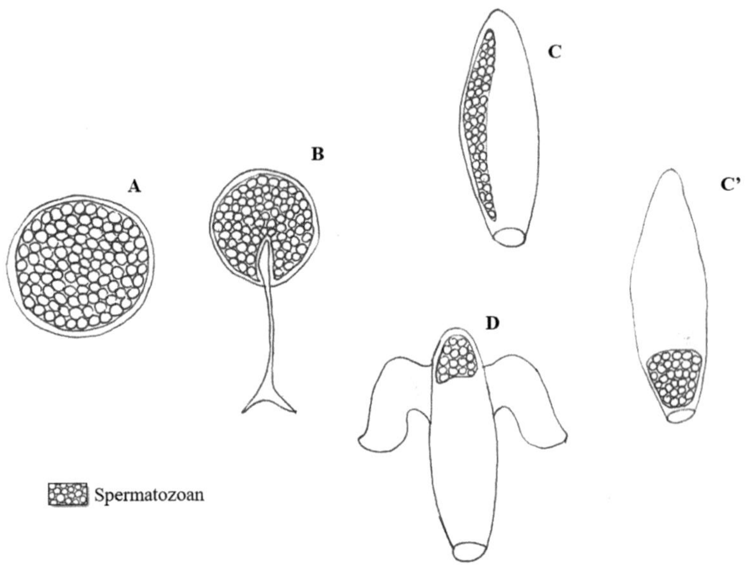

Fig. 10.7 Drawings showing the most representative kinds of crustacean spermatophores. (A) Symmetric and spherical spermatophore. (B) Stalk with a sperm ampule at one end and a glutinous pedestal at the other. (C) Non-winged spermatophore showing sperm along its inner periphery. (C′) Non-winged spermatophore with sperm at the base of the structure. (D) Winged spermatophore showing sperm at the end of the structure. Drawings are not at scale

1991; Leung-Trujillo and Lawrence 1991; Alfaro and Lozano 1993; Bauer and Min 1993; Heitzmann et al. 1993; Aline Staskowian et al. 2013; Benetti et al. 2013; Braga et al. 2013b).

In macrurans, spermatophores are set by a convoluted tube containing sperm embedded in a gelatinous matrix. Most female lobsters spawn several days after mating. Therefore, spermatophores have additional structures, such as an outer crust layer and a basal adhesive matrix. These structures provide protection and adherence for the epizoic fertilization. Spermatophores are simple, in female lobsters that spawn immediately after mating, and its gelatinous matrix disintegrates on contact with seawater, releasing a ribbon-like tube containing sperm (Subramoniam 1993; Robertson and Butler 2013; Fatihah et al. 2016; Nor Fatihah et al. 2017).

Female penaeids are divided into closed and open thelycum. Spermatophores that fertilize females with close thelycum consist of a simple spherical mass composed by few layers of viscous fluid around sperm and sometimes have a thick envelope, known as the sperm sac. In contrast, spermatophores that fertilize females with open thelycum contain a hollow, rigid, and tubular structure with its posterior portion open, some species contain a pair of thin extra structures known as "wings" at the anterior portion (Fig. 10.7). Spermatozoa suspended in seminal plasma are located inside the tubular structure, and several layers of adhesive mucoprotein cover the whole structure. Sperm location within this type of composed spermatophores can differ among species. Sperm could be found all along its inner periphery, at the base, or the end of the tubular structure (Fig. 10.7) (Chow et al. 1991; Heitzmann et al. 1993; Subramoniam 1993, 1995; Alfaro-Montoya 2010).

In summary, spermatophores are more than a sperm vessel. They also serve a mating plug, preventing subsequent copulation; and as storage container, keeping male gametes viable until ovulation or egg release takes place (Yano et al. 1988; Bauer and Min 1993; Hou et al. 2010; Fransozo et al. 2016).

10.7 Crustacean Spermatozoa

Crustacean spermatozoa lack structures typically found in other animal species, such as a motile flagellum, axoneme, and mitochondria. Crustacean spermatozoa are grouped into two broad categories: unistellate and multistellate. Unistellate spermatozoa display a single pointy structure known as a spike. In contrast, multistellate spermatozoa display several appendages. In unistellate spermatozoa, the spike is an extension of the acrosomal structure, but in multistellate spermatozoa, appendages could be composed of chromatin or microtubules (Kim et al. 2003; Klaus et al. 2009, 2013; Kurtz et al. 2009; Pitnick et al. 2009; Harlıoğlu et al. 2018; Krishnamoorthy et al. 2018).

In general, crustacean spermatozoa is composed of one or more spikes, hemispherical hood, filamentous network, and cytoplasmic particles. In contrast, crustaceans spermatozoa display less nuclear condensation than those in mammalian (Alfaro et al. 2007; Barbas and Mascarenhas 2009; Niksirat et al. 2013; Braga et al. 2013a).

In unistellate spermatozoa, the spike is made of numerous parallel microfilaments. There are two triangular points of high electron density located at the base of the spike that act as binding structures, keeping the spike attached or as a platform from where microfilaments rise. It is thought that they actively participate in the spike retraction during acrosomal (Alfaro et al. 2007; Aungsuchawan et al. 2011).

The acrosome is a specialized organelle derived from Golgi apparatus containing a specific mixture of enzymes. In mammals and most crustaceans, acrosome is located at the anterior region of spermatozoon head, and its enzymes participate in exocytosis and inset of nuclear content into the oocyte during the fertilization process (Gwo 2000). However, in close thelycum species, acrosome is located at the opposite end, at the base of the spike (Alfaro et al. 2007; Aungsuchawan et al. 2011).

Acrosomal capacitation is an event required for spermatozoa maturation and consists of a series of essential changes that occur in specific regions. The acrosomal reaction has been described in few crustacean species. However, molecular mechanisms involved have not been determined, and some morphological changes occur only in spermatozoa who had contact with the thelycum (Clark et al. 1973; Wallis et al. 1981; Clark and Griffin 1988; Alfaro et al. 2003, 2007; Vanichviriyakit et al. 2004; del Río et al. 2007; Pattira et al. 2007; Pongtippatee et al. 2007; Kruevaisayawan et al. 2008; Alfaro-Montoya 2010; Zhang et al. 2010).

Acrosomal reaction comprises of a series of morphological and biochemical changes that are presumed to begin at a female's reproductive organ. These changes are mainly variations in the composition and having repercussions on changes in the fluidity of the membrane, as well as phosphorylation of tyrosine residues in proteins associated with flagellar structures (del Río et al. 2007). In crustaceans, the acrosomal reaction consists of two successive morphological changes: spike loss followed by nuclear eversion (Subramoniam 1993; Wang et al. 1995).

10.8 Sperm Collection

Crustacean sperm collection is grouped into lethal and nonlethal. In crabs, the procedure to collect sperm is lethal. Males are anesthetized with ice-cold water, carapace opened, and internal organs dissected to obtain the spermatophores from the testes. On the other hand, the procedure in shrimps is usually nonlethal. Spermatophores could be manually extruded. When correctly performed, spermatophores regenerate, and the procedure could be repeated several times (Arce et al. 2000; Cabrita et al. 2009; Gwo 2011).

10.9 Crustacean Sperm Cryopreservation State of the Art

Currently, less than half of dozen crustacean species have been cultivated worldwide. Pacific white shrimp, *P. vannamei*, has been the most cultured species, holding 53% of total production. Minor production is represented by red swamp crawfish, *Procambarus clarkii* (12%); Chinese mitten crab, *Eriocheir sinensis* (10%); giant tiger prawn, *Penaeus monodon* (9%); oriental river prawn, *Macrobrachium nipponense* (4%); giant river prawn, *Macrobrachium rosenbergii* (3%), and other crustaceans (10%) (FAO 2018a, b). The success of culturing Pacific white shrimp all over the world is due to its high capacity to adapt to a wide variety of cultivation systems and its tolerance to an extensive range of salinities (Martínez Córdova et al. 1999; FAO 2004, 2012a, b, 2014; Briggs et al. 2005). China holds as the leading aquaculture crustacean producer. In 2017 China produced 1.6 million tons, followed by Viet Nam (694 thousand tons) and Indonesia (644 thousand tons) (Lightner 2011; WorldBank 2013; FAO 2012b, 2014).

Crustacean sperm cryopreservation simplifies reproductive management routines and adds flexibility to breeding program design. Among other benefits, it allows mobilization over long distances, extends sperm half-life for several years, enables the use of selected males (even after death), and prevent the reduction of genetic variability caused by inbreeding or environmental disasters or other external factors (Martinez-Paramo et al. 2017).

Between 2000 and 2010, combined effects of early mortality syndrome (EMS) and climate change affected negatively shrimp worldwide production. Most of the countries that produce shrimp suspended all forms of crustacean production to content EMS. Consequently, profit decreased to 54% (FAO 2012b, 2013; Tran et al. 2013). Shrimp aquaculture crisis demonstrated that good management practices were not enough to face an emergency, but having adequate protocols to back up valuable genetic lines (e.g., sperm cryopreservation) before any disease or climate contingency take place is very important (Cock et al. 2009; Pullin and White 2011; Moss et al. 2012; Stentiford et al. 2012).

Despite *P. vannamei* importance and commercial success, procedures granting its genetic diversity protection are not fully developed or widespread. Also, we must stress that one of the main reason that limits the development and use of biotechnological tools to preserve crustaceans diversity is the distinctiveness of their reproductive biology (Yano et al. 1988; Alfaro and Lozano 1993; Alfaro 1994; Alfaro et al. 2003, 2007; Campos-Ramos et al. 2006; Alfaro-Montoya 2010).

Up to date, few works about crustacean sperm cryopreservation have been reported (Table 10.2). Briefly, most evaluated sperm viability through microscopic examination of membrane integrity, with only two reporting viability through the use of sperm in artificial insemination (Chow et al. 1985; Morales Ueno and Paniagua Chávez 2017), and the most used cryoprotectant was DMSO, followed by glycerol.

Most of the earlier cryopreservation protocols use specialized equipment to process the samples. Programmable freezers are expensive, bulky, and difficult to move, hindering its use in field situations. Therefore, the use of less complicated

Table 10.2 Summary, crustacean sperm cryopreservation

Common name	Scientific name	References	Method used	Sample	Cryoprotectant used	Freezing container	Final temperature (°C)	Viability evaluation
Giant river prawn	*Macrobrachium rosenbergii*	Chow et al. (1985)	Controlled-rate freezing	Spermatozoa	Fresh water + 10% glycerol	1-cm diameter glass test tube	−196	Light microscopy and artificial insemination
Ridgeback rock shrimp	*Sicyonia ingentis*	Anchordoguy et al. (1988)	Slow programmable freezing	Spermatozoa	Artificial sea water + 5% DMSO	0.1-mL cryovial	−196	In vitro acrosome reaction
Giant mud crab	*Scylla serrata*	Jeyalectumie and Subramoniam (1989)	Controlled rate freezing	Spermatozoa	DMSO + trehalose	0.5-mL straw	−196	Biochemical analysis and eosin-nigrosin staining
Giant mud crab	*Scylla serrata*	Bhavanishankar and Subramoniam (1997)	Slow programmable freezing	Spermatozoon and spermatophores	Ca free artificial sea water + 10% DMSO	0.5-mL straw	−40	Trypan blue and eosin–nigrosin staining, hypo-osmotic sensitivity test and ionophore-induced acrosome reaction
Giant mud crab	*Scylla serrata*	Guan et al. (2002)	Unknown	Spermatozoa	Calcium-free artificial sea water + 5% DMSO	Unknown	4	Trypan blue staining
Pacific white shrimp	*Penaeus vannamei*	Lezcano et al. (2004)	Slow programmable freezing	Spermatozoa	10% methanol + 10% egg yolk + 0.2 M sucrose + sterile sea water	1-mL cryovial	−196	Flow cytometry
Giant tiger shrimp	*Penaeus monodon*	Bart et al. (2006)	Slow programmable freezing	Spermatophores	Calcium-free saline solution (0.09% NaCl) + 5% DMSO	0.5-mL cryovial	−196	Eosin–nigrosin staining
Giant tiger shrimp	*Penaeus. monodon*	Vuthiphandchai et al. (2007)	Slow programmable freezing	Spermatophores	Calcium-free artificial sea water + 5% DMSO	1.8-mL cryovial	−196	Eosin–nigrosin staining

Giant tiger shrimp	*Penaeus. monodon*	Nimrat et al. (2008)	Slow programmable freezing	Spermatophores	Calcium-free artificial sea water + 5% DMSO	1.8-mL cryovial	−196	Modified eosin–nigrosin staining
Chinese mitten crab	*Eriocheir sinensis*	Kang et al. (2009)	Ultra-fast freezing	Spermatozoa	Calcium-free saline solution (0.09% NaCl) + 10% glycerol + 5% DMSO	1-mL cryovial	−196	Light microscopy
Banana shrimp	*Penaeus merguiensis*	Memon et al. (2012)	Unknown	Spermatophores	Calcium-free artificial sea water + 15% magnesium chloride	1.8-mL cryovial	−196	Light microscopy
White shrimp	*Litopenaeus schmitti*	Castelo-Branco Chaves et al. (2014)	Unknown	Spermatozoa	Calcium-free artificial sea water + 5% glycerol	0.50-mL straw	−196	Eosin–nigrosin staining
White shrimp	*Litopenaeus schmitti*	Fernandes et al. (2014)	Controlled-rate freezing	Spermatozoa and spermatophores (postmortem)	Calcium-free artificial sea water + 10% glycerol	0.50-mL straw and 2-mL cryovial	−196	Eosin–nigrosin staining
Pacific white shrimp	*Penaeus vannamei*	Uberti et al. (2014)	Controlled-rate freezing	Spermatozoa	10% DMSO + 10% egg yolk + 0.2 M sucrose + sterile sea water	1-mL cryovial	−196	Eosin–nigrosin staining and flow cytometry
Orange mud crab	*Scylla olivacea*	Ikhwanuddin et al. (2015)	Controlled-rate freezing	Spermatozoa	Calcium-free artificial sea water + 5% DMSO	1.8-mL cryovial	−196	Eosin–nigrosin staining
Mud spiny lobster	*Panulirus polyphagus*	Fatihah et al. (2016)	Controlled-rate freezing	Spermatozoa	Calcium-free artificial sea water + 10% glycine	2-mL cryovial	−196	Eosin–nigrosin staining
Giant freshwater prawn	*Macrobrachium rosenbergii*	Valentina claudet et al. (2016)	Controlled-rate freezing	Spermatophores	Sterile-filtered pond water + 10% DMSO + 10% propylene glycol	Cryovial (volume not reported)	−196	Ionophore-induced acrosome reaction and scanning electron microscopy
Cinnamon river shrimp	*Macrobrachium acanthurus*	Costa et al. (2017)	Slow programmable freezing	Spermatophores	Distilled water + 10% glycerol	2-mL plastic tubes	−196	Eosin–nigrosin staining

(continued)

Table 10.2 (continued)

Common name	Scientific name	References	Method used	Sample	Cryoprotectant used	Freezing container	Final temperature (°C)	Viability evaluation
Pacific white shrimp	*Penaeus vannamei*	Zhao et al. (2017)	Controlled-rate freezing	Spermatozoa	Sterilized natural seawater + 5% DMSO	0.5-mL cryovial	−196	Scanning electron microscopy and transmission electron microscopy
Indian white shrimp	*Fenneropenaeus indicus*	Narasimman et al. (2018)	Controlled rate freezing	Spermatophores (postmortem)	Calcium-free artificial sea water + 5% DMSO + 5% methanol	Vial (volume not reported)	−196	Modified eosin–nigrosin staining
Pacific white shrimp	*Penaeus vannamei*	Morales Ueno and Paniagua Chávez (2017)	Vitrification	Sperm mass	PCT/MX2017050012	Patent pending	−196	Artificial insemination and hatching
Indian white shrimp	*Fenneropenaeus indicus*	Nimrat et al. (2020)	Controlled-rate freezing	Spermatophores	Calcium-free artificial sea water + 5% DMSO	1.5-mL cryovial	−196	Modified eosin–nigrosin staining

techniques such as the brief exposure of the sample in the gaseous phase of liquid nitrogen, followed by its total immersion in the liquid phase to complete the cryopreservation process or production of pellets on a dry ice surface, have been used. Also, these techniques are less expensive and do not require specialized equipment or personnel (Vajta and Nagy 2006; Tucker and Liebermann 2007; Tiersch 2011).

One of the difficulties that arise in the development of cryopreservation protocols for sperm in crustaceans lies in the diverse spermatophore morphology, coupled to its different fertilization strategy, makes challenging the development of suitable cryopreservation protocols and slows down the development of sperm cryopreservation protocols using another crustacean as models (Giese et al. 1982; Pitnick et al. 2009).

Most of the studies on crustaceans developed extenders for spermatozoa counting, facilitating the study on morphological/physiological traits or short-term storage. From these studies, only one report the development of a short-term extender solution to specifically increase the number of crustacean females inseminated per male (Morales-Ueno et al. 2013). The development of more protocols focused on increasing sperm dosages combined with extending storage time of crustacean sperm would improve the reproductive management at the hatcheries and would allow additional flexibility in the design of their breeding programs.

Several studies have reported that the presence of calcium ions triggers the acrosomal reaction in crustacean spermatozoa. Therefore, Ca-free solutions have been used in most of the studies (Table 10.2). Although it seems to be a good strategy for most crustacean species, it does not seem to work in all, particularly in freshwater crustaceans, where acrosomal reaction mechanisms have not yet been sufficiently studied.

Previous works had addressed cryoprotectant toxic effects on spermatozoa. However, morphology and physiology of gametes among species make them impossible to use these references to select the concentration or combination of substances that could work on all crustaceans, although they are valid as a material to make a first approximation (Vuthiphandchai et al. 2007; Nimrat et al. 2008; Memon et al. 2012; Castelo-Branco Chaves et al. 2014; Uberti et al. 2014).

Dimethyl sulfoxide (DMSO) has previously been used successfully to carry on mammal sperm and has been successfully introduced in most of the crustaceans (Table 10.2). During the cryopreservation process, DMSO serves as an osmotic buffer and increase sample viscosity depleting growth and propagation of ice. DMSO is potentially cytotoxic, but it is less harmful when is combined with a non-permeable cryoprotectant, such as trehalose (Anchordoguy et al. 1987, 1988; Beattie et al. 1997; Bart et al. 2006; Daly and Tiersch 2012; Katkov et al. 2012).

The addition of non-permeable agents to the membrane (sugars) could have exerted several beneficial processes. First, it is known that extracellular sugars allow spermatozoa dehydration, with the additional benefit of being nontoxic for most crustaceans (Shaw et al. 1992; Aisen et al. 2000; Julca et al. 2012). Second, its addition stabilizes cellular proteins and biological membranes. It is known that many organisms that produce them naturally to induce cellular dehydration to survive adverse environmental conditions, including some crustaceans such as saline Artemia (Rudolph and Crowe 1985; Almeida et al. 2007; Rodrigues et al. 2008).

The freezing rate is essential for the success of cryopreservation protocols. Most cases reported for crustaceans have used relatively rapid freezing rates, but the use of the reported devices (mostly cryovials) has likely interfered in the heat diffusion of the sample, delaying its cooling and thereby producing ice crystals during the process. The smaller the device used to hold the sample, the better the heat diffusion that accelerates the net freezing rates of the samples. Not only must the device be of the appropriate size, but the material that forms it must have characteristics that allow the total recovery of the sample without causing any detriment and preferably simple to disinfect and reusable (Isachenko et al. 2005; Pegg 2007; Succu et al. 2007; Bagchi et al. 2008; Jenkins 2011).

The use of ad hoc devices would also facilitate the cryopreservation process. There is a patented device (MX 369254) that allows simple handling of small, amorphous, glutinous, and adhesive samples (all spermatophore characteristics), which unlike the use of tubes or other devices, not only improves the freezing rate, it also facilitates assembly processes, addition of cryoprotectants, storage, and recovery of samples (Paniagua Chávez and Morales Ueno 2019).

In most of economically important crustacean species listed above, detailed reproductive features (e.g., egg extraction, egg activation, changes during spermatozoon capacitation, spermatozoon acrosome reaction, or spermatozoon–egg binding mechanisms) have not been fully explained. In some important economical crustacean species (*P. clarkii* and *M. nipponense*) cryopreservation studies have not yet been reported. An essential requirement to develop any cryopreservation protocol is to have a good understanding of the organism's reproduction and cellular/molecular mechanisms that control gamete maturation and fertilization. Regardless most of the biological and morphological features are alike among crustaceans. Lack of detailed information has made impossible to extrapolate technics for sperm collection and develop cryopreservation protocols among close related or morphologically similar crustacean species (Alfaro et al. 2003; Braga et al. 2013a; Fransozo et al. 2016). Additionally, more importantly, we should consider the species' reproductive adaptation to confined environments and repeated handling (e.g., spermatophores manual collection) because these factors have a profound impact on successful cryopreservation rates (Alfaro-Montoya 2010; Braga et al. 2011).

In crustaceans with external fertilization spermatophores produce an adhesive substance to facilitate its attachment to females' thelycum. The addition of egg yolk to the cryoprotectant caused the complete loss of its adherent capacity, which is essential to carry out the insemination processes.

As mentioned above, some crustacean females keep sperm for some time before spawning/ovulation. Therefore, different approaches for sample handling and cryopreservation have to be used depending on their fertilization system. For instance, in open thelycum crustaceans, sperm enclosed in spermatophores should be attached to the ventral surface of the female during mating. Consequently, protocols should be developed not only to keep sperm viability but also to keep viscosity so they can be added to the ventral surface of the female.

Many things can happen to the sperm during a microscopic feasibility review. First, the sample must be suspended and homogenized within a suspension. This process requires manual pressure on the sperm mass, with expected mechanical damage caused to the gametes during the process. Consequently, the damages caused by the process can interfere with the actual state of the gametes after thawing. Currently, not all chemical and mechanical effects of the cryopreservation process to spermatozoa are yet known. Therefore, the golden rule when cells are cryopreserved is to assess viability.

More than half of the studies reported in Table 10.2 require manual disruption of spermatophores and the maceration. As mentioned before, the persistence of spermatophore's adherence is essential for the successful artificial insemination in crustaceans with external fertilization. Gamete fertilization in these species occurs only during the passage of the oocytes through the sperm mass, and the complexities inherent in the maturation processes of the crustacean oocytes do not allow their use in in vitro insemination processes (Lynn and Clark 1983; Leung-Trujillo and Lawrence 1991; Bauer and Min 1993; Arce et al. 2000; Alfaro et al. 2004; Bart et al. 2006; Peixoto et al. 2011; Beirão et al. 2019). Up to date, only one artificial insemination using the thawed sperm mass of *P. vannamei* has been successfully carried out, resulting in the production of nauplii. This protocol has been patented (Table 10.2).

The cryopreservation of liquid samples simplifies the sperm counting and cryopreservation process, but the loss of the sperm mass structure or its glutinous adhesive material makes it impossible to use it again on artificial insemination once thawed. To date, successful artificial insemination by transferring sperm suspension within female's spermatheca has not been reported.

The vast majority of studies use indirect viability evaluation (e.g., microscopy, flow cytometry) of cryopreserved–thawed samples. However, as mentioned before, to achieve this, it is necessary to separate spermatozoa from spermatophore and suspend individual cells. This procedure causes mechanical stress during separation and homogenization processes. Consequently, the cells are damaged (e.g., cell membrane disruption, spontaneous cytoplasm eversion), and this damage can be interfering with the real status of gametes (Tieleman et al. 2003; Sinha et al. 2011).

Up to date, not all interactions between gametes and substances to which they are exposed are known. Therefore, in cryopreservation research, the verification of the fertilizing capacity of the sperm, it is considered as a "gold test" regardless of the results that would have been obtained through the use of indirect feasibility tests (Cuevas-Uribe and Tiersch 2011).

In any case, the survival rate measured based on membrane staining does not guarantee high fertilizing ability (Akarasanon et al. 2004). Thereby, patented protocol, described above, evaluated sperm viability based on artificial insemination success.

References

AbdelHafez F, Xu J, Goldberg J, Desai N (2011) Vitrification in open and closed carriers at different cell stages: assessment of embryo survival, development, DNA integrity and stability during vapor phase storage for transport. BMC Biotechnol 11:29. https://doi.org/10.1186/1472-6750-11-29

Aisen EG, Alvarez HL, Venturino A, Garde JJ (2000) Effect of trehalose and EDTA on cryoprotective action of ram semen diluents. Theriogenology 53(5):1053–1061. https://doi.org/10.1016/S0093-691X(00)00251-X

Akarasanon K, Damrongphol P, Poolsanguan W (2004) Long-term cryopreservation of spermatophore of the giant freshwater prawn, Macrobrachium rosenbergii (de Man). Aquac Res 35 (15):1415–1420. https://doi.org/10.1111/j.1365-2109.2004.01163.x

Alfaro J (1994) Ultraestructura de la glándula androgénica, espermatogénesis y oogénesis de camarones marinos (Decapoda: Penaeidae). Rev Biol Trop 42(2):121–129

Alfaro J, Lozano X (1993) Development and deterioration of spermatophores in pond-reared Penaeus vannamei. J World Aquacult Soc 24(4):522–529. https://doi.org/10.1111/j.1749-7345.1993.tb00583.x

Alfaro J, Muñoz N, Vargas M, Komen J (2003) Induction of sperm activation in open and closed thelycum penaeoid shrimps. Aquaculture 216(1–4):371–381. https://doi.org/10.1016/s0044-8486(02)00514-8

Alfaro J, Zúñiga G, Komen J (2004) Induction of ovarian maturation and spawning by combined treatment of serotonin and a dopamine antagonist, spiperone in Litopenaeus stylirostris and Litopenaeus vannamei. Aquaculture 236(1–4):511–522. https://doi.org/10.1016/j.aquaculture.2003.09.020

Alfaro J, Ulate K, Vargas M (2007) Sperm maturation and capacitation in the open thelycum shrimp Litopenaeus (Crustacea: Decapoda: Penaeoidea). Aquaculture 270(1–4):436–442. https://doi.org/10.1016/j.aquaculture.2007.05.011

Alfaro-Montoya J (2010) The reproductive conditions of male shrimps, genus Penaeus, sub-genus Litopenaeus (open thelyca penaeoid shrimps): a review. Aquaculture 300(1–4):1–9. https://doi.org/10.1016/j.aquaculture.2009.12.008

Aline Staskowian B, Marcos T, Daniela Carvalho dos S, Christopher T (2013) Ultrastructure of the spermatophores and spermatozoa from the gecarcinid land crab, Cardisoma guanhumi Latreille, 1828, (Crustacea, Brachyura, Gecarcinidae) from the western Atlantic. Invertebr Reprod Dev 57 (1):61–69. https://doi.org/10.1080/07924259.2012.666506

Almeida AM, Cardoso LA, Santos DM, Torné JM, Fevereiro PS (2007) Trehalose and its applications in plant biotechnology. In Vitro Cell Dev Biol Plant 43(3):167–177. https://doi.org/10.1007/s11627-006-9024-3

Anchordoguy TJ, Rudolph AS, Carpenter JF, Crowe JH (1987) Modes of interaction of cryoprotectants with membrane phospholipids during freezing. Cryobiology 24(4):324–331. https://doi.org/10.1016/0011-2240(87)90036-8

Anchordoguy T, Crowe JH, Griffin FJ, Clark WH (1988) Cryopreservation of sperm from the marine shrimp *Sicyonia ingentis*. Cryobiology 25(3):238–243. https://doi.org/10.1016/0011-2240(88)90031-4

Arce SM, Moss SM, Argue BJ (2000) Artificial insemination and spawning of pacific white shrimp *Litopenaeus vannamei*: implications for a selective breeding program. UJNR Tech Rep 28:5–7

Aungsuchawan S, Browdy CL, Withyachumnarnkul B (2011) Sperm capacitation of the shrimp Litopenaeus vannamei. Aquac Res 42(2):188–195. https://doi.org/10.1111/j.1365-2109.2010.02579.x

Bagchi A, Woods EJ, Critser JK (2008) Cryopreservation and vitrification: recent advances in fertility preservation technologies. Expert Rev Med Devices 5(3):359–370. https://doi.org/10.1586/17434440.5.3.359

Barbas J, Mascarenhas R (2009) Cryopreservation of domestic animal sperm cells. Cell Tissue Bank 10(1):49–62. https://doi.org/10.1007/s10561-008-9081-4

Bart AN, Choosuk S, Thakur DP (2006) Spermatophore cryopreservation and artificial insemination of black tiger shrimp, Penaeus monodon (Fabricius). Aquac Res 37(5):523–528. https://doi.org/10.1111/j.1365-2109.2006.01460.x

Bauer RT (1976) Mating behaviour and spermatophore transfer in the shrimp Heptacarpus pictus (Stimpson) (Decapoda: Caridea: Hippolytidae). J Nat Hist 10(4):415–440. https://doi.org/10.1080/00222937600770321

Bauer RT, Min LJ (1993) Spermatophores and plug substance of the marine shrimp Trachypenaeus similis (Crustacea: Decapoda: Penaeidae): formation in the male reproductive tract and disposition in the inseminated female. Biol Bull 185(2):174–185

Beattie GM, Crowe JH, Lopez AD, Cirulli V, Ricordi C, Hayek A (1997) Trehalose: a cryoprotectant that enhances recovery and preserves function of human pancreatic islets after long-term storage. Diabetes 46(3):519–523. https://doi.org/10.2337/diab.46.3.519

Beirão J, Boulais M, Gallego V, O'Brien JK, Peixoto S, Robeck TR, Cabrita E (2019) Sperm handling in aquatic animals for artificial reproduction. Theriogenology 133:161–178. https://doi.org/10.1016/j.theriogenology.2019.05.004

Benetti AS, Tavares M, Carvalho dos Santos D, Tudge C (2013) Ultrastructure of the spermatophores and spermatozoa from the gecarcinid land crab, Cardisoma guanhumi Latreille, 1828, (Crustacea, Brachyura, Gecarcinidae) from the western Atlantic. Invertebr Reprod Dev 57(1):61–69. https://doi.org/10.1080/07924259.2012.666506

Bhavanishankar S, Subramoniam T (1997) Cryopreservation of spermatozoa of the edible mud crab Scylla serrata (Forskal). J Exp Zool 277(4):326–336. https://doi.org/10.1002/(sici)1097-010x(19970301)277:4<326::aid-jez6>3.0.co;2-r

Braga AL, Lopes DLA, Krummenauer D, Poersch LH, Wasielesky W (2011) A comparison of the reproductive performance of the wild pink shrimp species Farfantepenaeus paulensis and Farfantepenaeus brasiliensis in captivity. J Shellfish Res 30(3):963–967. https://doi.org/10.2983/035.030.0339

Braga A, Nakayama C, Poersch L, Wasielesky W Jr (2013a) Unistellate spermatozoa of decapods: comparative evaluation and evolution of the morphology. Zoomorphology 132(3):1–24. https://doi.org/10.1007/s00435-013-0187-2

Braga AL, Lopes DLA, Poersch LH, Wasielesky W (2013b) Spermatophore and sperm quality of the pink shrimp Farfantepenaeus paulensis fed with fresh food supplemented with pollen and paprika. Aquaculture 380–383:29–32. https://doi.org/10.1016/j.aquaculture.2012.11.030

Briggs M, Funge-Smith S, Subasinghe RP, Phillips M (2005) Introduction and movement of two penaeid shrimp species in Asia and the Pacific, vol 476. FAO, Rome

Cabrita E, Robles V, Chereguini O, Wallace JC, Herraez MP (2003) Effect of different cryoprotectants and vitrificant solutions on the hatching rate of turbot embryos (Scophthalmus maximus). Cryobiology 47(3):204–213

Cabrita E, Robles V, Herráez P (2009) Methods in reproductive aquaculture: marine and freshwater species. In: Marine biology, 1st edn. CRC Press, Boca Ratón

Campos-Ramos R, Garza-Torres R, Guerrero-Tortolero DA, Maeda-Martínez AM, Obregón-Barboza H (2006) Environmental sex determination, external sex differentiation and structure of the androgenic gland in the Pacific white shrimp Litopenaeus vannamei (Boone). Aquac Res 37(15):1583–1593. https://doi.org/10.1111/j.1365-2109.2006.01604.x

Castelo-Branco Chaves T, Bambozzi Fernandes A, Mello Bourg MR, Oshiro LMY (2014) Cryopreservation and sperm storage of the white shrimp. Bol Inst Pesca Sao Paulo 40:17–22

Ceballos-Vázquez BP, Aparicio-Simón B, Palacios E, Racotta IS (2004) Sperm quality over consecutive spermatophore regenerations in the Pacific white shrimp *Litopenaeus vannamei*. J World Aquacult Soc 35(2):178–188

Chow S, Ogasawara Y, Taki Y (1982) Male reproductive system and fertilization of the Palaemonid shrimp *Macrobrachium rosenbergii*. Bull Jpn Soc Sci Fish 48(2):177–183

Chow S, Taki Y, Ogasawara Y (1985) Cryopreservation of spermatophore of the fresh water shrimp, Macrobrachium rosenbergii. Biol Bull 168(3):471. https://doi.org/10.2307/1541526

Chow S, Dougherty MM, Dougherty WJ, Sandifer PA (1991) Spermatophore formation in the white shrimps Penaeus setiferus and P. vannamei. J Crustac Biol 11(2):201–216. https://doi.org/10.2307/1548358

Clark WH, Griffin FJ (1988) The morphology and physiology of the acrosome reaction in the sperm of the decapod, Sicyonia ingentis (acrosome reaction/shrimp sperm/intracellular pH/cell activation). Dev Growth Differ 30(5):451–462. https://doi.org/10.1111/j.1440-169X.1988.00451.x

Clark WH Jr, Talbot P, Neal RA, Mock CR, Salser BR (1973) In vitro fertilization with non-motile spermatozoa of the brown shrimp Penaeus aztecus. Mar Biol 22(4):353–354. https://doi.org/10.1007/bf00391393

Cloud JG (2011) Extender solutions for sperm of salmonid fishes. In: Tiersch TR, Green CC (eds) Cryopreservation in aquatic species, 2nd edn. World Aquaculture Society, Baton Rouge, p 698

Cock J, Gitterle T, Salazar M, Rye M (2009) Breeding for disease resistance of Penaeid shrimps. Aquaculture 286(1–2):1–11. https://doi.org/10.1016/j.aquaculture.2008.09.011

Costa TV, Oshiro LMY, Mattos LA, López Greco LS, Melo EP, Flor HDR (2017) Toxicity, freezing and cold storage tests of native species semen. Bol Inst Pesca 43(3):344–346. https://doi.org/10.20950/1678-2305.2017v43n3p334

Cuevas-Uribe R, Tiersch TR (2011) Estimation of fish sperm concentration by use of spectrophotometry. In: Tiersch TR, Green CC (eds) Cryopreservation in aquatic species, 2nd edn. World Aquaculture Society, Baton Rouge, pp 162–200

Daly J, Tiersch TR (2012) Sources of variation in flow cytometric analysis of aquatic species sperm: the effect of cryoprotectants on flow cytometry scatter plots and subsequent population gating. Aquaculture 370–371(0):179–188. https://doi.org/10.1016/j.aquaculture.2012.09.024

Day JG, Stacey GN (2007) Cryopreservation and freeze-drying protocols. In: Methods in molecular biology, 2nd edn. Humana Press Inc., Totowa

del Río MJ, Godoy A, Toro A, Orellana R, Cortés ME, Moreno RD, Vigil P (2007) La reacción acrosómica del espermatozoide: avances recientes. Revista Internacional de Andrología 5 (4):368–373

Dougherty WJ, Dougherty MM, Harris SG (1986) Ultrastructural and histochemical observations on electroejaculated spermatophores of the palaemonid shrimp, Macrobrachium rosenbergii. Tissue Cell 18(5):709–724. https://doi.org/10.1016/0040-8166(86)90072-8

El-Sherief SS (1991) Fine structure of the sperm and spermatophores of Portunus pelagicus (L.) (Decapoda, Brachyura). Crustaceana 61(3):271–279

Endo Y, Fujii Y, Shintani K, Seo M, Motoyama H, Funahashi H (2012) Simple vitrification for small numbers of human spermatozoa. Reprod Biomed Online 24(3):301–307. https://doi.org/10.1016/j.rbmo.2011.11.016

FAO (2004) Manejo sanitario y mantenimiento de la bioseguridad de los laboratorios de postlarvas de camarón blanco (*Penaeus vannamei*) en América Latina. Documento Técnico de Pesca, vol 450. FAO, Roma

FAO (2012a) Fishery and aquaculture statistics 2010. Yearbook. Food and Agriculture Organization of the United Nations, Rome

FAO (2012b) The state of world fisheries and aquaculture 2012. FAO, Rome

FAO (2013) FAO/MARD technical workshop on early mortality syndrome (EMS) or acute Hepatopancreatic necrosis syndrome (AHPNS) of cultured shrimp. FAO, Hanoi

FAO (2014) State of the world fisheries and aquaculture. Opportunities and challenges. FAO, Rome

FAO (2018a) GLOBEFISH Highlights 2018. A quaterly update on world seafood markets, vol 2. FAO, Rome

FAO (2018b) State of world fisheries and aquaculture 2018. Meeting the sustainable development goals. FAO, Rome

Fatihah SN, Jasmani S, Abol-Munafi AB, Noorbaiduri S, Muhd-Farouk H, Ikhwanuddin M (2016) Development of a sperm cryopreservation protocol for the mud spiny lobster, Panulirus polyphagus. Aquaculture 462:56–63. https://doi.org/10.1016/j.aquaculture.2016.04.025

Fernandes AB, Mattos LAD, Mello M, Oshiro LMY (2014) Post-mortem spermatophore and sperm cryopreservation of the white shrimp Litopenaeus schmitti. Bol Inst Pesca 40(1):49–60

Foote RH (2002) The history of artificial insemination: selected notes and notables. J Anim Sci 80 (E-Suppl_2):1–10

Fransozo V, Fernandes AB, López-Greco LS, Zara FJ, Santos DC (2016) Functional morphology of the male reproductive system of the white shrimp Litopenaeus schmitti (Burkenroad, 1936) (Crustacea, Penaeidea) compared to other Litopenaeus. Invertebr Reprod Dev 60(3):161–174. https://doi.org/10.1080/07924259.2016.1174158

Fuller BJ, Lane N, Benson EE (2004) Life in the frozen state, 1st edn. CRC Press, New York

Gaitán-Espitia JD, Martínez-Silva MA, Borrero CE, Ramírez L, Valencia JP (2013) Cryogenic preservation of sperm from lane snapper (Lutjanus synagris): testing the effects of extenders and freezing rates on sperm quality. Aquaculture 384–387(0):6–12. https://doi.org/10.1016/j.aquaculture.2012.12.002

Garza-Torres R, Maeda-Martinez AM, Guerrero-Tortolero DA, Obregon-Barboza H, Campos-Ramos R (2011) Description of meiosis in female and male Pacific white shrimp *Litopenaeus vannamei* (Decapoda: Penaeidae). J Crustac Biol 31(1):75–81. https://doi.org/10.1651/10-3316.1

Giese AC, Pearse JS, Pearse VB (1982) Reproduction of marine invertebrates. In: General aspects: seeking unity in diversity, vol IX. Blackwell Scientific Publications, Oxford

Glenn DW III, Lang RP, Tiersch TR (2011) Evaluation of extenders for refrigerated storage of koi carp and goldfish sperm. In: Tiersch TR, Green CC (eds) Cryopreservation in aquatic species, 2nd edn. World Aquaculture Society, Baton Rouge, p 107

Guan W, Wang G, Li S, Chen G (2002) Cryopreservation of spermatozoa of mud crab (Scylla serrata) and viability assay by biostain method. J Oceanogr TaiwanStrait 21(4):457–462 (In Chinese)

Gwo JC (2000) Cryopreservation of aquatic invertebrate semen: a review. Aquac Res 31 (3):259–271. https://doi.org/10.1046/j.1365-2109.2000.00462.x

Gwo J-C (2011) Cryopreservation of eggs and embryos from aquatic organisms. In: Tiersch TR, Green CC (eds) Cryopreservation in aquatic species, 2nd edn. World Aquaculture Society, Baton Rouge, pp 539–560

Hamawaki A, Kuwayama M, Hamano S (1999) Minimum volume cooling method for bovine blastocyst vitrification. Theriogenology 51(1):165

Harlıoğlu MM, Farhadi A, Gür S (2018) Determination of sperm quality in decapod crustaceans. Aquaculture 490:185–193. https://doi.org/10.1016/j.aquaculture.2018.02.031

He L, Li Q, Liu L, Wang Y, Xie J, Yang H, Wang Q (2015) A catalog of proteins expressed in the AG secreted fluid during the mature phase of the Chinese mitten crabs (Eriocheir sinensis). PLoS One 10(8):1–19. https://doi.org/10.1371/journal.pone.0136266

Heitzmann JC, Diter A, Aquacop (1993) Spermatophore formation in the white shrimp, Penaeus vannamei Boone 1931: dependence on the intermoult cycle. Aquaculture 116(1):91–98. https://doi.org/10.1016/0044-8486(93)90225-N

Hou X-L, Mao Q, He L, Gong Y-N, Qu D, Wang Q (2010) Accessory sex gland proteins affect spermatophore digestion rate and spermatozoa acrosin activity in Eriocheir Sinensis. J Crustac Biol 30(3):435–440. https://doi.org/10.1651/09-3177.1

Ikhwanuddin M, Yin TP, Menon AJ, Jasmani S, Abol-Munafi AB (2015) Effect of different cryoprotectants and sperm densities of orange mud crab, Scylla olivacea (Herbst, 1796) for long-term storage of spermatozoa. Sains Malays 44(9):1283–1288

Isachenko E, Isachenko V, Katkov II, Dessole S, Nawroth F (2003) Vitrification of mammalian spermatozoa in the absence of cryoprotectants: from past practical difficulties to present success. Reprod Biomed Online 6(2):191–200

Isachenko V, Isachenko E, Montag M, Zaeva V, Krivokharchenko I, Nawroth F, Dessole S, Katkov II, der Ven HV (2005) Clean technique for cryoprotectant-free vitrification of human spermatozoa. Reprod Biomed Online 10(3):350–354. https://doi.org/10.1016/s1472-6483(10)61795-6

Jenkins JA (2011) Minimizing microbial contamination of sperm sample. In: Tiersch TR, Green CC (eds) Cryopreservation in aquatic species, 2nd edn. World Aquaculture Society, Baton Rouge, pp 684–687

Jeyalectumie C, Subramoniam T (1989) Cryopreservation of spermatophores and seminal plasma of the edible crab Scylla serrata. Biol Bull 177(2):247. https://doi.org/10.2307/1541940

Julca I, Alaminos M, Gonzalez-Lopez J, Manzanera M (2012) Xeroprotectants for the stabilization of biomaterials. Biotechnol Adv 30(6):1641–1654. https://doi.org/10.1016/j.biotechadv.2012.07.002

Kang X, Li G, Mu S, Guo M, Ge S (2009) Acrosome reaction of Chinese mitten-handed crab Eriocheir sinensis (Crustacea: Decapoda) spermatozoa: promoted by long-term cryopreservation. Aquaculture 295(3–4):195–199. https://doi.org/10.1016/j.aquaculture.2009.06.041

Katkov II, Bolyukh VF, Chernetsov OA, Dudin PI, Grigoriev AY, Isachenko V, Isachenko E, Lulat AG-M, Moskovtsev SI, Petrushko MP, Pinyaev VI, Sokol KM, Sokol YI, Sushko AB, Yakhnenko I (2012) Kinetic vitrification of spermatozoa of vertebrates: what can we learn from nature? In: Katkov I (ed) Current frontiers in cryobiology. Intech, London

Kim DH, Jo Q, Choi JH, Yun SJ, Oh TY, Kim BK, Han C-H (2003) Sperm structure of the pandalid shrimp *Pandalopsis japonica* (decapoda, pandalidae). J Crustac Biol 23(1):23–32

Klaus S, Schubart CD, Brandis D (2009) Ultrastructure of spermatozoa and spermatophores of old world freshwater crabs (Brachyura: Potamoidea: Gecarcinucidae, Potamidae, and Potamonautidae). J Morphol 270(2):175–193. https://doi.org/10.1002/jmor.10678

Klaus S, Münzner S, Modenbach A-C, Streit B, Tudge CC (2013) Spermatophore formation and sperm ultrastructure ofSundathelphusa philippina(Crustacea: Brachyura: Gecarcinucidae). Acta Zool 94(3):267–272. https://doi.org/10.1111/j.1463-6395.2011.00550.x

Kopeika E, Kopeika J, Zhang T (2007) Cryopreservation of fish sperm. In: Day JG, Stacey GN (eds) Cryopreservation and freeze-drying protocols, vol Methods in molecular biology, 2nd edn. Humana Press Inc., Totowa, p 212

Krishnamoorthy D, Narasimman S, Natesan M (2018) Ultrastructure of spermatozoa in a freshwater fairy shrimp, Streptocephalus dichotomus (Crustacea: Anostraca). Proc Zool Soc 71 (4):376–381. https://doi.org/10.1007/s12595-017-0221-3

Kruevaisayawan H, Vanichviriyakit R, Weerachatyanukul W, Iamsaard S, Withyachumnarnkul B, Basak A, Tanphaichitr N, Sobhon P (2008) Induction of the acrosome reaction in black tiger shrimp (Penaeus monodon) requires sperm trypsin-like enzyme activity. Biol Reprod 79 (1):134–141. https://doi.org/10.1095/biolreprod.107.066316

Kurtz K, Ausio J, Chiva M (2009) Preliminary study of sperm chromatin characteristics of the brachyuran crab Maja brachydactyla. Histones and nucleosome-like structures in decapod crustacean sperm nuclei previously described without SNBPs. Tissue Cell 41(5):334–344. https://doi.org/10.1016/j.tice.2009.02.003

Kuwayama M, Vajta G, Ieda S, Kato O (2005a) Comparison of open and closed methods for vitrification of human embryos and the elimination of potential contamination. Reprod Biomed Online 11(5):608–614. https://doi.org/10.1016/S1472-6483(10)61169-8

Kuwayama M, Vajta G, Kato O, Leibo SP (2005b) Highly efficient vitrification method for cryopreservation of human oocytes. Reprod Biomed Online 11(3):300–308. https://doi.org/10.1016/S1472-6483(10)60837-1

Kwantong S, Bart AN (2003) Effect of cryoprotectants, extenders and freezing rates on the fertilization rate of frozen striped catfish, Pangasius hypophthalmus (Sauvage), sperm. Aquac Res 34(10):887–893. https://doi.org/10.1046/j.1365-2109.2003.00897.x

Leung-Trujillo JR, Lawrence AL (1991) Spermatophore generation times in *Penaeus setiferus, P. vannamei*, and *P. stylirostris*. J World Aquacult Soc 22(4):244–251

Lezcano M, Granja C, Salazar M (2004) The use of flow cytometry in the evaluation of cell viability of cryopreserved sperm of the marine shrimp (*Litopenaeus vannamei*). Cryobiology 48 (3):349–356

Lightner DV (2011) Virus diseases of farmed shrimp in the Western Hemisphere (the Americas): a review. J Invertebr Pathol 106(1):110–130. https://doi.org/10.1016/j.jip.2010.09.012

López Greco LS, Vazquez F, Rodríguez EM (2007) Morphology of the male reproductive system and spermatophore formation in the freshwater 'red claw' crayfish Cherax quadricarinatus (Von Martens, 1898) (Decapoda, Parastacidae). Acta Zool 88(3):223–229. https://doi.org/10.1111/j.1463-6395.2007.00269.x

Lynn JW, Clark WH (1983) The fine structure of the mature sperm of the freshwater prawn, Macrobrachium rosenbergii. Biol Bull 164(3):459–470

Malek SRA, Bawab FM (1974) The formation of the spermatophore in Penaeus Kerathurus (Forskal, 1775) (Decapoda, Penaeidae). Crustaceana 26(3):273–285. https://doi.org/10.1163/156854074X00677

Mansour N, Richardson GF, McNiven MA (2006) Effect of extender composition and freezing rate on post-thaw motility and fertility of Arctic char, Salvelinus alpinus (L.), spermatozoa. Aquac Res 37(9):862–868. https://doi.org/10.1111/j.1365-2109.2006.01503.x

Martínez Córdova LR, Villareal Colmenares H, Cortés Jacinto E (1999) Biología del camarón. In: Martínez Córdova LR (ed) Cultivo de camarones peneidos. Principios y prácticas. AGT, México, pp 1–20

Martinez-Paramo S, Horvath A, Labbe C, Zhang T, Robles V, Herraez P, Suquet M, Adams S, Viveiros A, Tiersch TR, Cabrita E (2017) Cryobanking of aquatic species. Aquaculture 472:156–177. https://doi.org/10.1016/j.aquaculture.2016.05.042

Mazur P (2004) Principles of cryobiology. In: Fuller BJ, Lane N, Benson EE (eds) Life in the frozen state, 1st edn. CRC Press, New York, pp 3–65

McLay CL, López Greco LS (2011) A hypothesis about the origin of sperm storage in the Eubrachyura, the effects of seminal receptacle structure on mating strategies and the evolution of crab diversity: how did a race to be first become a race to be last? Zool Anz 250(4):378–406. https://doi.org/10.1016/j.jcz.2011.06.002

Memon AJ, Talpur AD, Khan MI, Fariddudin MO, Safiah J, Abol-Munafi AB, Ikhwanuddin M (2012) Optimization of spermatophores cryopreservation protocol of Banana shrimp (*Penaeus merguiensis*) (De Man, 1888). J Anim Vet Adv 11(10):1688–1704

Moawad AR, Amarnath D, Choi I, Fisher P, Zhu J, Campbell KHS (2008) Vitrification of immature ovine oocytes with Cryoloop: effect on in vitro fertilization rates. Cryobiology 57(3):323. https://doi.org/10.1016/j.cryobiol.2008.10.040

Mochida K, Hasegawa A, Li MW, Fray MD, Kito S, Vallelunga JM, Lloyd KC, Yoshiki A, Obata Y, Ogura A (2013) High osmolality vitrification: a new method for the simple and temperature-permissive cryopreservation of mouse embryos. PLoS One 8(1):e49316. https://doi.org/10.1371/journal.pone.0049316

Morales Ueno K, Paniagua Chávez C (2017) Protocol for the cryopreservation of high viscosity biological samples. Patent pending PCT/MX2017050012

Morales-Ueno K, Montaldo HH, Ortega AM, Paniagua-Chavez CG, Castillo-Juarez H (2013) An extender solution for the short-term storage of Litopenaeus vannamei sperm to be used in artificial insemination. Aquac Res 44(8):1254–1258. https://doi.org/10.1111/j.1365-2109.2012.03126.x

Moss S, Moss D, Arce S, Lightner D, Lotz J (2012) The role of selective breeding and biosecurity in the prevention of disease in penaeid shrimp aquaculture. J Invertebr Pathol 110(2):247–250. https://doi.org/10.1016/j.jip.2012.01.013

Muchlisin ZA (2005) Current status of extenders and cryoprotectants on fish spermatozoa cryopreservation. Biodiversitas 6(1):66–69

Narasimman S, Krishnamoorthy D, Buelah CD, Natesan M (2018) Retrieval and cryopreservation of sperm in spermatophores from cadaveric Indian white shrimp, Fenneropenaeus indicus (H. Milne Edwards, 1837). Anim Reprod Sci 192:185–192. https://doi.org/10.1016/j.anireprosci.2018.03.008

Niksirat H, Kouba A, Psenicka M, Kuklina I, Kozak P (2013) Ultrastructure of spermatozoa from three genera of crayfish Orconectes, Procambarus and Astacus (Decapoda: Astacoidea): new findings and comparisons. Zool Anz 252(2):226–233. https://doi.org/10.1016/j.jcz.2012.06.002

Nimrat S, Bart AN, Keatsaksit A, Vuthiphandchai V (2008) Microbial flora of spermatophores from black tiger shrimp (Penaeus monodon) declines over long-term cryostorage. Aquaculture 274(2–4):247–253. https://doi.org/10.1016/j.aquaculture.2007.11.018

Nimrat S, Noppakun S, Sripuak K, Boonthai T, Vuthiphandchai V (2020) Cryopreservation of banana shrimp (Fenneropenaeus merguiensis) spermatophores with supplementation of medicinal plant extracts: development of a programmable controlled-rate method and a practical method. Aquaculture 515:734537. https://doi.org/10.1016/j.aquaculture.2019.734537

Nor Fatihah S, Muhd-Farouk H, Amin-Safwan A, Hafiz Mahsol H, Ikhwanuddin M (2017) Histological characteristics on the testes of mud spiny lobster, Panulirus polyphagus (Herbst, 1793). Pak J Biol Sci 20(7):365–371. https://doi.org/10.3923/pjbs.2017.365.371

O'Reilly P, Doyle R (2007) Live gene banking of endangered populations of Atlantic Salmon. In: Verspoor E, Stradmeyer L, Nielsen J (eds) The Atlantic salmon: genetics, conservation and management. Blackwell Publishing Ltd, Hoboken, pp 425–469

Pacey AA, Tomlinson MJ (2009) The history of sperm cryopreservation. In: Sperm banking: theory and practice. Cambridge University Press, Cambridge

Paniagua Chávez C, Morales Ueno K (2019) Sistema de Manejo de muestras biológicas de alta viscosidad. Mexico Patent MX369254

Park C, Chapman FA (2005) An extender solution for the short-term storage of sturgeon semen. N Am J Aquac 67:52–57

Pattira P, Rapeepan V, Jittipan C, Pornthep P, Boonyarath P, Prasert S, Boonsirm W (2007) Acrosome reaction in the sperm of the black tiger shrimp Penaeus monodon (Decapoda, Penaeidae). Aquac Res 38. https://doi.org/10.1111/j.1365-2109.2007.01824.x

Pegg DE (2007) Principles of cryopreservation. In: Cryopreservation and freeze-drying protocols. Humana Press Inc., Totowa, pp 39–57

Peixoto S, Wasielesky W, Cavalli RO (2011) Broodstock maturation and reproduction of the indigenous pink shrimp Farfantepenaeus paulensis in Brazil: an updated review on research and development. Aquaculture 315(1-2):9–15. https://doi.org/10.1016/j.aquaculture.2010.04.009

Peralta Martínez MÁ, Martínez Morales I, Unzueta Bustamante ML, Montaldo HH, Castillo-Juárez H (2013) Morfología del sistema reproductor y del espermatóforo de Litopenaeus vannamei, camarón blanco del Pacífico. Hidrobiologica 23(3):313–319

Pitnick S, Hosken DJ, Birkhead TR (2009) Sperm morphological diversity. In: Tim RB, David JH, Scott Pitnick A, Birkhead DJH TR, Scott P (eds) Sperm biology. Academic Press, London, pp 69–149. https://doi.org/10.1016/b978-0-12-372568-4.00003-3

Pongtippatee P, Vanichviriyakit R, Chavadej J, Plodpai P, Pratoomchart B, Sobhon P, Withyachumnarnkul B (2007) Acrosome reaction in the sperm of the black tiger shrimp Penaeus monodon (Decapoda, Penaeidae). Aquac Res 38(15):1635–1644. https://doi.org/10.1111/j.1365-2109.2007.01824.x

Pullin R, White P (2011) Climate change and aquatic genetic resources for food and agriculture: state of knowledge, risks and opportunities. FAO, Rome

Robertson DN, Butler MJ (2013) Mate choice and sperm limitation in the spotted spiny lobster, Panulirus guttatus. Mar Biol Res 9(1):69–76. https://doi.org/10.1080/17451000.2012.727429

Rodrigues JP, Paraguassú-Braga FH, Carvalho L, Abdelhay E, Bouzas LF, Porto LC (2008) Evaluation of trehalose and sucrose as cryoprotectants for hematopoietic stem cells of umbilical cord blood. Cryobiology 56(2):144–151. https://doi.org/10.1016/j.cryobiol.2008.01.003

Rudolph AS, Crowe JH (1985) Membrane stabilization during freezing: the role of two natural cryoprotectants, trehalose and proline. Cryobiology 22(4):367–377. https://doi.org/10.1016/0011-2240(85)90184-1

Sal Moyano MP, Gavio MA, Cuartas EI (2009) Morphology and function of the reproductive tract of the spider crab Libinia spinosa (Crustacea, Brachyura, Majoidea): pattern of sperm storage. Helgol Mar Res 64(3):213–221. https://doi.org/10.1007/s10152-009-0180-9

Salazar M, Lezcano M, Granja C (2009) Protocol for cryopreservation of *Penaeus vannamei* sperm cells. In: Cabrita E, Robles V, Herráez P (eds) Methods in reproductive aquaculture: marine and freshwater species, vol Marine biology, 1st edn. CRC Press, Boca Ratón, p 505

Sandifer PA, Lawrence AL, Harris SG, Chamberlain GW, Stokes AD, Bray WA (1984) Electrical stimulation of spermatophore expulsion in marine shrimp, Penaeus spp. Aquaculture 41 (2):181–187. https://doi.org/10.1016/0044-8486(84)90096-6

Saragusty J, Arav A (2011) Current progress in oocyte and embryo cryopreservation by slow freezing and vitrification. Reproduction 141(1):1–19. https://doi.org/10.1530/REP-10-0236

Shaw PW, Bernard AG, Fuller BJ, Hunter JH, Shaw RW (1992) Vitrification of mouse oocytes using short cryoprotectant exposure: effects of varying exposure times on survival. Mol Reprod Dev 33(2):210–214. https://doi.org/10.1002/mrd.1080330214

Simeo CG, Ribes E, Rotllant G (2009) Internal anatomy and ultrastructure of the male reproductive system of the spider crab Maja brachydactyla (Decapoda: Brachyura). Tissue Cell 41 (5):345–361. https://doi.org/10.1016/j.tice.2009.02.002

Sinha B, Köster D, Ruez R, Gonnord P, Bastiani M, Abankwa D, Stan RV, Butler-Browne G, Vedie B, Johannes L, Morone N, Parton RG, Raposo G, Sens P, Lamaze C, Nassoy P (2011) Cells respond to mechanical stress by rapid disassembly of Caveolae. Cell 144(3):402–413. https://doi.org/10.1016/j.cell.2010.12.031

Stentiford GD, Neil DM, Peeler EJ, Shields JD, Small HJ, Flegel TW, Vlak JM, Jones B, Morado F, Moss S, Lotz J, Bartholomay L, Behringer DC, Hauton C, Lightner DV (2012) Disease will limit future food supply from the global crustacean fishery and aquaculture sectors. J Invertebr Pathol 110(2):141–157. https://doi.org/10.1016/j.jip.2012.03.013

Subramoniam T (1993) Spermatophores and sperm transfer in marine crustaceans. In: Blaxter JHS, Southward AJ (eds) Advances in marine biology, vol 29. Academic, pp 129–214. https://doi.org/10.1016/S0065-2881(08)60131-5

Subramoniam T (1995) Light and electron microscopic studies on the seminal secretions and the vas deferens of the penaeiodean shrimp, *Sicyonia ingentis*. J Biosci (Penang) 20(5):691–706

Succu S, Leoni GG, Bebbere D, Berlinguer F, Mossa F, Bogliolo L, Madeddu M, Ledda S, Naitana S (2007) Vitrification devices affect structural and molecular status of in vitro matured ovine oocytes. Mol Reprod Dev 74(10):1337–1344. https://doi.org/10.1002/mrd.20693

Tieleman DP, Leontiadou H, Mark AE, Marrink S-J (2003) Simulation of pore formation in lipid bilayers by mechanical stress and electric fields. J Am Chem Soc 125(21):6382–6383. https://doi.org/10.1021/ja029504i

Tiersch TR (2011) Process pathways for cryopreservation research, application and commercialization. In: Cryopreservation in aquatic species, 2nd edn. World Aquaculture Society, Baton Rouge, pp 646–671

Tiersch TR, Green CC (eds) (2011) Cryopreservation in aquatic species, 2nd edn. World Aquaculture Society, Baton Rouge

Tran L, Nunan L, Redman R, Mohney L, Pantoja C, Fitzsimmons K, Lightner D (2013) Determination of the infectious nature of the agent of acute hepatopancreatic necrosis syndrome affecting penaeid shrimp. Dis Aquat Org 105(1):45–55. https://doi.org/10.3354/dao02621

Tucker MJ, Liebermann J (2007) Vitrification in assisted reproduction: a user's manual and troubleshooting guide. Reproductive medicine and assisted reproductive techniques series. CRC Press

Uberti MF, Vieira FN, Salência HR, Vieira GS, Vinatea LA (2014) Assessment of viability of sperm cells of Litopenaeus vannamei on cryopreservation. Braz Arch Biol Technol 57 (3):374–380. https://doi.org/10.1590/s1516-89132014005000013

Ulate K, Alfaro-Montoya J (2010) Sperm–egg incompatibility in inter-specific inseminations of Penaeus (Litopenaeus) occidentalis, P. (Litopenaeus) stylirostris and P. (Litopenaeus) vannamei. Aquaculture 309(1–4):290–292. https://doi.org/10.1016/j.aquaculture.2010.08.029

Vajta G, Nagy ZP (2006) Are programmable freezers still needed in the embryo laboratory? Review on vitrification. Reprod Biomed Online 12(6):779–796. https://doi.org/10.1016/S1472-6483(10)61091-7

Valentina Claudet P, Narasimman S, Natesan M (2016) Effect of cryoprotectants and cooling rates on fertility potential of sperm in the giant freshwater prawn, Macrobrachium rosenbergii (De Man). Anim Reprod Sci 171:49–57. https://doi.org/10.1016/j.anireprosci.2016.05.013

Vanichviriyakit R, Kruevaisayawan H, Weerachatyanukul W, Tawipreeda P, Withyachumnarnkul B, Pratoomchat B, Chavadej J, Sobhon P (2004) Molecular modification of Penaeus monodon sperm in female thelycum and its consequent responses. Mol Reprod Dev 69(3):356–363. https://doi.org/10.1002/mrd.20138

Vuthiphandchai V, Nimrat S, Kotcharat S, Bart AN (2007) Development of a cryopreservation protocol for long-term storage of black tiger shrimp (Penaeus monodon) spermatophores. Theriogenology 68(8):1192–1199. https://doi.org/10.1016/j.theriogenology.2007.08.024

Wallis HC, Maurice GK, Ashley IY (1981) An acrosome reaction in natantian sperm. J Exp Zool 218. https://doi.org/10.1002/jez.1402180220

Wang Q, Misamore M, Jiang CQ, Browdy CL (1995) Egg water induced reaction and biostain assay of sperm from marine shrimp Penaeus vannamei: dietary effects on sperm quality. J World Aquacult Soc 26(3):261–271. https://doi.org/10.1111/j.1749-7345.1995.tb00254.x

Wang Y-L, Sun W-J, He L, Li Q, Wang Q (2015) Morphological alterations of all stages of spermatogenesis and acrosome reaction in Chinese mitten crab Eriocheir sinensis. Cell Tissue Res 360(2):401–412. https://doi.org/10.1007/s00441-014-2092-5

Wayman WR, Figiel CR Jr, Tiersch TR (2011) Simple extenders for refrigerated storage and cryopreservation of channel catfish sperm. In: Tiersch TR, Green CC (eds) Cryopreservation in aquatic species, 2nd edn. World Aquaculture Society, Baton Rouge, p 692

WorldBank (2013) FISH TO 2030: prospects for fisheries and aquaculture. International Bank for Reconstruction and Development, Washington DC

Yano I, Kanna RA, Oyama RN, Wyban JA (1988) Mating behaviour in the penaeid shrimp Penaeus vannamei. Mar Biol 97(2):171–175. https://doi.org/10.1007/BF00391299

Zhang Z, Cheng H, Wang Y, Wang S, Xie F, Li S (2010) Acrosome reaction of sperm in the mud crab Scylla serrata as a sensitive toxicity test for metal exposures. Arch Environ Contam Toxicol 58(1):96–104. https://doi.org/10.1007/s00244-009-9326-7

Zhao Y, Yang C, Chen X, Peng M, Chen X, Zeng D (2017) Effects of cryopreservation on ultrastructural morphology of white shrimp (Litopenaeus Vannamei) sperm. CryoLetters 38 (5):357–363

State of the Art in Cryopreservation of Bivalve Spermatozoa

11

Demoy-Schneider Marina, Schmitt Nelly, Suquet Marc, and Cosson Jacky

Abstract

In the current context of decreasing, shrinking or disappearing of ecosystems, species, wild populations, local plant and animal varieties and local breeds of domestic animals, cryo-conservation is a powerful way to safeguard the highly endangered biodiversity. In this chapter, we review the present knowledge on the reproductive biology of different bivalve species and on the characteristics and quality of their spermatozoa, including their structure, motility, and energy metabolism. Moreover, we summarized the various ways to collect sperm prior to cryo-conservation and the different methodologies used for conservation of bivalve sperm: cold storage and the key steps of the cryo-conservation techniques commonly used to preserve the ability to move of the spermatozoa of these species, such as cryoprotectant selection and equilibration, packing of samples for freezing, cooling rate and freezing, and finally thawing. Taking in account the quality of sperm prior to freezing and understanding of the successive steps involved in the cryo-conservation process constitute an unescapable way to improve the success of this valuable technique for oysters and marine mollusk species sperm long-term preservation.

Keywords

Cryopreservation · Scallop · Oyster · Clam · Spermatozoa

D.-S. Marina (✉) · S. Nelly
UMR EIO 241, Université de la Polynésie Française, Tahiti, French Polynesia
e-mail: marina.demoy-schneider@upf.pf

S. Marc
Ifremer, UMR 6539 Lemar, Site expérimental d'Argenton, Argenton en Landunvez, France

C. Jacky
Faculty of Fisheries and Protection of Waters, Research Institute of Fish Culture and Hydrobiology, University of South Bohemia in České Budějovice, Vodňany, Czech Republic

J. Betsy, S. Kumar (eds.), *Cryopreservation of Fish Gametes*,
https://doi.org/10.1007/978-981-15-4025-7_11

Abbreviations

ATP	Adenosine triphosphate
BLPO	Black-lip pearl oyster
DMSO	Dimethyl sulfoxide
EG	Ethylene glycol
EGTA	Ethylene glycol-bis(β-aminoethyl ether)-N,N,N′,N′-tetraacetic acid
EO	European oyster
GS	Great scallop
JS	Japanese scallop
LN	Liquid nitrogen
MC	Manila clam
OP	Osmotic pressure
OXPHOS	Oxidative phosphorylation
PEG	Polyethylene glycol
PO	Pacific oyster
SW	Sea water
VAP	Velocity of the average path

11.1 Introduction

Cryopreservation allows an optimal management of the commercial and biological risks that are associated with selective breeding: for example, economic and market changes in desirable traits or animal loss due to disease. In hatcheries, cryopreservation of oocytes, sperm, or larvae would enable spat production year-round without the animals conditioning constraint needed for out-of-season production (Adams et al. 2011).

Beyond the benefits of the gamete cryopreservation reported by Suquet et al. (2000) in case of fish: synchronization of gamete availability of both sexes, use of the total volume of available semen, simplification of broodstock maintenance, transportation of gamete, prevention of sperm aging, experimental programs for genetic studies, conservation of genetic variability in domesticated populations and in the current context of decreasing; shrinking; or disappearance in ecosystems; species; wild populations; local plant; and animal varieties and local breeds of domestic animals (IPBES 2019), this technique is a powerful way to safeguard the highly endangered biodiversity. In the past years, this technique has been improved in case of many marine mammals, fish, and bivalve species which constitute more than 30% among one million species threatened with extinction. In this field, there is still moderate knowledge about bivalve sperm manipulation techniques compared to the situation in fish. This book chapter intends to describe in detail the present awareness on the biological features of different bivalve species and on their spermatozoa characteristics, including their structure, motility, energy metabolism, and the cryo-conservation techniques commonly used to preserve these species.

11.2 Structure and Composition of Sperm in Bivalve Species

11.2.1 Presentation of the Main Species Used in Aquaculture

11.2.1.1 Pacific Oyster: *Crassostrea gigas*

Distribution: *C. gigas* (Thunberg 1793) (Fig. 11.1a) is a mollusk valued for its meat and produced by aquaculture in many countries. This bivalve is native from the northwest of Pacific Ocean but because it has a good ecological adaptability, it has been widely distributed around the world (Fig. 11.1b). It was introduced first in Europe during the 1960s for aquaculture purpose (Drinkwaard 1999). Nowadays, *C. gigas* can be considered, besides its economic interest, as an invasive species

Fig. 11.1 (**a**) General morphology of the Pacific oyster (*C. gigas*): on left, after upper shell removal with the male gonad visible at the tip of the metal needle; on right, the PO in its natural environment. (**b**) *C. gigas* producing countries (from FAO 2003a, Fisheries and Aquaculture Department). (**c**) *C. gigas* in the spawning process with a cloud of sperm emitted from the gonopore (from Demoy-Schneider et al. 2014)

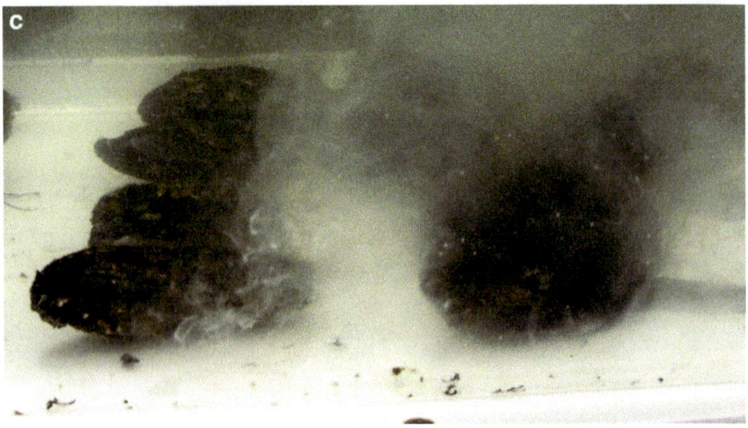

Fig. 11.1 (continued)

(Anglès d'Auriac et al. 2017). Currently this species is spreading widely in northern Europe and in the rest of the world. This species is the most widely farmed aquaculture species worldwide.

Production: The larvae are picked up after spawning with collectors on which they attach themselves. They will stay on until the age of 18 months. Then breeding will start using four techniques: suspension under a table, flat on the foreshore, in deep water, and raising in pockets on tables. The refinement process of oysters improves their organoleptic qualities, weight, and color. This stage occurs in ripening pools called "claires." The oyster diet is then enriched with microalgae (Van Houcke et al. 2017). Worldwide production (Fig 11.1b) was estimated in 2016 by FAO at 4864 tons (FAO 2018).

Tolerance to environmental factors: *C. gigas* installation in a seawater area is dependent on a temperature range. It is the main parameter which affects growth, development, and larval survey. It is known to have a high-temperature tolerance which explains its wide geographic location (Clark et al. 2013). Gametes are released between 16 and 30 °C (Ruiz et al. 1992). Adult oysters can survive between 1.8 and 35 °C. Optimal range salinity is between 20 and 25%, but can tolerate salinity under 10% and upon 35% (FAO 2008).

Reproduction: *C. gigas* is a dioecy protandrous hermaphrodite. In this species, sex is genetically determined and may be influenced by environmental factors (Guo et al. 1998). Shpigel (1989) reported a sexual asynchrony in maturation with the males becoming ripe earlier than the females. Sexual maturity is usually reached between 12 and 18 months. Male and female gametes might be released in seawater at the same time (Fig. 11.1c) (Dinamani 1987).

11.2.1.2 Black-Lip Pearl Oyster: *Pinctada margaritifera*

Distribution: *P. margaritifera* (Fig. 11.2a) is a benthic bivalve widely ranged from the Gulf of California, Mexico to the Eastern Mediterranean Sea, and is largely distributed in the Indo-Pacific region, particularly abundant in the South Pacific, New Guinea, Hawaiian Islands, and Polynesia (Fig. 11.2b) (Wada and Temkin 2008; Le Pennec et al. 2010; Van Dyk 2011). Under natural conditions, this species lives near the bottom in coral reef waters, while cultivated pearl oysters are hung from down lines suspended on subsurface long lines, ~10 m from the surface. Its economic value is due to the pearl production. But the first interest in the past, especially in French Polynesia, has been the shaping of button from the shell mother of pearl. Then in the 1960s pearl farms began to develop in the Tuamotu atolls. (Le Pennec et al. 2010).

Production: The black-lip pearl oyster culture for pearl production is widespread in the islands of French Polynesia, the Cook, the Marshall, and Salmon islands and

Fig. 11.2 (**a**) General morphology of the black-lip pearl oyster (*P. margaritifera*): on left after upper shell removal with the gonad visible in the middle at the top; on right, the BLPO after collection from its natural environment. (**b**) World distribution of *P. margaritifera* (from Van Dyk 2011)

also in southern China, northern and western Australia, in Seychelles and Sudan. Farming of *P. margaritifera* started in the early 1960s and presently it represents the main aquacultural activity in the lagoons of French Polynesia. This activity plays a major economic role, thanks to the production of black pearls, the largest export industry in this Polynesian area (Soyez et al. 2011). However, the production relies almost exclusively on the collection of wild spat of the black-lip pearl oyster, *P. margaritifera*, which makes the aquaculture and marketing highly dependent on natural resources. Hatchery practice was only developed in experimental pearl farms (Hui et al. 2011). Harvesting of juveniles is carried out on collectors in the lagoon. Two breeding techniques are used in pearl farms: platform and line or leash, the latter being the most widely used. The whole breeding process takes place in the lagoon.

Tolerance to environmental factors: Wild black-lip pearl oysters live mainly in lagoons with maximum abundance at 20–30 m depth. Temperatures allowing growth and reproduction are between 26 and 30 °C (Pouvreau and Prasil 2001; Joubert et al. 2014). The salinity must be for good living conditions between 28 and 32% (Douroudi et al. 2001).

Reproduction: *P. margaritifera* is a protandrous hermaphrodite (Tranter 1958). This species reaches sexual maturity at the end of the first year (length around 40 mm) (Pouvreau et al. 2000a, b). This species is sexually male during the first 2 years of life and later becomes female over several years to reach a sex ratio close to 1:1 in populations older than 8 years (Thielley et al. 1993). Male and female gametes may be released in seawater at the same time. Oocytes are activated prior spawning (Tranter 1958). Spermatozoa motility is activated at spawning when they get in contact with seawater, usually due to alkaline pH (Demoy-Schneider et al. 2012). Reproduction occurs at different period around the world depending of the seasons. Spawning period occurs during the hot season, with a rising of seawater temperature in country with seasonality and year-round in French Polynesia where the seawater temperature does not vary outside of the possible limits (Hwang 2007).

11.2.1.3 European Flat Oyster: *Ostrea edulis*

Distribution: The European flat oyster (EO), *O. edulis* (Fig. 11.3a) is native in Europe. It has been imported to North America in the 1940s and 1950s. Nowadays, they are scattered from the North Sea to the Mediterranean and Black Sea (Fig. 11.3b) (Jaziri 1990).

Production: *O. edulis* is consumed by human since a long period of time. It is a native species that has been overexploited in the past and is nowadays classified as an endangered species in Europe. The production of this species declined drastically in the 1960s due to the impact of parasitic epizooites. In 2010, the world production was about 4000 tons (Helm et al. 2004; FAO 2012). Juveniles are obtained by collecting wild born or hatchery produced specimen. The oyster farmers use two breeding techniques: heightening where oysters are kept flottered by various techniques (rafts, dies, plastic baskets...) and flat: oyster spat is directly sown by boat on a sub tidal bottom at a density of 50–100 kg/ha. Oysters are harvested at their

Fig. 11.3 (**a**) *O. edulis* (from Pouvreau 2017). (**b**) Distribution of *O. edulis.* (from Jaziri 1990)

young stage (2 years) to be marketed to avoid mortality. Currently, oysters may survive 3–4 years because of Bonamiosis and Marteilia disease (FAO 2008). The EO is now considered as a threatened species by OSPAR commission whose protection must be increased by several approaches including the study of its reproduction success (Haelters and Kerckhof 2009).

Tolerance to environmental factors: It often occurs in large beds on muddy sand, muddy gravel and rocks. Oysters filter phytoplankton and other particulate material present in seawater. *O. edulis* can be found in estuaries and tolerates salinity values from 15 to 23% and even 35% in Brittany. This species survives temperature ranging from −2 to 32 °C (Lapègue et al. 2006). A major problem for this species is its sensitivity to Bonamiosis. Current research is presently oriented towards the search of strains genetically resisting this disease (FAO 2008).

Reproduction: *O. edulis* is a protandric hermaphrodite, changing sex generally twice in a single reproductive season. Oysters function first as males early in the spawning season and later change to females before changing to males again. The particularity of this oyster is that eggs are fertilized in the pallial cavity by external spermatozoa and then incubated during 8–10 days before being dropped in seawater (Lapègue et al. 2006).

11.2.1.4 Manila Clam: *Ruditapes philippinarum*

Distribution: The Manila clam (MC) (Fig. 11.4a) was first located in the Indo-Pacific region. This is a subtropical to low boreal species of the western Pacific and distributed in temperate areas in Europe. The natural populations are distributed in the Philippines, the Southern and Eastern China Seas, Yellow Sea, Sea of Japan, and around the Southern Kuril Islands (Fig. 11.4b) (Scarlato 1981; Goulletquer and Héral 1997). Nowadays, the distribution of *R. philippinarum* is broad, it is found for example in Ireland where it has been introduced in 1982 for aquaculture (Britton 1991; Humphreys et al. 2015). This species is characterized by a sediment dwelling (Huo et al. 2017).

Production: This mollusk species has a considerable economic value so its production is widely represented. The worldwide production was of 4229 tons in 2016 (FAO 2018). China is the largest producer (Goulletquer 1997). Juveniles are collected as wild- or hatchery-born. Shell culture is based on intertidal on-bottom culture. The culture can initiate when shells reach 2–3 mm size. Clams may take 2–4 years to reach marketable size (FAO 2018).

Tolerance to environmental factors: Temperature range survival and reproduction is between 12 and 25 °C. Gonad activity is possible down from 8 °C but gametes maturation does not occur below 12 °C (Laing et al. 1987; Laing and Utting 1994). This species tolerates salinity down to 15% so it is able to extend into estuaries (Jensen et al. 2005). Clams are susceptible to different diseases like viruses, bacteria, rickettsia, and protozoan parasites (FAO 2018).

Reproduction: Clams are sequential hermaphrodites, either male or female with a sexual distribution of approximately 1:1 (Jones et al. 1993). Lee et al. reported a sex reversal rate of 19%. The male to female sex reversal ratio was 21.1%, which was significantly higher than the female to male sex reversal ratio of 12.9% (Lee et al. 2013). Temperature is the most significant factor that influenced gametogenetic cycles and spawning. 14 °C is the lower temperature limit for spawning which usually occurs between 20 and 25 °C. There are generally two spawning

Fig. 11.4 (**a**) A spawning female Manila clam (from Laing et al. 2004). (**b**) Main producer countries of *R. philippinarum* (Goulletquer 2015)

periods in the same reproductive season separated by 2–3 months of rest (Mann 1979).

11.2.1.5 Great Scallop: *Pecten maximus*

Distribution: *P. maximus* (Fig. 11.5a) is distributed along shores of northern Norway to North Africa (Fig. 11.5b) (Beaumont and Gjedrem 2007). Common names of this species are "great scallop" or "king scallop"; it usually lives on sand and gravel bottom but can be found in mud as well. Scallops have specific habitat preferences, living according to aggregated distributions and are quite motionless; therefore, they are easily detected and collected for commercial fisheries or recreational purposes (Brand 1991).

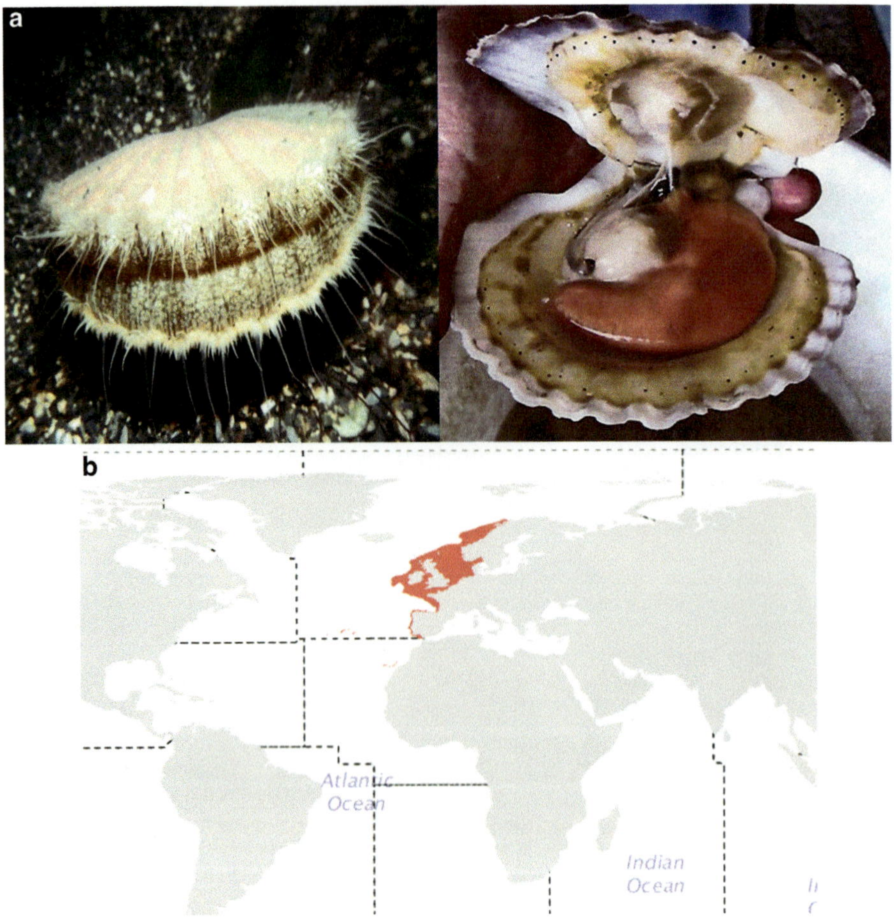

Fig. 11.5 (**a**) Great scallop *P. maximus*: on left, whole animal (from Howarth and Stewart 2014); on right, after opening of the two shells, visualization of the pinkish gonad. (**b**) *P. maximus* producing countries (from FAO 2003b, Fisheries and Aquaculture Department)

Production: The trading of this species relies partly on the natural stocks harvesting and as a complement, on the aquaculture. In 2013, about 65,632 tons were marketed (Brugère and De Young 2015). The main harvest method uses dredging on natural beds with sea ranching supplementation implemented in specific locations to complement the natural production, as mainly occurring in the Bay of Brest, France. In this country, this species is protected by severe collection regulations: aquaculture represents an alternative way to protect wild stocks. Aquaculture of this species requires the collection of wild spat, using bags filled with suitable "cultch" that are suspended in the water column. An alternative way is the use of scallop hatcheries production, with the advantage of genitor selection and application of genetic manipulation. The growing stage management of scallops makes use of two systems: either hanging or bottom culture (Beaumont and Gjedrem 2007).

Tolerance to environmental factors: Living conditions of *P. maximus* are adapted to temperate and cold regions. The natural habitat is subtidal where temperature, salinity, and food levels are relatively constant (Brand 1991).

Reproduction: *P. maximus* is a permanent functional hermaphrodite. Reproductive maturity is reached at 2–3 years. The breeding season is from April to September. Reproductive location is in water column. Fully developed male or female gonads contain sex cells which are not spawned simultaneously when mature, thus the risk of self-fertilization is excluded (Barnabé 2003; Spencer 2008). Artificial reproduction is poorly controlled in hatcheries (Andersen et al. 2011).

11.2.1.6 Japanese Oyster: *Patinopecten yessoensis*

Distribution: Alternative names for this mollusk are Japanese scallop (JS) or Yezo scallop (Fig. 11.6a). It is an economically important shellfish species worldwide. *P. yessoensis* is a cold-water (4–8 °C) species widely distributed in the northwest part of the Pacific Ocean (Fig. 11.6b) (FAO 2006).

Production: China and Japan are the major producers with 1.1 million tons in 2003. The production relies almost entirely on wild spat collection and growth in suspended or bottom-sowing culture systems. Adult scallops are harvested when the shell reaches about 100 mm size, after 2 or 3 years of breeding (FAO 2006).

Tolerance to environmental factors: Optimal temperature for growth is 4–8 °C but the tolerance range extends from −2 to 26 °C. JS can survive for 16 h in anoxic conditions and at least 2 days in hypoxic conditions with oxygen saturation of 20%. Scallops died within 24 h if the salinity decreased to 18 PSU or less and within a few hours if the salinity dropped to 10 PSU or less. Regular reproduction occurs at salinities ranging 32–34% (Chang et al. 1985; Sakurai et al. 2000).

Reproduction: Even if *P. yessoensis* is a protandric hermaphrodite species (Yamamoto 1964), the sex ratio varies between populations. However, this species is not strict protandric because some young individuals are hermaphrodite and then, some scallops become females without going through a male stage. Moreover, some specimen can be transiently both male and female within a large age range. The age of sex switch depends on the age structure of the population and is also socially controlled: males change sex earlier in the presence of younger males, which leads to support successful egg fertilization (Silina 2018). The sexual maturity is reached at 2 years age. From this age, scallops become dioecious, with a low incidence of hermaphrodites. The spawning season starts in March when water temperature is close to 7–12 °C and ends in May. Fertilization occurs in sea water where sperms and eggs are released (Bregman 1979; Chang et al. 1985; Yamamoto 1964).

Fig. 11.6 (**a**) The Japanese scallop, *P. yessoensis* (photo credit Harum Koh). (**b**) *P. yessoensis* producing countries (FAO 2003c, Fisheries and Aquaculture Department)

11.2.2 Sperm Morphology and Structure

11.2.2.1 Sperm Morphology

In the studied species, there are some variations in the sperm morphology.

C. gigas: Pacific oyster (GS) spermatozoon head is 2–2.6 μm long and 2.3–2.5 μm wide. The intermediate piece is very short and the flagellum is 45 μm long and 0.3 μm wide. The shape of the head is about spherical with a diameter of 2.2 ± 1.0 μm (Faure 1996; Boulais et al. 2019).

P. margaritifera: In the BLPO spermatozoon, the head is pear-shaped measuring 1.7 μm width (Le Pennec et al. 2010). According to live images obtained by phase contrast microscopy, the total length of this cell is 49 ± 2 μm, among which, the flagellar length represents 47 μm. An additional thin terminal filament, about 3 μm long, was observed in some spermatozoa (Demoy-Schneider et al. 2012).

O. edulis: The EO spermatozoa are not free but gathered in particular aggregates called "spermatozeugmata," acellular structures with an average diameter of 60 μm (O'Foighil 1989; Suquet et al. 2018) in which the spermatozoa heads are embedded. In these structures, each sperm flagellum moves according to gentle asynchronous beats but remain immotile until their release into seawater (Suquet et al. 2018).

R. philippinarum: In MC, the morphology of the mature spermatozoon is of the primitive type and is somewhat different from those of other bivalves with an elongated pyramid-like head. The spermatozoon is approximately 48–51 μm long. The head is about 6.2 μm diameter, including an elongated and curved nucleus (about 3.4 μm) and a long acrosome (about 2.4 μm long) (Dai et al. 2004; Kim et al. 2013).

P. maximus: The great scallop (GS) spermatozoon is 45–50 μm long. It can be considered as a primitive type with a pear-shape head overlapped with a short conical acrosome deprived of axial rod. The plasma membrane looks slightly corrugated. The flagellum length ranges 45 μm including a thinner part of 6 μm length (Dorange and Le Pennec 1989; Suquet et al. 2016).

P. yessoensis: In the JS, the mature spermatozoon is composed of a cone-shaped acrosome, an elongated nucleus, a short midpiece, and a single flagellum. The head is 3.9 μm long and the flagellum has a total length of 50.2 μm. Fresh sperm appears covered by smooth plasma membrane (Li et al. 2000a, b).

11.2.2.2 Fine Structure

In bivalves, the male gamete consists of a head, a midpiece, and a flagellum (Fig. 11.7) (Sastry 1979; Mackie 1984; Treen et al. 2012). The head contains the nucleus with DNA material and an acrosome located at the anterior head region. The acrosome is a Golgi-derived secretory vesicle that contains enzymes involved in lysing the oocyte membrane during fertilization. Mitochondria and centrioles are located in the sperm midpiece. Generally four mitochondria produce energy required

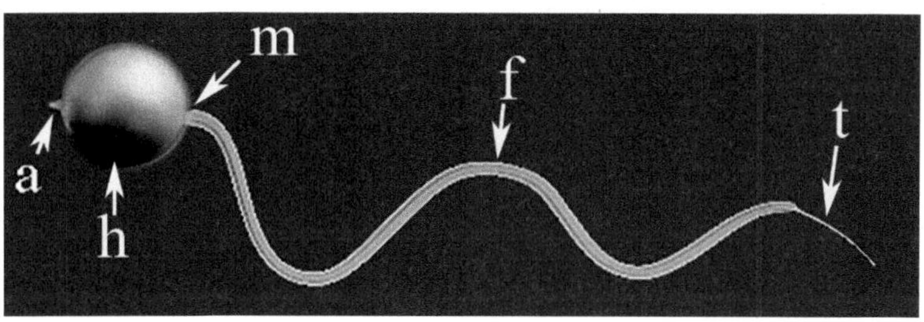

Fig. 11.7 Three-dimensional schematic drawing of a typical bivalve spermatozoon. *a* acrosome, *h* head of spherical shape in most oyster species, *m* mid-piece including mitochondria and centriolar complex, *f* flagellum containing the motor part called axoneme, *t* terminal filament. Head and tail are not at scale

for the motility apparatus of spermatozoa, also called axoneme. The distal centriole forms the basal body of the axoneme with a structure of 9 + 2 microtubules surrounded by a plasma membrane as described in Bondarenko and Cosson (2019). Dynein arms and radial spokes are involved in spermatozoa motility. Fueled by ATP hydrolysis, dynein arms generate force and alternative movement of microtubules, regulating flagellar waveform and velocity (Morse and Zardus 1997; Dong et al. 2005a, b, c). It is worth to note that axonemal structure is highly conserved through evolution (Inaba 2007; Drozdov et al. 2009). Spermatozoa motility is initiated at spawning after their release from the reproductive tract into the aquatic environment. Various chemical signals, including pH, ions, and cyclic nucleotides, are involved in controlling sperm motility in bivalves.

11.2.3 Seminal Plasma Characteristics

Several studies reported the ionic composition and pH of seminal fluid in some species of bivalves. Testicular fluid composition in Na^+, K^+, and Ca^{2+} was detailed and Na^+ concentration was found slightly lower in the studied species than in artificial seawater (ASW), whereas K^+ concentration was higher in bivalve testicular fluid than in ASW. Ca^{2+} concentration was similar in testicular fluid and ASW. At last, it was observed that pH in testicular fluid is acidic compared to pH in ASW, with a value of 8.2 (Faure et al. 1994; Alavi et al. 2014; Azeredo 2015; Boulais et al. 2018, 2019).

11.3 Collection of Bivalve Sperm

11.3.1 Spontaneous Spawning

In most oyster species, natural spawning depends on several environmental factors including temperature, salinity, currents, and food availability.

C. gigas: In the PO, spawning is triggered by factors such as temperature, phytoplanton bloom, or tides (Grangeré et al. 2009). Natural spawning occurs when seawater temperature rises from 22 to 26 °C, during the summer months, either in the morning or in the evening, during high tides (Shpigel 1989; Bernard et al. 2016). Moreover, Ubertini et al. (2017) reported that full and dark moon in combination with high temperature enhance spawning events. Oocytes are released (10–50 millions per female) in seawater and fertilization successfully happens after contact between spermatozoa and oocytes for at least a 10 min period at a minimal ratio of 500 spermatozoa per oocyte to be successful (Boulais et al. 2019).

P. margaritifera: In the black-lip pearl oyster, spawning may occur throughout the year with a peak during the warm season (December to April). Natural shedding is often linked to temperature or abrupt changes in some environmental conditions, including a reduction in salinity, food availability, changes in currents, calm seas, crowding, and other stressors such as handling and exposure to the air. Gametes are released into seawater where fertilization takes place (Thielley et al. 1993; Sims 1994).

O. edulis: In Europe, spawning of the EO occurs during summer and there may be some periodicity in spawning with peaks during full and new moon periods (Korringa 1947; Ubertini et al. 2017). Shpigel (1989) reported that, in Israel, the EO produces gametes from March to May when water temperature ranges from 18 to 22 °C, with a resting period during the summer months. Loosanoff (1962) showed that in the waters of Maine, active spring gametogenesis begins in may and general spawning begins approximately during the second or third week of July and continues until about the end of August. Males produce spermatozeugmata, where spermatozoa are gathered and after their liberation in seawater, spermatozeugmata dissociate and release free swimming spermatozoa which fertilize oocytes (up to one million eggs per spawning) into the female pallial cavity (O'Foighil 1989; Lapègue et al. 2006; Suquet et al. 2018).

R. philippinarum: For the MC, natural spawning is influenced by temperature with a minimum of 14 °C; the spawning season depends on the geographic location and occurs from May to October with a peak in September (Jones et al. 1993; Drummond et al. 2006). In Portugal, Moura et al. (2018) described an extensive spawning period from May until November–December. In Japan, MC has only one spawning period: from June to October with a peak period from mid-July to early-August (Abe et al. 2019). The exposure to the spawn of another clam triggers also natural spawning. The sperm is released into the water column from the exhalent siphon as a thick, cloudy, and white stream. The spawned eggs are fertilized externally to the parent animals (Jones et al. 1993).

P. maximus: In 2005, Slater reviewed spawning data on a more than 20 years period and reported that, in Europe, the GS major spawning periods occurred between April and August when temperature ranges 10–14 °C and lasts at this level until September. It has been described that spawning occurred during spring when water temperature was increasing and spawning in the fall occurred when water temperature was decreasing (Stanley 1967). Moreover, in 2006, Barber and Blake reported that various environmental factors such as a rapid change in temperature, disturbances as winds and tides, presence of gametes of the opposite sex, and lunar periodicity initiate spawning. They also mentioned that some neurosecretions such as serotonin and dopamine or similar influenced the synchronization of spawning in the GS population (2006). Le Pennec (2003) estimated that 15–21 million oocytes are released per each three-year old adult female emission. External fertilization occurs after gamete release into the water column (Mason 1983).

P. yessoensis: For the Japanese oyster, natural spawning occurs from late March to late May at seawater temperatures between 7 and 11 °C in Korea (Uddin et al. 2007) and fertilization is external (Meng et al. 2012). According to Chang et al. (1985), in Japan, spawning season varies from May to July depending on the localization and has been associated with a sudden increase of the seawater temperature (Yamamoto 1950, 1951a, b) but was not related with lunar phase (Yamamoto 1952).

11.3.2 Induced Spawning

Different techniques are used to induce bivalve spawning (Velasco et al. 2007). In 2013, Suquet et al. reported some of them such as gamete stripping, gonad scarification, thermal manipulation, addition of oocytes in seawater surrounding animals, intragonadal serotonin or prostaglandin injection, collection after sperm shedding. These spawning induction techniques are used in a variable way, depending on the authors, for different species.

11.3.2.1 Gametes Stripping

When gametes are fully mature, they can be stripped or removed: shedding can be obtained after manual mincing of gonad, using a Pasteur pipette or an automatic pipette fitted with a 200 μL empty tip and exerting a gentle suction as reported by Suquet et al. in the flat oyster *O. edulis* (2018) or by Demoy-Schneider et al. in *P. margaritifera* (2012, 2014). This technique was also used for the mussel *Mytilus edulis* (Everett et al. 2004; Jha et al. 2007) or the PO *C. gigas* (Cosson et al. 2008a; Suquet et al. 2010, 2012).

11.3.2.2 Gonad Scarification

Gonad scarification consists in the application of incisions to the gonad with a scalpel and collection of the exuding gametes. This alternative sperm collection procedure was used for the GS *P. maximus* (Suquet et al. 2013), but also for the EO *O. edulis* (Suquet et al. 2017, 2018). Right after the incision, sperm is collected using a micropipette fitted with a 200 μL tip. Ky et al. (2015) also obtained sperm using this technique in the case of the black-lip pearl oyster *P. margaritifera.*

11.3.2.3 Serotonin, Aminergic Neurohormones, and Prostaglandin Injection

Artificial induction of spawning has often been obtained using intragonadal serotonin (hydroxytryptamine; 5-HT) injection in many bivalve species: *P. yessoensis* (Matsutani and Nomura, 1982, 1987), *Argopecten irradians*, *Crassostrea virginica, Spisula solidissima*, (Gibbons and Castagna 1984, 1985; Hirai et al. 1988). In *Pecten albicans, Mercenaria mercenaria* and *Artica islandica*, injection of serotonin into the anterior adductor muscle also induced spawning (Gibbons et al. 1983; Gibbons and Castagna 1984). Falese et al. (2011) showed that male clams (*Nutricola confusa* and *N. tantilla*) released spermatozeugmata when exposed to fluvoxamine, a serotonin reuptake inhibitor.

11.3.2.4 Thermal Manipulation

Thermal stimulation is a key factor for spawning induction in many bivalve mollusk species and was described by Loosanoff and Davis in 1950. Indeed, this stimulation has been used efficiently for many years in different species and a rapid temperature increase induces sperm release in American oyster (*C. virginica*) (Galstoff 1940), in Caribbean scallops *Argopecten nucleus* and *Nodipecten nodosus* (Velasco et al. 2007). In the GS, or in the PO, artificial spawning can also be induced by application of a thermal shock (Devauchelle et al. 1994). For the tropical scallop, *Pecten ziczac*, spawning was induced using a rapid rise of the water temperature from 20 to 29 °C and Velez et al. (1990) reported that a thermal shock is also efficient in association with serotonin intragonadal injection for spermatozoa release but not for oocytes spawning. In the BLPO spawning is also induced using a thermal shock consisting in the exposure of animals to an overnight low water temperature (22 °C) followed by a transfer at high temperature (32–33 °C) where spawning occurs (Ky et al. 2015). For MC, for which best growth temperature is around 25 °C, spawning is induced by decreasing the temperature to 15–17 °C for 8 h in the shade (Zhang and Yan 2006). For the JS, Yamamoto (1964) reported that a rapid increase of the critical temperature of 8.0–8.5 °C induces the spawning.

11.3.2.5 Addition of Oocytes in Seawater

In *Ostrea virginica*, Galstoff observed the sperm release after the addition of oocytes in seawater surrounding animals (1964). Jones et al. reported that the addition of gonadal extract (eggs or sperm) in a 25–30 °C seawater will normally trigger spawning of the mature MC, *R. philippinarum*; if not, seawater temperature was concomitantly decreased to 10 °C prior to the addition of eggs or sperm (1993).

11.4 Spermatozoa Motility in Bivalves

In the testis, bivalve spermatozoa are immotile and unable to fertilize (Faure et al. 1993; Alavi et al. 2014). Among bivalve species, spermatozoa ability to initiate motility varies after dilution in SW; activated spermatozoa have various movement characteristics (Table 11.1). Motility parameters of spermatozoa can be automatically measured thanks to CASA systems (computer-assisted sperm analyzer) (Wilson-Leedy and Ingermann 2007).

11.4.1 Spermatozoa Motility in Normal Seawater

In the PO, motility of spermatozoa is initiated in SW up to 57% (Alavi et al. 2014) to 100% (Suquet et al. 2010), motility lasts for several hours up to 24 h (Suquet et al. 2010; Boulais et al. 2015). There is a time-dependent reduction of percentage of motile spermatozoa to 10% at 30 min (Alavi et al. 2014) or at 24 h (Suquet et al. 2010; Boulais et al. 2015). Dong et al. (2002) reported a high variability (39–82%) in the PO spermatozoa motility while Riesco et al. (2017) reported values of 45–85% in the Portuguese oyster, *Crassostrea angulata*, with movement lasting several days with a similar percentage of motile cells. After release from spermatozeugmata (sperm clumps) of the EO, spermatozoa are motile for about 10 min (Dai et al. 2004). Motility of 75% of BLPO spermatozoa is triggered in SW (Lyons et al. 2005). However, spermatozoa motility is not fully initiated in SW (Acosta-Salmón et al. 2007) with a high interindividual variation. Actually, spermatozoa motility is not immediately initiated in SW (Acosta-Salmón et al. 2007; Demoy-Schneider et al.

Table 11.1 Morphology and motility kinetics of spermatozoa in most studied bivalve species (adapted from Boulais et al. 2019)

Species	Black-lip pearl oyster	Pacific oyster	Japanese scallop	Manila clam
Head length (μm)		2–2.6	3.9	6.2
Head width (μm)		2.3–2.5		
Number of mitochondria	4	4	4	4
Flagellar length (μm)	47	37–41	50	42–45
Motility—SW (%)	0, 20–53	55–73	78	43
Motility—5-HT, Caffeine, $DCSB_4$ (%)	85	77	85	95
VCL—SW (μm s^{-1})			255	100
VSL—SW (μm s^{-1})	67			
VAP—SW (μm s^{-1})	166	60–70		
Waves velocity (μm s^{-1})	403			
Beat frequency (Hz)	49			
Wave amplitude (μm)	5.4			
Wave length (μm)	15.9			
Local curvature or bend angle	From −69 to +65			
Number of curvatures	3.1			
Diameter of circular head tracks (μm)	347			
ATP content (nmol 10^8 spz)	3.4	4.5, 18.6		

Black-lip pearl oyster, *P. margaritifera*; Japanese scallop, *P. yessoensis*; Manila clam, *R. philippinarum*; Pacific oyster, *C. gigas*

2018). Motility of 50–85% of spermatozoa was triggered after dilution in a saline solution (Demoy-Schneider et al. 2012, 2018).

When movement is triggered in SW, 6% of spermatozoa are motile in the Atlantic surf clam *S. solidissima* (Kadam and Koide 1990). In the MC, sperm motility lasts for up to 7 days (Alavi et al. 2014). At 1 min post dilution in SW, motility is triggered in 43% of spermatozoa but reaches 72% at 1 h. Following dilution in SW, motility of GS and JS spermatozoa is initiated in about 75% of cells, but decreases down to 45% at 30 min post-activation in the GS and to <10% at 1-day post-dilution in the JS (Faure et al. 1994; Alavi et al. 2014). Faure et al. (1994) reported a motility duration of 6 h after dilution in SW in GS.

11.4.2 Serotonin-Induced Spermatozoa by Serotonin (5-HT)

The involvement of serotonin (5-HT) in spawning regulation was reported in the JS (Matsutani and Nomura 1982), but not its metabolite 5-hydroxyindoleacetic acid (5-HIAA) (Kadam and Koide 1990) in the SC. In various bivalve species including the MC, JS, and PO, 5-HT is capable of inducing spermatozoa hypermotility while the percentage of motile spermatozoa increases after dilution of sperm in SW containing 5-HT: about 90% of MC spermatozoa become motile within 3–5 min post-activation in the presence of 5-HT (Alavi et al. 2014), 85% in the JS at 2 h post-activation and 40% after 1 day. Active concentration of 5-HT differs among species: 10–3, 5.10–6, 10–6, and 10–7 M in JS, SC, MC, and PO spermatozoa, respectively. Furthermore, 5-HT-induced spermatozoa hypermotility is time-dependent, probably mediated by a 5-HT receptor. In mammals (hamster and human) 5-HT induces hyper-motility (Fujinoki 2011; Jiménez-Trejo et al. 2012) a process known as “hyper-activation.” Furthermore, 5-HT hyperpolarizes membrane potential in 5-HT-induced hypermotile bivalve spermatozoa (unpublished). Spermatozoa velocity of the JS and MC is higher after motility activation with 5-HT than SW (Alavi et al. 2014), suggesting that 5-HT stimulates flagellar beating as reported in sperm of the blue mussel and SC (Gwo et al. 2002).

In bivalves, 5-HT receptors were cloned in the JS, 5-HTpy (Bockaert et al. 2010), and the pearl oyster, *Pinctada fucata*, 5-HTpf (Tanabe et al. 2010; Wang and He 2014). The situation in bivalves is quite similar to that of vertebrates (Stefano et al. 1976). Kadam and Koide (1990) observed that motility is initiated in 80, 70, and 76% of spermatozoa in the presence of 5 μM of 8-OH-DPAT (agonist of 5-HT1 receptor), 5-MT (agonist of the 5-HT4 receptor), and 2-methyl-5-HT (a nonselective agonist of 5-HT1 receptor), respectively; and that motility is not triggered when applying RU24969 (agonist of the 5-HT1B receptor).

Overall, these studies demonstrate the involvement of 5-HT receptors in spermatozoa motility signaling and suggest the presence of a serotoninergic system in the reproductive tract of male that might regulate sperm physiology and biochemistry. It is likely that 5-HT-induced hypermotility in bivalve spermatozoa includes intracellular regulatory signals for axonemal beating and is mediated by 5-HT receptors.

11.4.3 Control of Sperm Motility in Bivalves: pH, Osmolarity, Ions (K^+, Na^+, $Ca2^+$), cAMP

11.4.3.1 Osmolality-Independent Activation of Spermatozoa Motility

Dilution of JS, MC, and PO spermatozoa in nonionic media, such as 1000 mM sucrose, 20 mM Tris, pH 8.0 ± 0.1 containing 5-HT or 10 mM Ca^{2+} with an osmolality similar to that of SW, does not activate their motility (Alavi et al. 2014). The OP (Osmotic Pressure) of testicular fluid is 1061 mOsmol L^{-1}, a value similar to that of SW (Boulais et al. 2018). These results indicate that spermatozoa motility initiation in bivalves is an osmolality-independent mechanism. An OP below 900 and above 1100 mOsmol kg^{-1} prevents spermatozoa motility initiation (Dong et al. 2002).

11.4.3.2 pH Regulation of Spermatozoa Motility Initiation

In bivalves, pH of the testicular fluid is lower than that of SW: this inhibits spermatozoa motility activation (Faure et al. 1994; Dai et al. 2004; Demoy-Schneider et al. 2012; Alavi et al. 2014; Boulais et al. 2018) (Table 11.2) as bivalves

Table 11.2 pH and ionic constituents in testicular fluid (TF) and blood plasma (BP) of bivalves compared to seawater (adapted from Boulais et al. 2019)

		Na^+ mM	K^+ mM	Ca^{2+} mM	pH	Osmolality mOsmol kg^{-1}
Artificial seawater		469	11	10	8.2	1000
Pacific oyster	TF	360	18	11	5.5–6.5	
	BF	418	20	14	ND	
	TF	427	32	9	5.82	1061
Japanese scallop	TF	351	41	7	7.0	
	BF	523	11	9	ND	
Great scallop	TF				6.5	
Black-lip pearl oyster	TF				6.6	

ND not determined. Black-lip pearl oyster *P. margaritifera*; Great scallop, *P. maximus*; Japanese scallop, *P. yessoensis*; Pacific oyster, *C. gigas*

spermatozoa require an alkaline medium to become motile. Gonadal pH of the PO is 5.82 and spermatozoa motility is not triggered in SW if the pH value is below 6.0 (Boulais et al. 2018). At pH 6.0–7.0, the percentage of motility is $<10\%$ and optimal pH values for spermatozoa motility ranges 7.5–9.5 (Boulais et al. 2018). The percentage of motile spermatozoa is higher in the presence of 5-HT for a same pH value. Motility is triggered within 15 min when PO spermatozoa are incubated in artificial SW at pH 7.0 containing 5-HT (Alavi et al. 2014). Spermatozoa velocity (VAP: velocity of the average path) also increased in a pH-dependent manner (Boulais et al. 2018).

In the BLPO, pH values of the male gonad and gonadal tract were 6.6 and 7.4, respectively (Demoy-Schneider et al. 2012). A specific solution called DCSB4 (pH 8.2) acts as a motility activator for BLPO spermatozoa. DCSB4 is a saline solution composed of 19.5 g L^{-1} NaCl, 6.25 g L^{-1} glycine, 0.15 g L^{-1} $CaCl_2$, 0.19 g L^{-1} $MgSO_4$, 2.42 g L^{-1} Tris, pH 8.2. Following transfer in DCSB4, spermatozoa need a few minutes to reach fully activated motility; then, spermatozoa remain motile for at least 10 min. Motility stops at more acidic pH values (Demoy-Schneider et al. 2012).

Motility of GS spermatozoa is not initiated if SW pH is value below 5.0, and fully triggered at pH above 8.0 (Faure et al. 1994). In the JS, spermatozoa motility does not occur in pH 4 SW, but is fully activated at pH 6.0 and above (Alavi et al. 2014). Movement of MC spermatozoa is fully activated in SW at pH value above 8.0 (Alavi et al. 2014). Inhibition of spermatozoa motility at low external pH ([pH]e) is reversible since increasing the activation medium pH or washing spermatozoa with alkaline SW results in motility triggering (Faure et al. 1994). After dilution in SW at high pH values ≥ 9 compared to <9, spermatozoa become motile within a shorter period of time (Alavi et al. 2014; Boulais et al. 2018). As an example, at pH 8 and pH 9 within 15- and 3-min post-activation in SW, respectively, MC sperm become fully motile. Similarly, percentage of motile sperm PO is higher at 5 min post-activation compared to 1 min post-activation in SW if pH is <8.5. It appears that [pH]e mediates intracellular pH ([pH]i) (Parisi et al. 1984). Modification of [pH] i results from NH_3 or NH_4Cl that rapidly alkalinize sperm [pH]i. Motility of Japanese PO and MC spermatozoa is immediately and fully triggered in artificial SW including 5-HT 2 mM NH_3 (Meizel and Turner 1983; Alavi et al. 2014). Similarly, in the PO, a threefold increase of the percentage of motile sperm is observed in NH_4Cl-SW (Boulais et al. 2018).

An increase in [pH]i is hypothetically required for optimal dynein ATPase activity and thus axonemal beating (Christen et al. 1982; Parisi et al. 1984; Young and Laing 1990). Optimal pH of dynein ATPase activity ranges values from 7.4 to 8.6, while activity stops at pH <7.2 (Nakajima et al. 2005; Ohta et al. 2007). Acidic pH of testicular fluid is thus a factor maintaining spermatozoa in a quiescent state in the testis.

11.4.3.3 K^+ Regulation of Spermatozoa Motility Initiation

Concentration of K^+ ions in testicular fluid of the PO and JS is higher than in SW (Alavi et al. 2014; Boulais et al. 2018) (Table 11.2). Increasing extracellular $[K^+]$e to the value of K^+ in the testicular fluid reduces JS spermatozoa and decreases sperm velocity in SW $\pm$5-HT (Alavi et al. 2014). However, percentage of motile spermatozoa is higher in SW containing 10 mM K^+: this concentration is optimal for initiation and maintenance of sperm motility in bivalves.

The percentage of motile spermatozoa was measured in K^+-free SW with or without 5-HT (Alavi et al. 2014; Boulais et al. 2018). Motility of JS spermatozoa is triggered in K^+-free SW with a lower percentage than in SW. Similarly, spermatozoa of the PO become fully motile after dilution in K^+-free SW (Boulais et al. 2018), thus K^+ influx is not required to trigger sperm motility. Sperm motility and velocity decrease in presence of a voltage-dependent K^+ channel inhibitor (4-AP). K^+-inhibited spermatozoa motility initiation is also reported in salmonid and sturgeon (Boitano and Omoto 1991; Cosson et al. 1999).

11.4.3.4 Na^+ Regulation of Spermatozoa Motility Initiation

Bivalves testicular fluid contains sodium as a main constituent (Alavi et al. 2014; Boulais et al. 2018) (Table 11.2). Sperm motility of JS, MC, and PO is not triggered in a Na^+–SW, even supplemented with 10 mM Ca^{2+} A23187 (a Ca^{2+} ionophore) (Alavi et al. 2014). In the GS, JS, MC, and PO spermatozoa motility is not initiated in Na^+-free artificial SW (Faure et al. 1994; Alavi et al. 2014; Boulais et al. 2018) but, nevertheless, $[Na^+]$e is needed for motility: SW containing >150 mM Na^+ is required to trigger spermatozoa motility in the GS and inhibition of sperm motility in Na^+-free SW is reversible (Faure et al. 1994).

In SW-containing 5-HT, $[Na^+]$e requirement for spermatozoa motility initiation seems to be species specific. Chelating extracellular $Ca2^+$ ions ($[Ca^{2+}]$e) results in full suppression of spermatozoa motility activation in Na^+-free 5-HT SW (Alavi et al. 2014). The role of Ca^{2+} ion was also investigated using flunarizine, a $Na^+/Ca2^+$ exchanger (NCX) blocker (Alavi and Cosson 2006; Morisawa 2008), which resulted in inhibition of spermatozoa motility associated with a decrease in spermatozoa velocity (Alavi et al. 2014). Overall, $[Na^+]$e is essential to spermatozoa motility initiation, which requires Na^+ influx mediated by regulation of $[Ca^{2+}]$i.

11.4.3.5 Ca^{2+} Regulation of Spermatozoa Motility Initiation

Ca^{2+} ions concentration in the testicular fluid is similar to that in SW (Alavi et al. 2014; Boulais et al. 2018). The physiological roles of Ca^{2+} ions in bivalve sperm motility vary among species. In the MC, chelating $[Ca^{2+}]e$ by EGTA totally suppresses sperm movement in SW (±5-HT). In the JS, sperm motility decreases in SW ±5-HT containing 2.5–5 mM EGTA and suppressed at 10 mM. However, EGTA in 5-HT SW has no effects on sperm motility compared to 5-HT SW. In the PO, 1–10 mM EGTA reduces the percentage of motile sperm in a dose-dependent manner (Alavi et al. 2014). In contrast, 1 mM EGTA increases the percentage of motile spermatozoa compared to SW in the PO (Boulais et al. 2018). Similarly, other authors reported that Ca^{2+}-free SW increases motility of PO sperm (Dong et al. 2002). Altogether, bivalves spermatozoa may require lower $[Ca^{2+}]e$ to trigger initiation of motility in the presence of 5-HT compared to SW (Alavi et al. 2014), However Ca^{2+} itself is not the primary factor triggering spermatozoa motility. It was suggested that the higher percentage of motile spermatozoa in Ca^{2+}-free artificial SW compared to SW is associated with changes in $[Na^{+}]i$ concentration (Boulais et al. 2018).

It was observed that spermatozoa motility in the JS decreases following activation in SW containing Ca^{2+} channel blockers including 50 μM mibefradil, 200 μM verapamil, and 200 μM nifedipine at 1, 120, and 15 min post-activation, respectively. In SW containing 5-HT, spermatozoa motility is decreased in a shorter period of time compared to SW by mibefradil, verapamil, and nifedipine at 1, 1, and 5 min post-activation, respectively. Spermatozoa velocity in the JS also decreases in SW with or without 5-HT containing aforementioned Ca^{2+} channel blockers at 1 min post-activation. The observed decrease in both spermatozoa motility and velocity following activation in SW or 5-HT containing Ca^{2+} channel blockers suggests that Ca^{2+} influx is mediated by voltage-dependent Ca^{2+} channels (Alavi et al. 2014).

To better understand role of Ca^{2+} in spermatozoa motility initiation, changes in $[Ca^{2+]}i$ were investigated in MC spermatozoa following activation in SW with or without 5-HT containing 10 mM EGTA or 50 μM mibefradil. In this species, 5-HT triggers spermatozoa motility in a time-dependent manner. Spermatozoa motility initiation is associated with time-dependent increase and oscillation in $[Ca^{2+}]i$. In SW, $[Ca^{2+}]i$ increases and Ca^{2+} oscillations are similar to those observed in 5-HT. In the presence of EGTA, no $[Ca^{2+}]i$ increase and oscillations are observed in SW with or without 5-HT. In the presence of mibefradil, Ca^{2+} oscillations are observed, but $[Ca^{2+}]i$ decrease in 5-HT activated spermatozoa throughout the activation period. These results confirm that $[Ca^{2+}]e$ play an important role in spermatozoa motility activation. Further experiments have shown that spermatozoa motility in the MC and PO is suppressed in artificial SW and 5-HT containing 100 or 200 μM W-7 (an inhibitor for Ca^{2+}–calmodulin (CaM) dependent phosphodiesterase), which is explained by $[Ca^{2+}]e$ requirement for Ca^{2+}/CaM-dependent flagellar beating. Taken together, these results indicate that $[Ca^{2+}]e$ is required for spermatozoa motility initiation in bivalves. It is suggested that Ca^{2+} influx, mediated by voltage-dependent

Ca^{2+} channels, is required to activate Ca^{2+}/CaM proteins in the axoneme leading to flagellar beating. Investigating the roles of NCX, [Ca^{2+}]i store, and their physiological interactions in controlling spermatozoa motility initiation would be of interest for future studies. Moreover, considering the presence of Ca^{2+} in the testicular fluid, Ca^{2+} is not an inhibitory factor maintaining sperm in the quiescent state in the testis (Alavi et al. 2014).

11.4.3.6 Cyclic Nucleotide Regulation of Sperm Motility Initiation

Intracellular concentration of cAMP (Correia et al. 2015) and cGMP levels (Kuroda et al. 1999) are modulated by various cation channels modulate in spermatozoa of marine organisms. Activation of spermatozoa motility may occur via cAMP-dependent phosphorylation of axonemal proteins. In the case of PO, cAMP activation is not related with control neither by K^+ nor by Ca^{2+}. Intracellular cAMP rise could be controlled by 5-HT in bivalve spermatozoa.

In the Asian clam, *Corbicula fluminea*, a freshwater species, spermatozoa include two flagella with equal size, each having a typical "9 + 2" axoneme and normal beating (Sanchez et al. 2001; Ho and Suarez 2003). Spermatozoa of the AC are immotile in the gonad and in freshwater. Spermatozoa motility initiation is triggered by cGMP, cAMP, and cAMP analogs. A selective cAMP-dependent protein kinase (PKA) inhibitor (H-89 at 1 μM) inhibits spermatozoa motility. Involvement of PKA in sperm motility initiation is downstream of cAMP, but not of cGMP while PKG is also involved but differentially (Sanchez et al. 2001).

11.4.3.7 Signaling Pathway Controlling Spermatozoa Motility Initiation in Bivalves

The K^+ ions concentration and acidic pH of testicular fluid are the main factors inhibiting spermatozoa motility in bivalve testis according to an osmolality-independent mechanism. Alkalization of spermatozoa [pH]i is a mandatory condition to activate flagellar beating. During spawning, motility is triggered by a Na^+-dependent [pH]i alkalization probably mediated by a Na^+/H^+ exchanger (Fig. 11.8). Consequently, the primary factor controlling spermatozoa motility activation is [pH] I, not Na^+. Sperm motility initiation is induced by hyperpolarization of the membrane potential via a 5-HT rise in the testis (Fig. 11.8). A K^+ efflux via voltage-dependent K^+ channels and associated with opening of voltage-dependent Ca^{2+} channels under alkaline conditions also mediate membrane hyperpolarization and possibly a release of Ca^{2+} from intracellular stores which triggers Ca^{2+}/CaM-dependent flagellar beating. Ca^{2+} exchanges via NCX are also probably regulated by [Na^+] e. In addition 5-HT is probably involved in controlling intracellular cAMP, leading to cAMP-dependent phosphorylation of axonemal proteins (Boulais et al. 2019).

11.4.4 Flagellar Beating Characteristics of Bivalve Spermatozoa

At contact with SW $\pm$5-HT, flagellum beats according to asymmetric wave pattern (Fig. 11.9) which displays a circular movement of spermatozoa (Fig. 11.10).

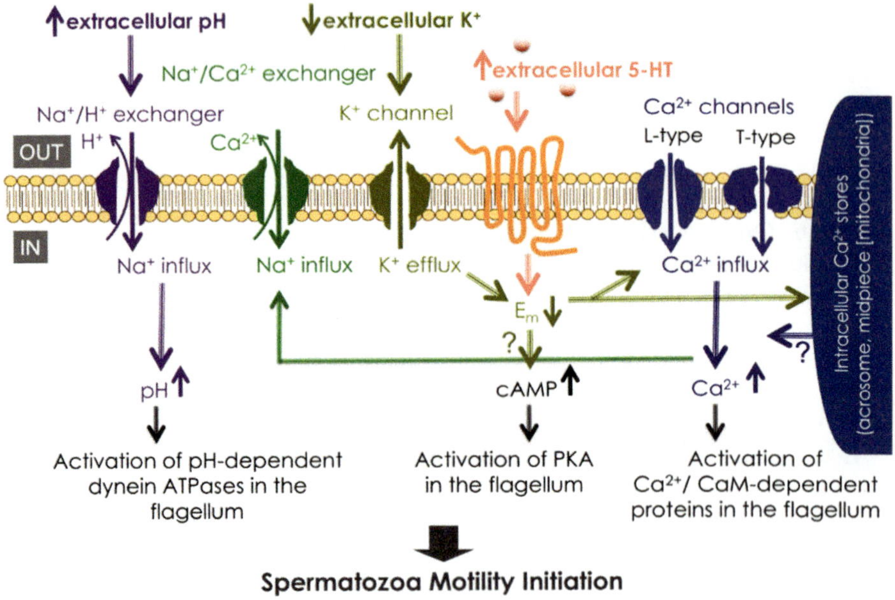

Fig. 11.8 Current known signaling pathway of spermatozoa motility in bivalves. During spawning, spermatozoa motility is triggered by a Na^+-dependent alkalization of internal pH mediated by a Na^+/H^+ exchanger. In addition, decrease of extracellular K^+ or presence of extracellular 5-hydroxytryptamine creatinine sulfate (5-HT) induces hyperpolarization of spermatozoa membrane potential (Em) mediated by K^+ efflux via voltage-dependent K^+ channels associated with opening of voltage-dependent Ca^{2+} channels under alkaline conditions. Subsequent increase in intracellular Ca^{2+} content ($[Ca^{2+}]_i$) triggers Ca^{2+}/calmodulin (CaM)-dependent flagellar beating. Release of Ca^{2+} from intracellular stores may also contribute to increase $[Ca^{2+}]_i$. Na^+ influx probably regulates Ca^{2+} exchange via Na^+/Ca^{2+} exchanger. The mechanism through which 5-HT induces spermatozoa hypermotility is unclear. It may be possible that 5-HT is involved in controlling intracellular cAMP rise in bivalve spermatozoa, leading to cAMP-dependent protein kinase (PKA) phosphorylation in the flagellum. It is also possible that 5-HT-induced hyperpolarization control rise in $[Ca^{2+}]_i$ concentrations via Ca^{2+} influx or Ca^{2+} release from $[Ca^{2+}]_i$ stores. (From Boulais et al. 2019)

Different flagellar wave characteristics were reported (Demoy-Schneider et al. 2012), including "full," "twitching," and "declining" propagation of wave. Propagating full wave means that flagellar waves are developed along flagella, with planar waves of sinusoidal shape from head to distal tip (Figs. 11.11, 11.12 and 11.13). Wave amplitude (5–6 μm) is nearly constant along the flagellum length. Sperm tracks are mostly circular with arcs interspersed with short linear segments (Fig. 11.10). Tracks are according to a clockwise direction when close to the cover slip surface and counter-clockwise in the vicinity of glass slide surface (Cosson et al. 2003). Sperm heads alternate both sides of the mean track (Boulais et al. 2019).

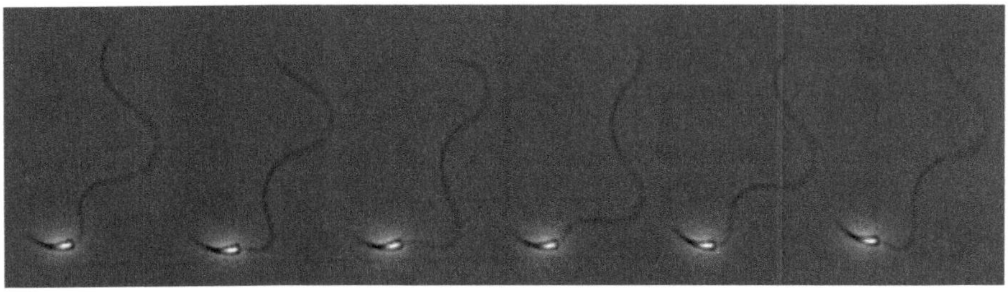

Fig. 11.9 Successive frames of flagellar beating of a Manila clam, *R. philippinarum*, spermatozoon. Motility of spermatozoa was recorded using a high-speed CCD camera (HAS-220; Ditect) under a phase contrast microscope (Olympus BX51, 20X magnification) with an acquisition rate of 200 frames per second. Some parts of the flagellum are in focus while others are slightly out of focus illustrating the 3D distortion of the flagellar plane of beating. To activate motility, sperm was diluted 2000 folds in artificial seawater containing 10^{-5} M 5-hydroxytryptamine creatinine sulfate (5-HT) and BSA (0.1% w:v) was added to prevent sperm adhering to the glass slide, no cover slip was used (from Boulais et al. 2019)

11.5 Energy Metabolism in Bivalve Sperm

11.5.1 Mitochondrial Respiration

The respiration rate of oyster spermatozoa can be easily measured because they have a long motility duration, which contrasts with fish spermatozoa that present a short motility, period (30 s–20 min). Activation of sperm motility is closely related to mitochondrial respiration: sperm oxygen consumption strongly increases during motility activation. Oxygen consumption of BLPO sperm ranges from 0 prior to motility activation to 29 nmol O_2 min^{-1} per 10^9 sperm after activation in alkaline SW (Demoy-Schneider et al. 2012). A maximum respiration rate value is obtained by dissipation of the proton gradient across the inner mitochondrial membrane by application of an oxidative phosphorylation "uncoupler" such as CCCP (Carbonyl cyanide m-chlorophenyl hydrazine), highlighting that mitochondrial respiratory chain activity can be greatly increased in PO spermatozoa.

During a short period after transfer in SW appears a "twitching" behavior of sperm: flagella develop one local bend without sinusoidal waves, leading to a non-efficient movement (Fig. 11.13). Before reaching the behavior of fully activated, spermatozoa go through a "half activation" state (around 50% of motile spermatozoa) where they alternatively twitch and stop several times (Fig. 11.14) (Boulais et al. 2019).

11.5.2 Intracellular ATP Concentration

ATP is required to fuel spermatozoa movement phase because flagellar movement occurs through the flagellar dynein-ATPases (Konishi et al. 1998). ATP content

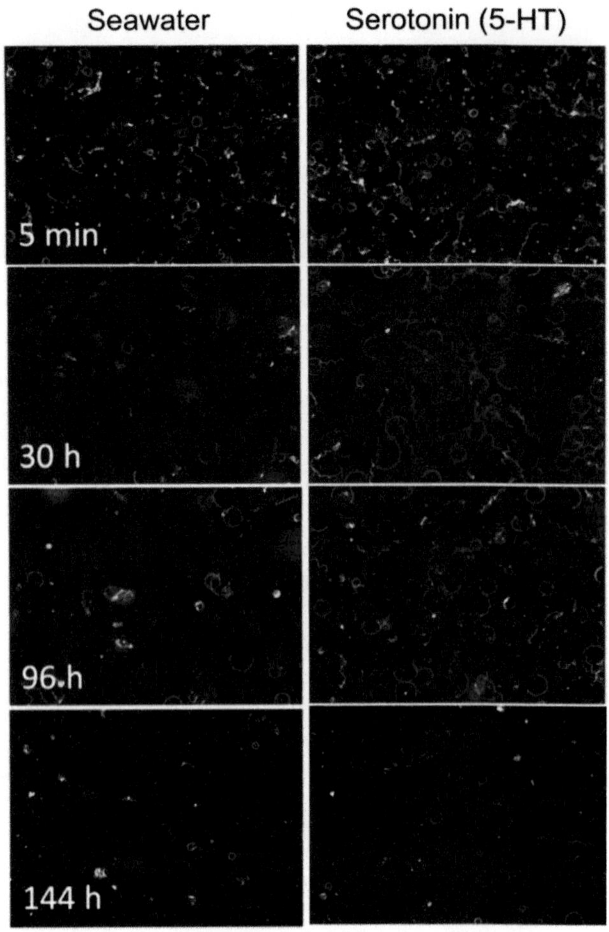

Fig. 11.10 Head trajectories in motile spermatozoa of the Manila clam, *R. philippinarum*. Motility of spermatozoa was activated in artificial seawater or artificial seawater containing 10^{-5} M 5-hydroxytryptamine creatinine sulfate (5-HT) at 30, 96, and 144 h post-activation, and recorded under a phase-contrast microscope (Olympus BX51) connected to a digital CCD camera (Hamamatsu Photonics) at 50 frames per second. Spermatozoa movement trajectories were analyzed using semen motility analysis system (SMAS). To activate spermatozoa motility, sperm was diluted 500–1000 folds in artificial seawater with or without 5-HT. BSA (0.1% w: v) was added to prevent sperm adhering to the glass slide and no cover slip was used (from Boulais et al. 2019)

declines within tens of seconds to few minutes depending on species leading to arrest of spermatozoa movement (Kraemer et al. 1986). Changes in ATP content during the movement phase of spermatozoa have been little documented in bivalves. In the GS, sperm stop their movement after 10 h of motility, due to the exhaustion of [ATPi] (Deguchi and Osanai 1994). In BLPO, the [ATPi] is 700 or 543 nmol per 10^9 spermatozoa at 2 min after motility activation in SW or in DCSB4, respectively (Demoy-Schneider et al. 2012). An [ATPi] content of 157.2 nmol 10^{-9} spermatozoa at 1 h after activation in SW was reported in the PO (Howard et al. 2004). A stable [ATP_i] of 185.6 nmol per 10^9 spermatozoa and 140.1 nmol per 10^9 spermatozoa at 90 min and 9 h after activation, respectively in this species (Suquet et al. 2010).

Arrest of spermatozoa movement is not explained in the PO by a low [ATPi] (75–94% of initial content) (Suquet et al. 2010; Boulais et al. 2015) nor by the death of spermatozoa as 94% remain motile at the end of the motility phase (24 h) (Boulais et al. 2015). One hypothesis to the arrest of movement could be the disruption of

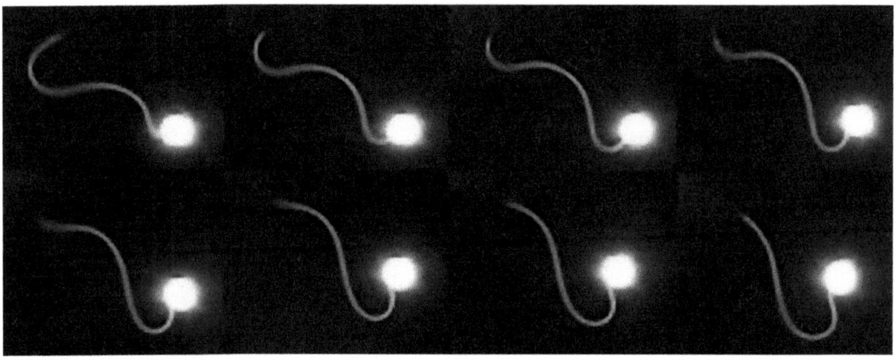

Fig. 11.11 Successive frames of flagellar beating of a Pacific oyster, *C. gigas*, spermatozoon in seawater. Spermatozoa motility was activated in seawater and observed using a phase contrast microscope (dark-field, Olympus BX51) microscope, connected to a high-speed camera at 1.000 images s^{-1} (interval between two frames is 1 ms. Some parts of the flagellum are in focus while others are slightly out of focus, illustrating the 3D distortion of the flagellar plane of beating. To activate spermatozoa motility, sperm was diluted 500 folds in seawater and pluronic acid (0.1% w: v) was added to prevent sperm adhering to the glass slide, no cover slip was used (from Boulais et al. 2019)

ATP transport from mitochondria to the flagellum, possibly suggesting the involvement of phosphagens.

11.5.3 Metabolic Pathways for ATP Synthesis

Two metabolic pathways lead to ATP production: glycolysis and OXPHOS, the latter being the most efficient way to produce ATP. In sea urchins, spermatozoa do not undergo glycolysis and rely entirely on OXPHOS (Cook and Babcock 1993; Cosson et al. 2008b). The long period of sperm movement in PO compared to other marine invertebrates (Suquet et al. 2010) was studied (Boulais et al. 2015) and demonstrated that OXPHOS produces a stable throughout 24 h. A steady decrease of [ATPi] during the first 9 h could stimulate ATP synthesis via OXPHOS, thus compensating for ATP hydrolysis during the last part of the movement phase (Boulais et al. 2015). Sperm of OP are motile during the first 2.5 h of movement when mitochondrial ATP synthesis is inhibited by 1 μM CCCP, an uncoupler (Boulais et al. 2015). These and other results suggest that ATP-sustaining flagellar movement can originate from alternative metabolic sources during the first 2 h of motility such as glycolysis, stored ATP, or phosphagens. High-energetic compounds, such as arginine- or creatine-phosphate are phosphagens that serve as [ATPi] shuttles from mitochondria to dynein-ATPases and during periods of high-energy demand (Suquet et al. 2013).

Overall, these specificities of PO sperm energetics reveal a specific strategy developed by this species, reflecting the adaptation of PO to sessile life, thus

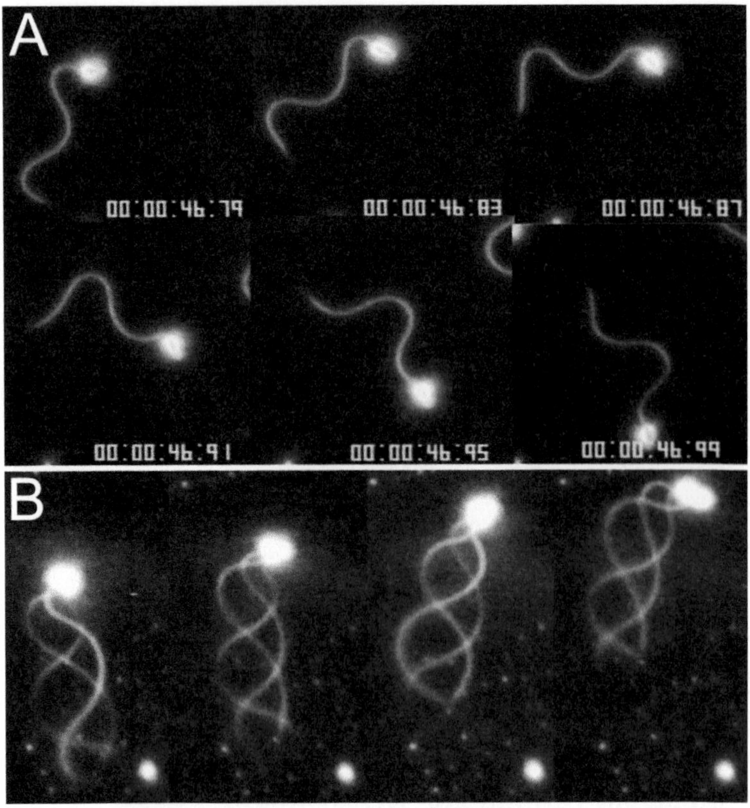

Fig. 11.12 Successive frames of flagellar beating of a Pacific oyster, *C. gigas*, spermatozoon in seawater at 1 mM 5-HT. Spermatozoa movement was observed using a phase contrast microscope (dark-field, Olympus BX51) supplied by stroboscopic illumination, and connected to a video camera at 50 frames s^{-1} at 150 Hz (interval between two frames is 20 ms). A: Each frame shows one position of the same spermatozoon B: Each frame includes three successive positions of the same spermatozoon. To activate spermatozoa motility, sperm was diluted 500 folds in seawater and pluronic acid (0.1% w:v) was added to prevent sperm adhering to the glass slide, no cover slip was used (from Boulais et al. 2019)

improving oocyte fertilization success (Boulais et al. 2017) and contributing to the dispersion of PO in the wild (Mita and Nakamura 1993).

11.6 Methodologies Used for Conservation of Bivalve Sperm

11.6.1 Cold Storage

Sperm storage at 4 °C is an alternative method for conservation when sperm cannot be cryopreserved immediately after collection. In the Australian flat oyster *Ostrea angasi,* when spermatozoa are stored at 4 °C, they retain potent motility during

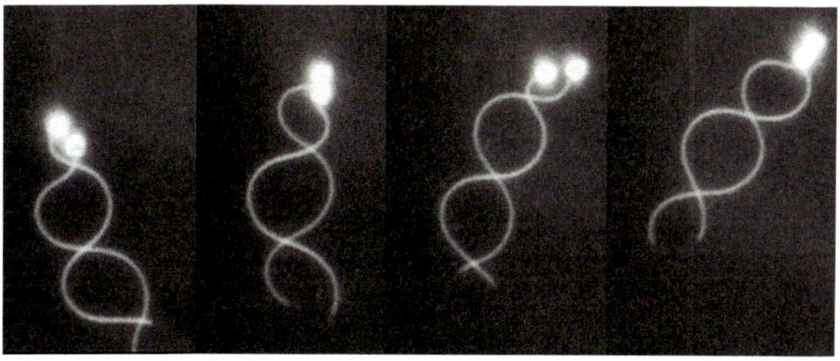

Fig. 11.13 Swimming spermatozoon of *P. margaritifera*. From left to right, successive video-images are every 20 ms; each image represents two positions of the same spermatozoon separated by 10 ms. Dark field microscopy with 40× objective lens and stroboscopic illumination

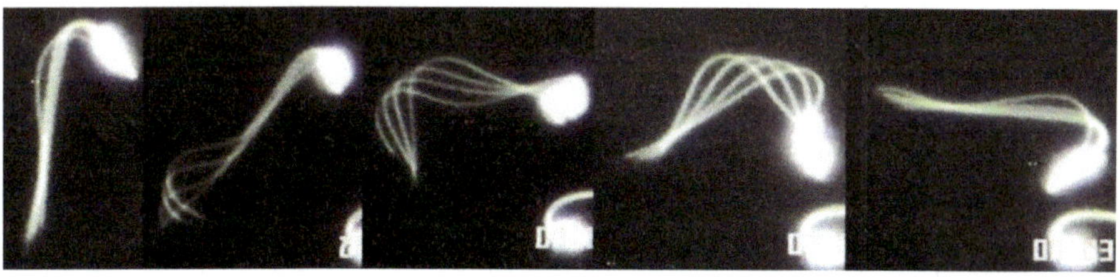

Fig. 11.14 Successive frames of "twitching" behavior in a Pacific oyster, *C. gigas*, spermatozoon. Spermatozoon shows only one local bending point along the flagellum, and thus no flagellar sinusoidal waves. Spermatozoa motility was observed at the beginning of movement activation, prior to full activation of spermatozoa motility in seawater, using a phase contrast microscope (dark field, Olympus BX51) supplied by stroboscopic illumination, and connected to a video camera at 50 frames s^{-1} (interval between two frames is 20 ms) at 150 Hz. Each frame includes three successive positions of the same spermatozoon. To activate spermatozoa motility, sperm was diluted 500 folds in seawater and pluronic acid (0.1% w:v) was added to prevent sperm adhering to the glass slide, no cover slip was used (from Boulais et al. 2019)

5 days in natural SW and the equilibration in a 5% DMSO (dimethyl sulfoxide) + 0.45 M trehalose mixture extended the potent motility up to 7 days (Hassan et al. 2017a, b). Similar results were obtained for the PO spermatozoa (Vanderhost et al. 1985; Dong et al. 2005c). In *P. margaritifera*, spermatozoa held in regular seawater (pH 7.8) at 4 °C retain potent motility for several days and can be subsequently activated by an alkaline swimming medium. During this storage period, spermatozoa saved their motility behavior and characteristics, even if the motility performances progressively declined. Mitochondrial respiration rates and ATP utilization are closely bound to motility activation. The respiration rate is strongly increased and ATP is more rapidly consumed immediately after motility activation (Demoy-Schneider et al. 2012, 2014). Gourtay et al. (2016) tested short-term storage for *P. maximus* spermatozoa and showed that spermatozoa can retain potent motility up to 64 h at 4 °C in a *P. maximus* specific diluent or in natural SW.

11.6.2 Cryo-conservation Techniques

11.6.2.1 Cryoprotectant Selection and Equilibration

Cryoprotectants are selected according to their ability to protect sperm from the damage caused to the gametes and their cytotoxicity. Common cryo-protecting agents (CPA) for oysters cryopreservation include dimethyl sulfoxide (DMSO), N-dimethyl acetamide, glycerol, propylene glycol (PG), ethylene glycol (EG), and methanol (Gwo 2000; Heres et al. 2019). CPAs are usually prepared in SW at a wide concentration range between 0.5 and 3 M. Most of the time, glycine, sucrose, trehalose, or glucose are added in one step or progressively in the CPA mixture at various concentrations, depending on the species (Paredes 2015; Liu et al. 2015; Hassan et al. 2015). The equilibration duration depends on the species, the chemical species of CPA used and its concentration. The CPA type is also very important for sperm survival. In most cases for oysters, permeating CPAs as glycine, DMSO or trehalose, are used to prevent membrane injury, in combination with non-permeating CPAs like sugars or polymers which are stabilizing agents of the spermatozoa membrane (Hassan et al. 2015).

11.6.2.2 Packing of Samples for Freezing

With respect to cryopreservation techniques, packing of samples for freezing and storing is a critical point to assure sample identification and to standardize cooling and thawing rates particularly if liquid nitrogen (LN) is used for cryo-conservation. Because of the small volume of sperm available, standard "French straws" (250–500 μL) are considered as best compromise for sperm packaging. This kind of straws offers the advantages of efficient and reliable sample identification by a permanent bar code printed on straws, sample safety by complete sealing of the straw tips, and standardization of the cooling and thawing processes. Straws are manually filled with sperm diluted with CPA using a micropipette or by aspiration and then sealed with modeling clay or heat application (Hassan et al. 2015).

11.6.2.3 Cooling Rate and Freezing

According to Tiersch et al. (2007), cooling rate is a primordial key step for sperm cryopreservation. Theoretically, it should be fast enough to minimize the exposure time of sperm to concentrated extracellular solutions and yet it should be slow enough to avoid, by cellular dehydration, the formation of intracellular ice crystals below a damaging level. The optimum cooling rates vary according to cryoprotectants and species (Paredes 2015). Nevertheless, samples should not remain at room temperature more than 10 min between the dilution in CPA and the beginning of the cooling process. Cooling generally starts by diluting the samples in the chilled CPA at 4 °C, then by setting the samples on a styrofoam tray floating in the LN vapors on the LN surface for 3–20 min (Paniagua-Chavez et al. 2006; Hui et al. 2011; Paredes 2015; Liu et al. 2015). For most of the bivalve species, freezing is performed by direct immersion in LN but if available, the cooling process being controlled by a programmable freezer including one, two, or multiple-step cooling rate as reported by Liu et al. (2018) and Paredes (2015). Various cooling rates have been reported, from 2.5 °C min^{-1} (Paniagua-Chavez et al. 2006) to 50 °C min^{-1} (Li et al. 2000b). Combination of slow and fast cooling rates has also been reported in some oyster species by Hassan et al. (2015). Whatever the programmable method, samples are progressively cooled from 4 to −80 °C and then immersed in LN for long-term storage (Hassan et al. 2015).

11.6.3 Thawing

It has been reported that thawing procedure is another critical step due to the risk of recrystallization of water. Thawing is generally performed either at room temperature or in a warm temperature-controlled water-bath. The thawing rate is faster than the cooling rate. The thawing duration varies between 2 s and 2 min and optimal thawing temperatures are ranging 16–75 °C (Hassan et al. 2015; Paredes 2015). Liu et al. (2015) detailed the different temperatures used to thaw frozen oyster spermatozoa: three ranges are categorized, low (<29 °C), medium, (30 °C–49 °C) and high (>50 °C). For the PO and abalone, the three ranges of temperatures are used with a satisfying rate of fertilization after thawing whereas for black-lip pear oysters, scallops, Japanese oysters and mussels, thawing temperatures are lower for highest fertilization.

11.6.4 Overview of the Published Results

The most relevant results concerning bivalve cryopreservation for the studied species are gathered in Table 11.3. Most of the studies reported in the reviews written by Hassan et al. (2015), Liu et al. (2015) or Paredes (2015) list the different cryopreservation protocols used to optimize spermatozoa long-term storage.

11.7 Conclusion

Sperm cryo-conservation is a technique that is practiced for many years in case of marine animal species and especially in mollusks and nowadays the cryopreservation steps are quite well-established, leading to high-level results for most of the bivalve species. Previously published studies reported that the choice of the extender depends on the mollusk species and that DMSO is the common suitable cryoprotective agent for most of the mollusk species (Liu et al., 2015) but also highlighted that sperm quality is an unavoidable factor to guaranty successful cryo-conservation. All these points emphasize the necessity of additional research efforts to improve the knowledge of the different parameters, such as ultrastructure, movement characteristics, energy stocks, and metabolism, involved in the quality of each sperm species prior to cryo-conservation and thus obtain higher levels of fertilization rate.

Table 11.3 List of sperm cryopreservation studies performed in bivalves (adapted from Hassan et al. 2015; Liu et al. 2015; Paredes 2015)

Species	CPA	Packaging	Cooling rate and freezing	Fertilization rate	References
	DMSO 20% in SW	1 mL ampoules	2 min in LN vapor and plunged into LN	Up to 10% fertilization	Lannan (1971)
	DMSO 3.3–20% in SW		Immersion in LN	79% fertilization	Hwang and Chen (1973)
	DMSO 20% + Gly 0.6%		At -5 °C min^{-1} or at -30 °C min^{-1}	36.3% fertilization	Staeger (1974)
	DMSO 10%		5 cm over LN vapor for 3 min	92% fertilization	Bougrier and Robemanana (1986)
	DMSO 10%		Held 7 cm over LN	30% motility	Korukura et al. (1990)
	DMSO 10% in SW	1.8 mL cryotubes	At -4.7 °C min^{-1} down to -70 °C and plunged into LN	48–93% fertilization	Yankson and Moyse (1991)
	Concentrated polysaccharide in distilled water + DMSO 5%	0.25, 0.5, and 2.5 mL straws	At -50 °C min^{-1} and plunged into LN	0–100% fertilization	Smith et al. (2001)
C. gigas	DMSO 10% in SW	1.5 mL tubes	From room temperature to -30 °C at -15 °C min^{-1} and plunged into LN	40% fertilization	Gwo et al. (2003)
	EG 10% in SW	2 mL containers	From 26 °C to -70 °C at -6 °C min^{-1} and plunged into LN	58.9% fertilization	Ieropoli et al. (2004)
	2.5–15% in distilled water	0.25 mL straws	From 0 °C to -80 °C at -50 °C min^{-1}, held at-80 °C for 10 min and plunged into LN	80% fertilization	Adams et al. (2004)
	DMSO 5% + 0.55 M trehalose	4.5 mL vials	Held in methanol bath -75 °C for 10 min and then plunged in LN	80% fertilization	Adams et al. (2004)
	DMSO 5% + 0.55 M trehalose	4.5 mL vials	3 cm above LN for 10 min and then plunged in LN	80% fertilization	Adams et al. (2004)

	DMSO 5% + 0.55 M trehalose	4.5 mL vials	Directly plunged in LN	10–60% fertilization	Adams et al. (2004)
	PG 5% + DMSO 8% in Ca^{++} – free HBSS	0.5 mL straws	At –16 °C min^{-1}, down to –140 °C and plunged into LN	0–96% fertilization	Dong et al. (2005b)
	PG 2% + methanol 6% in Ca^{++} – free HBSS	0.5 mL straws	At –5 °C min^{-1}, down to –30 °C and plunged into LN	98% fertilization	Dong et al. (2005c)
	PEG 6% + PG 4% or PEG 6% + DMSO 4% in Ca^{++} – free HBSS	0.25 or 0.5 mL straws	At –5 °C min^{-1}, down to –30 °C and plunged into LN and at –45 °C min^{-1}, down to –80 °C and plunged into LN	21% fertilization	Dong et al. (2006)
	Methanol 10% + DMSO 12% in Ca^{++} – free HBSS	0.5 mL straws	At –5 °C min^{-1}, down to –30 °C and plunged into LN and At –45 °C min^{-1}, down to –80 °C and plunged into LN	96% fertilization	Dong et al. (2007a)
	Methanol 2% + PEG 4% in Ca^{++} – free HBSS	0.5 mL straws	At –16 °C min^{-1}, down to –140 °C and plunged into LN	42% fertilization	Dong et al. (2007b)
	DMSO 5% + 0.54 M trehalose	4.5 mL vials	Frozen with dry ice down to –75 °C and plunged into LN	90% fertilization	Adams et al. (2008)
	DMSO 5% + 1 M trehalose		Held over LN vapor for 10 min	86% motility	Lyons et al. (2005)
	DMSO +0.45 M trehalose	0.25 mL straws	6 cm above LN for 10 min	35% motility	Acosta-Salmon et al. (2007)
P. margaritifera	DMSO 0.8 M + trehalose 0.7 M + oyster haemolymph 10%	0.5 mL straws	3 cm above LN for 10 min	80% motility	Hui et al. (2011)
			At –5 °C min^{-1}, down to –30 °C and plunged into LN	50% motility	Vitiello et al. (2011)
	DMSO +0.45 M trehalose	0.25 ml straws	3 cm above LN for 10 min	54% motility	Demoy-Schneider et al. (2018)

(continued)

Table 11.3 (continued)

Species	CPA	Packaging	Cooling rate and freezing	Fertilization rate	References
O. edulis	DMSO 10% in Ca^{++} – free HBSS	0.5 mL straws	LN vapors for 3 min and plunged into LN	8% motility	Horváth et al. (2012)
P. maximus	PEG 20% in DCSB4	0.5 mL straws	5.5 cm above LN for 10 min and plunged into LN	20% motility	Suquet et al. (2016)
P. yessoensis			20 cm above LN for 3 min	45% motility	Yang et al. (2007)
			3 cm above LN for 3 min	26% motility	Yang et al. (2008)

Acknowledgements Marina Demoy-Schneider work was supported by the French *Ministère des Outremers* and the research program "La cryo-conservation des spermatozoïdes de l'huître perlière *Pinctada margaritifera*, un atout pour une perliculture durable et la sauvegarde de sa biodiversité." Jacky Cosson work was supported by the Ministry of Education, Youth and Sports of the Czech Republic (CENAKVA: CZ.1.05/2.1.00/01.0024 and CENAKVA II: LO1205 under the NPU I program, Biodiverzita: CZ.02.1.01/0.0/0.0/16 025/0007370), by the Grant Agency of the University of South Bohemia in Ceske Budejovice (125/2016/Z) and a grant from European Union's Horizon 2020 research and innovation program (No 652831 AQUAEXCEL2020).

References

Abe H, Hamaguchi M, Kajihara N, Taniai Y, Oshino A, Moriyama A, Kamiyama T (2019) Population dynamics of the Manila clam *Ruditapes philipinnarum* and the implications of the 2011 tsunami impact in two-shallow semi-enclosed bays in northeastern Japan. In: Komatzu T et al (eds) Oceanography challenges to future earth: human and natural impact on our seas, vol 29. Springer Nature, Basel, pp 365–386

Acosta-Salmón H, Jerry DJ, Southgate PC (2007) Effects of cryoprotectant agents and freezing protocol on motility of black-lip pearl oyster (*Pinctada margaritifera* L.) spermatozoa. Cryobiology 54:13–18

Adams SL, Smith JF, Roberts RD, Janke A, Kaspar HF, Tervit HR, Pugh PA, Webb SC, King N (2004) Cryopreservation of sperm of the pacific oyster (*Crassostrea gigas*): development of a practical method for commercial spat production. Aquaculture 242(1–4):271–282

Adams SL, Smith JF, Roberts RD, Janke AR, King N, Tervit HR, Webb SC (2008) Application of sperm cryopreservation in selective breeding of the Pacific oyster, *Crassostrea gigas* (Thunberg). Aquacult Res 39(13):1434–1442. https://doi.org/10.1111/j.1365-2109.2008.02013.x

Adams SL, Tervit HR, Salinas-Flores L, Smith JF, McGowan LT, Roberts RD, Janke A, King NG, Webb SC, Gale SL (2011) Cryopreservation of Pacific oyster oocytes. In: Tiersch TR, Green CC (eds) Cryopreservation in aquatic species, 2nd edn. World Aquaculture Society, Baton Rouge, pp 616–623

Alavi SMH, Matsumura N, Shiba K, Itoh N, Takahashi KG, Inaba K, Osada M (2014) Roles of extracellular ions and pH in 5-HT-induced sperm motility in marine bivalve. Reproduction 147:331–345. https://doi.org/10.1530/REP-13-0418

Alavi SMH, Cosson J (2006) Sperm motility in fishes: (II) effects of ions and osmotic pressure. Cell Biol Int 30:1–14. https://doi.org/10.1016/j.cellbi.2005.06.004

Andersen S, Christophersen G, Magnesen T (2011) Spat production of the great scallop (*Pecten maximus*): a roller coaster1. Can J Zool 89(7):579–598

Anglès d'Auriac M, Rinde E, Norling P, Lapègue S, Staalstrom A, Hjermann D, Thaulow J (2017) Rapid expansion of the invasive oyster *Crassostrea gigas* at its northern distribution limit in Europe: naturally dispersed or introduced? PLoS One 12(5):e0177481. https://doi.org/10.1371/journal.pone.0177481

Azeredo FF (2015) Development and standardization of a protocol for sperm cryopreservation of two important commercial oyster species. Thesis for Master of Science Degree, Faculty of Sciences and Technology, University of Algarve, pp 65

Barber BJ, Blake NJ (2006) Reproductive physiology. In: Shumway SE, Parsons GJ (eds) Scallops: biology ecology and aquaculture, Developments in aquaculture and fisheries, vol 6. Science Elsevier Press, Amsterdam, pp 357–416. https://doi.org/10.1016/S0167-9309(06)80033-5

Barnabé G (ed) (2003) Aquaculture: biology and ecology of cultured species, 1st edn. CRC Press Book, Boca Raton

Beaumont A, Gjedrem T (2007) Scallops—*Pecten maximus* and *P. Jacobaeus.* In: Svasand T, et al. (ed.) Genetic impact of aquaculture activities on native populations. Genimpact Final Scientific Report (EU Contract n. RICA-CT-2005-022802), pp 83–90

Bernard I, Massabuau JC, Ciret P, Sottolichio A, Pouvreau S, Tran D (2016) In situ spawning in a marine broadcast spawner, the Pacific oyster *Crassostrea gigas*: timing and environmental triggers. J Limnol Oceanogr 61:635–647

Bockaert J, Claeysen S, Dumuis A, Marin P (2010) Classification and signaling characteristics of 5-HT receptors. In: Müller CP, Jacobs BL (eds) Handbook of the behavioral neurobiology of serotonin, handbook of behavioral neuroscience, vol 21. Elsevier, Amsterdam, pp 103–121

Boitano S, Omoto CK (1991) Membrane hyperpolarization activates trout sperm without an increase in intracellular pH. J Cell Sci 98:343–349

Bondarenko V, Cosson J (2019) Theriogeneology structure and beating behavior of the sperm motility apparatus of aquatic animals. Theriogenology 135:152–163. https://doi.org/10.1016/j.theriogenology.2019.06.005

Boulais M, Soudant P, Le Goïc N, Quéré C, Boudry P, Suquet M (2015) Involvement of mitochondrial activity and OXPHOS in ATP synthesis during the motility phase of spermatozoa in the Pacific oyster, *Crassostrea gigas*. Biol Reprod 93:1–7. https://doi.org/10.1095/biolreprod.115.128538

Boulais M, Soudant P, Le Goïc N, Quéré C, Boudry P, Suquet M (2017) ATP content and viability of spermatozoa drive variability of fertilization success in the Pacific oyster (*Crassostrea gigas*). Aquaculture 479:114–119. https://doi.org/10.1016/j.aquaculture.2017.05.035

Boulais M, Suquet M, Arsenault-Pernet EJ, Malo F, Queau I, Pignet P, Ratiskol D, Le Grand J, Huber M, Cosson J (2018) pH controls spermatozoa motility in the Pacific oyster (*Crassostrea gigas*). Biol Open 7:bio031427. https://doi.org/10.1242/bio.031427

Boulais M, Demoy-Schneider M, Alavi SMH, Cosson J (2019) Spermatozoa motility in bivalves: signaling, flagellar beating behavior, and energetics. Theriogenology 136:15–27. https://doi.org/10.1016/j.theriogenology.2019.06.025

Bougrier S, Robemanana LD (1986) Cryopreservation of the spermatozoa of the Japanese oyster, *Crassostrea gigas*. Aquaculture 58:277–280

Brand AR (1991) Scallop ecology: distributions and behavior. In: Shumway SE, Parsons GJ (eds) Scallops: biology ecology and aquaculture, Developments in aquaculture and fisheries, vol 21. Science Elsevier Press, Amsterdam, pp 517–584

Bregman YE (1979) Population-genetic structure of bivalve Patinopecten yessoensis. Izvestiya TINRO 103:66–78. (In Russian with English abstract)

Britton W (1991) Clam cultivation manual. Aquaculture explained, vol 8. Aqua. Tech. Sec. Bord Iascaigh Mhara, Dun Laoghaire, Co., Dublin, Ireland

Brugère C, De Young C (2015) Assessing climate change vulnerability in fisheries and aquaculture. Available methodologies and their relevance for the sector. In: FAO (ed) Fisheries and aquaculture technical paper 597, Rome

Chang YJ, Mori K, Nomura T (1985) Studies on the scallop, *Patinopecten yessoensis*, in sowing cultures in Abashiri waters-reproductive periodicity. Tokohu J Agric Res 35(2–4):91–105

Christen RW, Schackmann RW, Shapiro BM (1982) Elevation of intracellular pH activates respiration and motility of sperm of the sea urchin *Strongylocentrotus purpuratus*. J Biol Chem 257:14881–14890

Clark MS, Husmann G, Thorne MAS, Burns G, Truebano M, Peck LS, Abele D, Philipp EE (2013) Hypoxia impacts large adults first: consequences in a warming world. Glob Change Biol 19:2215–2263

Cook SP, Babcock DF (1993) Selective modulation by cGMP of the K channel activated by speract. J Biol Chem 268(30):22402–22407

Correia J, Michelangeli F, Publicover S (2015) Regulation and roles of Ca^{2+} stores in human sperm. Reproduction 150:65–76. https://doi.org/10.1530/REP-15-0102

Cosson J, Billard R, Cibert C, Dreanno C, Suquet M (1999) Ionic factors regulating the motility of fish sperm. In: Gagnon C (ed) The male gamete: from basic to clinical applications. Cache River Press, Vienna, pp 161–186

Cosson J, Huitorel P, Gagnon C (2003) How spermatozoa come to be confined to surfaces. Cell Motil Cytoskel 54(1):56–63

Cosson J, Faure C, Devauchelle N, Suquet M (2008a) Activation of oyster (*Crassostrea gigas*) sperm motility. In: Physiomar 08. Brest, France, 1–4 September 2008. Book of abstracts. http://archimer.ifremer.fr/doc/2008/acte-4600.pdf

Cosson J, Dreanno C, Fauvel C, Groison AL, Suquet M, Billard R (2008b) Marine fish spermatozoa: racing ephemeral swimmers. Reproduction 136:277–294. https://doi.org/10.1530/REP-07-0522

Dai PF, Rao XZ, Chen YS (2004) Ultrastructural studies on spermatozoon and spermatogenesis of *Ruditapes philippinarum*. Chin J Zool 39:26–32

Deguchi R, Osanai K (1994) Repetitive intracellular Ca^{2+} increases at fertilization and the role of Ca^{2+} in meiosis reinitiation from the first metaphase in oocytes of marine bivalves. Dev Biol 163:162–174. https://doi.org/10.1006/dbio.1994.1132

Demoy-Schneider M, Levêque A, Schmitt N, Le Pennec M, Cosson J (2012) Motility activation and metabolism characteristics of spermatozoa of the black-lip-pearl oyster *Pinctada margaritifera var: cumingii* (Jameson 1901). Theriogenology 77:53–64. https://doi.org/10.1016/j.theriogenology.2011.07.014

Demoy-Schneider M, Schmitt N, Suquet M, Labbé C, Boulais M, Prokopchuk G, Cosson J (2014) Biological characteristics of sperm in two oyster species: the Pacific oyster, *Crassostrea gigas,* and the black-lip-pearl oyster, *Pinctada Margaritifera*. In: Erickson BT (ed) Spermatozoa, biology, motility and function and chromosomal abnormalities. Nova Science Publishers Inc., New York, pp 15–75

Demoy-Schneider M, Schmitt N, Le Pennec G, Suquet M, Cosson J (2018) Quality assessment of cryopreserved black-lip pearl oyster *Pinctada margaritifera* spermatozoa. Aquaculture 97:278–286. https://doi.org/10.1016/j.aquaculture.2018.07.067

Devauchelle N, Faure C, Girard JP (1994) The quality of sperm of two bivalves: the oyster, *Crassostrea gigas* and the scallop, *Pecten maximus*. In: Muir J, Sevila F (eds.) Measures for success, vol 21. Eur. Aquac. Soc., Spec. Publ., pp 214–216

Dinamani P (1987) Gametogenic patterns in populations of Pacific oyster, *Crassostrea gigas*, in Northland, New Zealand. Aquaculture 64:65–76. https://doi.org/10.1016/0044-8486(87)90206-7

Dong Q, Eudeline B, Allen SK Jr, Tiersch TR (2002) Factors affecting sperm motility of tetraploid Pacific oysters. J Shellfish Res 21:719–723

Dong Q, Huang C, Tiersch TR (2005a) Spermatozoal ultrastructure of diploid and tetraploid Pacific oysters. Aquaculture 249:487–496. https://doi.org/10.1016/j.aquaculture.2005.03.009

Dong Q, Eudeline B, Huang C, Allen SK Jr, Tiersch TR (2005b) Commercial-scale sperm cryopreservation of diploid and tetraploid Pacific oysters, Crassostrea gigas. Cryobiology 50 (1):1–16

Dong Q, Huang C, Eudeline B, Tiersch TR (2005c) Systematic factor optimization for cryopreservation of shipped sperm samples of diploid Pacific oysters, *Crassostrea gigas*. Cryobiology 51 (2):176–197

Dong Q, Huang C, Eudeline B, Allen SK Jr, Tiersch TR (2006) Systematic factor optimization for sperm cryoconservation of tetraploïd Pacific oysters, *Crassostrea gigas*. Theriogenology 66:387–403

Dong Q, Huang C, Tiersch TR (2007a) Control of sperm concentration is necessary for standardization of sperm cryopreservation in aquatic species: evidence from sperm agglutination in oysters. Cryobiology 54(1):87–98. https://doi.org/10.1016/j.cryobiol.2006.11.007

Dong Q, Huang C, Eudeline B, Tiersch TR (2007b) Cryoprotectant optimization for sperm of diploïd Pacific oysters by use of commercial dairy sperm freezing facilities. Aquaculture 271:537–545

Dorange G, Le Pennec M (1989) Ultrastructural characteristics of spermatogenesis in Pecten maximus (Mollusca, Bivalvia). Invertebrate Reprod Dev 15:109–117. https://doi.org/10.1080/07924259.1989.9672031

Douroudi MS, Southgate PC, Mayer RJ (2001) The combined effects of temperature and salinity on embryos and larvae of the black-lip pearl oyster, *Pinctada margaritifera* (L.). Aquacult Res 30 (4):271–277. https://doi.org/10.1046/j.1365-2109.1999.00324.x

Drinkwaard AC (1999) Introductions and developments of oysters in the North Sea area: a review. Helgol Mar Res 52:301–308

Drozdov AL, Sharina SN, Tyurin SA (2009) Sperm ultrastructure in representatives of six bivalve families from Peter the Great Bay, Sea of Japan. Russ J Mar Biol 35:236–241

Drummond L, Mulcahy M, Culloty SC (2006) The reproductive biology of the Manila clam, *Ruditapes philippinarum*, from the north-west of Ireland. Aquaculture 254(1–4):326–340. https://doi.org/10.1016/j.aquaculture.2005.10.052

Everett EM, Williams PJ, Gibson G, Stewart DT (2004) Mitochondrial DNA polymorphisms and sperm motility in *Mytilus edulis* (Bivalvia : Mytilidae). J Exp Zool 301A:906–910

Falese LE, Russel MP, Dollahon N (2011) Spermcasting of spermatozeugmata by the bivalves *Nutricola confusa* and *N. tantilla*. Invertebr Biol 130:334–343

FAO (Food and Agriculture Organization of the United Nations) (2003a) *Crassostrea gigas* (Thunberg, 1793) (Ostreidae). In: Cultured aquatic species fact sheets. Fisheries and Aquaculture Department

FAO (Food and Agriculture Organization of the United Nations) (2003b) *Pecten maximus* (Linnaeus, 1758). In: Cultured aquatic species fact sheets. Fisheries and Aquaculture Department

FAO (Food and Agriculture Organization of the United Nations) (2003c) *Patinopecten yessoensis* (Jay, 1857) Cultured aquatic species fact sheets. Fisheries and Aquaculture Department

FAO (Food and Agricultural Organization of the United Nations) (2006) The state of world fisheries and aquaculture. Rome, pp 162

FAO (Food and Agricultural Organization of the United Nations) (2008) The state of world fisheries and aquaculture. Rome, pp 176

FAO (Food and Agricultural Organization of the United Nations) (2012) The state of world fisheries and aquaculture. Rome, pp 209

FAO (Food and Agricultural Organization of the United Nations) (2018) The state of world fisheries and aquaculture. Meeting the sustainable development goals. Rome. Licence: CC BY-NC-SA 3.0 IGO. pp 210

Faure C (1996) Paramètres physiologiques de l'émission et de l'activation des gametes mâles de deux mollusques bivalves, la coquille Saint Jacques Pecten maximus (L.) et l'huître creuse Crassostrea gigas (Thunberg). Ph.D. Thesis, Fac Diderot, Univ. Paris VII, Paris, pp 255

Faure C, Devauchelle N, Girard JP, Cosson J (1993) The quality of *Pecten maximus* sperm. Proceedings of the 9th international pectinid workshop, J fish res board can

Faure C, Devauchelle N, Girard JP (1994) Ionic factors affecting motility, respiration and fertilization of the sperm of the bivalve *Pecten maximus* (L.). J Comp Physiol Part A 164:444–450. https://doi.org/10.1007/BF00714581

Fujinoki M (2011) Serotonin-enhanced hyperactivation of hamster sperm. Reproduction 142 (2):255–266

Galstoff PS (1940) Physiology of reproduction of *Ostrea virginica*. III; Stimulation of spawning in the male oyster. Biol Bull 78:117–135

Galstoff PS (1964) The American Oyster Crassostrea virginica Gmelin, vol 64. Fishery Bulletin of the Fish and Wildlife Service, Washington, pp 219–238

Gibbons M, Goodsell JG, Castagna M, Luz R (1983) Chemical stimulation of spawning by serotonin in the ocean quahog *Artica islandica* (Linné). J Shellfish Res 3:203–205

Gibbons M, Castagna M (1984) Serotonin as an inducer of spawning in six bivalve species. Aquaculture 40:189–191

Gibbons M, Castagna M (1985) Responses of hard clam *Mercenaria mercenaria* (Linné) to induction of spawning by serotonin. J Shellfish Res 5:65–67

Goulletquer P (1997) A bibliography of the Manila clam *Tapes philippinarum*. IFREMER, RIDRV-97.02/RA/LA, La Tremblade, pp 122

Goulletquer P (2015) *Ruditapes philippinarum* (Adams & Reeve, 1850). In: FAO (ed) Cultured aquatic species information programme, 2005–2015. FAO Fisheries and Aquaculture Department, Rome

Goulletquer P, Héral M (1997) Marine molluscan production trends in France: from fisheries to aquaculture. *NOAA Technical Report NMFS*, vol 129, pp 137–164

Gourtay C, Pignet P, Mingant C, Ratiskol D, Suquet M (2016) Setting of a protocol for short-term storage of great scallop (Pecten maximus) sperm. SEB, Brighton, June 2016. https://doi.org/10.13140/RG.2.2.15920.97283

Grangeré K, Ménesguen A, Lefebvre S, Bacher C, Pouvreau S (2009) Modelling the influence of environmental factors on the physiological status of the Pacific oyster Crassostrea gigas in an estuarine embayment: The Baie des Veys (France). J Sea Res 62(2–3):147–158. https://doi.org/10.1016/j.seares.2009.02.002

Guo X, Hedgecock D, Hershberger WK, Cooper K, Allen SK Jr (1998) Genetic determinants of protandric sex in the Pacific oyster, *Crassostrea gigas* (Thunberg). Evolution 52(2):394–402

Gwo JC (2000) Cryoconservation of aquatic invertebrate semen: a review. Aquacult Res 31:259–271

Gwo JC, Yang WT, Sheu YT, Cheng HY (2002) Spermatozoan morphology of four species of bivalve (*Heterodonta, veneridae*) from Taiwan. Tissue Cell 34(1):39–43. https://doi.org/10.1054/tice.2001.0222

Gwo JC, Wu CY, Chang WS, Cheng HY (2003) Evaluation of damage in Pacific oyster (*Crassostrea gigas*) spermatooa before and after cryopreservation using comet assay. Cryo Letters 24:171–180

Haelters J, Kerckhof F (2009) Background document Ostrea edulis and Ostrea edulis beds. OSPAR Commission, pp 21

Hassan MM, Qin JG, Li X (2015) Sperm cryopreservation in oysters: a review of its current status and potentials for future application in aquaculture. Aquaculture 438:24–32

Hassan MM, Li X, Liu Y, Qin JG (2017a) Sperm cryopreservation in the spermcasting Australian flat oyster *Ostrea angasi* by a programmable freezing method. Cryobiology 76:119–124

Hassan MM, Li X, Qin JG (2017b) Improvement of post-thaw sperm survivals using liquid nitrogen vapor in a spermcasting oyster *Ostrea angasi*. Cryobiology 78:1–7

Helm MM, Bourne N, Lovatelli A (2004) Hatchery culture of bivalves. A practical manual. FAO, Fisheries Technical Paper N°471, Rome, p 200

Heres P, Rodriguez-Riveiro R, Troncoso J, Paredes E (2019) Toxicity tests of cryoprotecting agents for *Mytilus galloprovincialis* (Lamark, 1819) early developmental stages. Cryobiology 86:40. https://doi.org/10.1016/j.cryobiol.2019.01.001

Hirai S, Kishimoto T, Kadan AL, Kanatani H, Koide SS (1988) Induction of spawning and oocyte maturation by 5-hydroxytryptamine in the surf clam. J Exp Zool 245:318–321. https://doi.org/10.1002/jez.1402450312

Ho HC, Suarez SS (2003) Characterization of the intracellular calcium store at the base of the sperm flagellum that regulates hyperactivated motility. Biol Reprod 68:1590–1596. https://doi.org/10.1095/biolreprod.102.011320

Howard DR, Trantow CM, Thaler CD (2004) Motility of a biflagellate sperm: waveform analysis and cyclic nucleotide activation. Cell Motil Cytoskeleton 59:120–130. https://doi.org/10.1002/cm.20027

Howarth LM, Stewart BD (2014) The dredge fishery for scallops in the United Kingdom (UK): effects on marine ecosystems and proposals for future management. Report to the Sustainable Inshore Fisheries Trust. Marine Ecosystem Management Report no. 5, University of York, pp 54

Horváth Ā, Bubalo A, Čučević A, Bartulović V, Kotrik L, Urbányi B, Glamuzina B (2012) Cryopreservation of sperm and larvae of of the European flat oyster (*Ostrea edulis*). J Appl Ichthyol 28:948–951

Hui B, Vonau V, Moriceau J, Tetumu R, Vanaa V, Demoy-Schneider M, Suquet M, Le Moullac G (2011) Hatchery-sclae trials using cryopreserved spermatozoa of black-lip pearl oyster, *Pinctada margaritifera*. Aquat Living Resour 24:219–223

Humphreys J, Harris MRC, Herbert RJH, Farrell P, Jensen A, Cragg SM (2015) Introduction, dispersal and naturalization of the Manila clam Ruditapes philippinarum in British estuaries, 1980–2010. J Mar Biol Assoc U K 95(6):1163–1172. https://doi.org/10.1017/S0025315415000132

Huo Z, Li Y, Zhang X, Yan X, Yang F (2017) Growth improvement of shell length in the orange strain of Manila clam, *Ruditapes philippinarum*. J World Aquacult Society 48(6):860–866. https://doi.org/10.1111/jwas.12392

Hwang JJ (2007) Reproductive cycles of the pearl oysters, *Pinctada fucata* (Gould) and *Pinctada margaritifera* (Linnaeus) (Bivalvia: Pteriidae) in southwestern Taiwan waters. J Mar Sci Technol 15(2):67–75

Hwang SW, Chen HP (1973) Fertility of male oyster gametes after freeze-thawing. Chin Am J Commun Rural Reconstr Fish Ser 15:1–5

Ieropoli S, Masullo P, Santo M, Sansone G (2004) Effects of extender composition, cooling rate and freezing on the fertilization viability of spermatozoa of the Pacific oyster (*Crassostrea gigas*). Cryobiology 49:250–257

Inaba K (2007) Molecular basis of sperm flagellar axonemes: structural and evolutionary aspects. Ann N Y Acad Sci 1101:506–526. https://doi.org/10.1196/annals.1389.017

Intergovernmental Platform on Biodiversity and Ecosystem Services (IPBES) Summary for policymarkers of the global assessment report on biodiversity and ecosystem services of the Intergovernmental Science-policy platform on Biodiversity and Ecosystem Services. Advance unedited version, 6 May 2019, pp 39

Jaziri H (1990) Variations génétiques et structuration biogéographique chez un bivalve marin: l'huître plate *Ostrea edulis* L. PhD dissertation, University of Montpellier, France

Jensen A, Humphreys J, Caldow R, Cesar C (2005) The Manila clam in Poole harbour. In: Humphreys J, May V (eds) The ecology of Poole harbour, vol 13, pp 163–173

Jha M, Côté J, Hoeh WR, Blier PU, Stewart DT (2007) Sperm motility in *Mytilus edulis* in relation to mitochondrial DNA polymorphisms : implications for the evolution of doubly uniparental inheritance in bivalves. Evolution 62:99–106

Jiménez-Trejo F, Tapia-Rodriguez M, Cerbon M, Kuhn DM, Manjarrez-Gutiérrez G, Mendoza-Rodriguez CA, Picazo O (2012) Evidence of 5-HT components in human sperm: implications for protein tyrosine phosphorylation and the physiology of motility. Reproduction 144:677–685. https://doi.org/10.1530/REP-12-0145

Jones GG, Sanford CL and Jones BL (1993) Manila clams: Hatchery and Nursery Methods Innovative Aquaculture Products Ltd and Science Council of British Columbia, Skerry Bay, Lasqueti Island, Victoria, BC, Canada, VOR 2JO, pp 73

Joubert C, Linard C, Le Moullac G, Soyez C, Saulnier D, Teaniniuraitemoana V, Ky CL, Gueguen Y (2014) Temperature and food influence shell growth and mantle gene expression of shell matrix proteins in the pearl oyster *Pinctada margaritifera*. PLoS One 9(8):e103944. https://doi.org/10.1371/journal.pone.0103944

Kadam AL, Koide SS (1990) Stimulation of *Spisula* sperm motility by 5-hydroxytryptamine analogs invertebrate. Reprod Dev 17:33–37. https://doi.org/10.1080/07924259.1990.9672085

Kim JH, Chung JS, Lee KY (2013) Ultrastructural characteristics of the testis, spermatogenesis and taxonomic values of sperm morphology in male *Ruditapes philippinarum* in Western Korea. Dev Reprod 17:121–132. https://doi.org/10.12717/DR.2013.17.2.121

Konishi K, Kawamura K, Furuita H, Komaru A (1998) Spermatogenesis of the freshwater clam *Corbicula aff. fluminea* Mueller (Bivalvia: Corbiculidae). J Shellfish Res 17:185–189

Korringa P (1947) Relations between the moon and periodicity in the breeding of marine animals. Ecol Monogr 17:347–381

Korukura H, Namba K, Ishikawa T (1990) Lesions of spermatozoa by cryopreservation in oyster *Crassostrea gigas*. Nippon Suisan Gakkaishi 56:1803–1806

Kraemer LR, Swanson C, Galloway M, Kraemer R (1986) Biological basis of behavior in *Corbicula fluminea*. II. Functional morphology of reproduction and development and review of evidence for self-fertilization. In: Proceedings of the second international *Corbicula* symposium. Am Malacol Bull, p 193–201

Kuroda Y, Kaneko S, Yoshimura Y, Nozawa S, Mikoshiba K (1999) Are there inositol 1,4,5-triphosphate (IP3) receptors in human sperm? Life Sci 65:135–143. https://doi.org/10.1016/S0024-3205(99)00230-1

Ky CL, Lau C, Sham Koua M, Lo C (2015) Growth performance comparison of *Pinctada margaritifera* juveniles produced by thermal shock or gonad scarification spawning procedures. J Shellfish Res 34(3):811–817. https://doi.org/10.2983/035.034.0310

Laing ISD, Utting SD, Kilada RWS (1987) Interactive effect of diet and temperature on the growth of juvenile clams. J Exp Mar Biol Ecol 113:23–28

Laing ISD, Utting SD (1994) The physiologuy and biochemistry of diploid and triploid clams. J Exp Mar Biol Ecol 184:159–169

Laing ISD, Lees D, Page DJ, Henshilwood K (2004) Research on shellfish cultivation, vol 122. CEFAS, Lowestoft, p 68

Lannan JE (1971) Experimental self-fertilization of the pacific oyster, *Crassostrea gigas*, utilizing cryopreserved sperm. Genetics 68:599–601

Lapègue S, Beaumont A, Boudry P, Goulletquer P (2006) European flat oyster - Ostrea edulis. In: Crosetti D, Lapègue S, Olesen SI, Svaasand T (eds).Genetic effects of domestication, culture and breeding of fish and shellfish, and their impacts on wild populations. Genimpact project: evaluation of genetic impact of aquaculture activities on native populations. A European network. WP1 workshop "genetics of domestication, breeding and enhancement of performance of fish and shellfish", Viterbo, Italy, 12–17th June, 2006, pp 6

Lee JS, Park JS, Shin YK, Yeon Gyu Lee YG, Park JJ (2013) Sequential hermaphroditism in Manila clam *Ruditapes philippinarum* (*Bivalvia: Veneridae*). Invertebrate Reprod Dev 57 (3):185–188. https://doi.org/10.1080/07924259.2012.717109

Le Pennec M (2003) The pelagic life of the pectinid *Pecten maximus*-a review. ICES J Mar Sci 60 (2):211–223. https://doi.org/10.1016/S1054-3139(02)00270-9

Le Pennec M, Anastas M, Bichet H, Buestel D, Cochard JC, Cochennec-Loreau N, Coeroli M, Conte E, Correia P, Fougerouse-Tsing A, Langy S, Le Moullac G, Lo C, Peltzer L, Pham A (2010) Huître perlière et perle de Tahiti. Université de la Polynésie Française, pp 204

Li Q, Osada M, Kashihara M, Hirohashi K, Kijima A (2000a) Effects of ultraviolet irradiation on genetical inactivation and morphological structure of sperm of the Japanese scallop, *Patinopecten yessoensis*. Aquaculture 186:233–242. https://doi.org/10.1016/S0044-8486(99)00384-1

Li C, Li J, Xue QZ (2000b) Cryopreservation of the spermatozoa of *Chlamys* (*Azumapecten*) *farreri*. Mar Fish Res 21:57–62

Liu Y, Li X, Robinson N, Qin JG (2015) Cryoconservation in marine mollusk: a review. Aquacult Int 23:1505–1524

Liu Y, Liu S, Liu B, Li X (2018) Cryoconservation of strip spawned sperm using programmable freezing technique in the blue mussel, *Mytilus galloprovincialis*. J Oceanol Limnol 36 (6):2351–2357

Loosanoff VL (1962) Gametogenesis and spawning of the European oyster, *O. edulis*, in waters of Maine. Biol Bull 122(1):86–94

Loosanoff VL, Davis H (1950) Rearing of bivalve molluscs. Adv Mar Biol 1:1–136

Lyons L, Jerry DR, Southgate PC (2005) Cryopreservation of black-lip pearl oyster (*Pinctada margaritifera*, l.) spermatozoa: effects of cryoprotectants on spermatozoa motility. J Shellfish Res 24:1187–1190. https://doi.org/10.2983/0730-8000(2005)24[1187:COBPOP]2.0.CO.2

Mackie GL (1984) Bivalves. In: Tompa AS, Verdonk NH, van den Biggelaar JAM (eds) The Mollusca, Reproduction, vol 7. New York, Academic, pp 351–418

Mann R (1979) The effect of temperature on growth, physiology and gametogenesis in the Manila clam Tapes Philippinarum (Adam & Reeve, 1850). J Exp Mar Biol Ecol 38:121–133

Mason J (1983) Scallop: an queen fisheries in the British isles. Fishing New Books, Farnham

Matsutani T, Nomura T (1982) Induction of spawning by serotonin in the scallop *Patinopecten yessoensis* (Jay). Mar Biol Lett 3:353–358

Matsutani T, Nomura T (1987) *In vitro* effects of serotonin and prostaglandins in release of eggs from the ovary of the scallop *Patinopecten yessoensis*. Gen Comp Endocrinol 67:111–118

Meizel S, Turner KO (1983) Serotonin or its agonist 5-methoxytryptamine can stimulate hamster sperm acrosome reactions in a more direct manner than catecholamines. J Exp Zool 226:171–174. https://doi.org/10.1002/jez.1402260120

Meng Q, Bao Z, Wang Z, Wang S, Hu J, Hu X, Huang X (2012) Growth and reproductive performance of triploid Yesso scallops (*Patinopecten yessoensis*) induced by hypotonic shock. J Shellfish Res 31(4):1113–1122. https://doi.org/10.2983/035.031.0422

Mita M, Nakamura M (1993) Phosphatidylcholine is an endogenous substrate for energy metabolism in spermatozoa of sea urchins of the order Echinoidea. Zool Sci 10:73–83

Morisawa M (2008) Adaptation and strategy for fertilization in the sperm of teleost fishes. J Appl Ichthyol 24:362–370. https://doi.org/10.1111/j.1439-0426.2008.01126.x

Morse MP, Zardus JD (1997) Bivalvia. In: Harrison FW, Kohn AJ (eds) Microscopic anatomy of invertebrates, vol 6A. Wiley, Mollusca II, New York, pp 7–118

Moura P, Vasconcelos P, Pereira F, Chainho P (2018) Reproductive cycle of the Manila clam (*Ruditapes philippinarum*): an intensively harvested invasive species in the Tagus Estuary (Portugal). J Mar Biol Assoc U K 98(7):1645–1657. https://doi.org/10.1017/S0025315417001382

Nakajima A, Morita M, Takemura A, Kamimura S, Okuno M (2005) Increase in intracellular pH induces phosphorylation of axonemal proteins for flagellar motility activation in starfish sperm. J Exp Biol 208:4411–4418. https://doi.org/10.1242/jeb.01906

O'Foighil D (1989) Role of spermatozeugmata in the spawning ecology of the brooding oyster *Ostrea edulis*. Gamete Res 24:219–228. https://doi.org/10.1002/mrd.1120240209

Ohta H, Kawamoto T, Isowa K, Aoki H, Hayashi M, Narita T, Komaru A (2007) Motility of spermatozoa obtained from testes of Japanese pearl oyster *Pinctada fucata martensii*. Fish Sci 73:107–111. https://doi.org/10.1111/j.1444-2906.2007.01308.x

Paniagua-Chávez CG, Jenkins J, Segovia M, Tiersch TR (2006) Assessment of gamete quality for the eastern oyster (Crassostrea virginica) by use of fluorescent dyes. Cryobiology 53 (1):128–138

Paredes E (2015) Exploring the evolution of marine invertebrate cryopreservation-landmarks, state of the art and future lines of research. Cryobiology 71(2):198–209. https://doi.org/10.1016/j.cryobiol.2015.08.011

Parisi E, De Prisco P, Capasso A, del Prete M (1984) Serotonin and sperm motility. Cell Biol Int Rep 8:95. https://doi.org/10.1016/0309-1651(84)90075-4

Pouvreau S (2017) Underwater images of the last native oysters beds in Brittany (France) IFREMER. https://image.ifremer.fr/data/00377/48842/

Pouvreau S, Tiapari J, Gangnery A, Lagarde F, Garnier M, Teissier H, Haumani G, Buestel D, Bodoy A (2000a) Growth of the black-lip pearl oyster, *Pinctada margaritifera*, in suspended culture under hydrobiological conditions of Takapoto Lagoon (French Polynesia). Aquaculture 184:133–154

Pouvreau S, Gangnery A, Tiapari J, Lagarde F, Garnier M, Bodoy A (2000b) Gametogenic cycleand reproductive effort of the tropical black-lip pearl oyster, *Pinctada margaritifera* (Bivalvia: Pteridae), cultivated in Takapoto atoll (French Polynesia). Aquat Living Resour 13:37–48

Pouvreau S, Prasil V (2001) Growth of the black-lip pearl oyster, *Pinctada margaritifera*, at nine culture sites of French Polynesia synthesis of several sampling designs conducted between 1994 and 1999. Aquat Living Resour 14:155–163

Riesco MF, Félix F, Matias D, Joaquim S, Suquet M, Cabrita E (2017) First study in cryopreserved *Crassostrea angulata* sperm. Gen Comp Endocrinol 245:108–115. https://doi.org/10.1016/j.ygcen.2016.05.003

Ruiz C, Abad M, Sedano F, Garcia-Martin LO, Sanchez Lopez JL (1992) Influence of seasonal environmental changes on the gamete production and biochemical composition of *Crassostrea gigas* (Thunberg) in suspended culture in El Grove, Galicia, Spain. J Exp Mar Biol Ecol 155:249–262

Sakurai I, Seto M, Makiguchi N, Ogata T (2000) Tolerance of the Japanese scallop, *Patinopecten yessoensis*, to low concentrations of dissolved oxygen and salinity. Aquacult Sci 48 (1):137–138. https://doi.org/10.11233/aquaculturesci1953.48.137

Sanchez D, Labarca P, Darszon A (2001) Sea urchin sperm cation-selective channels directly modulated by cAMP. FEBS Lett 503:111–115. https://doi.org/10.1016/S0014-5793(01) 02713-2

Sastry AN (1979) *Pelecypoda* (excluding Ostreidae). In: Giese AC, Pearse JS (eds) Reproduction of marine invertebrates, Molluscs: Pelecypods and lesser classes, vol V. Academic, New York, pp 113–292

Scarlato OA (1981) Bivalves of temperate waters of the Northwestern part of the Pacific ocean. Nauka Press, Leningrad, p 408

Shpigel M (1989) Gametogenesis of the European flat oyster (*Ostrea edulis*) and Pacific oyster (*Crassostrea gigas*) in fann water in Israel. Aquaculture 80:343–349

Silina AV (2018) Sex change in scallop *Patinopecten yessoensis*: response to population composition. PeerJ 6:e5240. https://doi.org/10.7717/peerj.5240

Sims NA (1994) Growth of wild and cultured black-lip pearl oysters *Pinctada margaritifera* (L.) (Pteriidae Bivalvia) in the Cook islands. Aquaculture 122:181–191

Slater JW (2005) Spawning of king scallops, *Pecten maximus* (L.) in Mulroy Bay and the relationship with spatfall intensity. J Shellfish Res 24:951–958. https://doi.org/10.2983/0730-8000 (2005)24[951:SOKSPM]2.0.CO

Smith JF, Pugh PA, Tervit HR, Roberts RD, Janke AR, Kaspar HF, Adams SL (2001) Crypreservation of shellfish sperm, eggs and embryos. Proc NZ Soc Anim Prod 61:31–34

Stanley CA (1967) The commercial scallop Pecten maximus in Northern irish waters. PhD thesis, Queen's University of Ireland, Belfast, pp 111

Stefano GB, Catapane J, Aiello E (1976) Dopaminergic agents: influence on serotonin in the molluscan nervous system. Science 194:539–541

Spencer BE (2008) Molluscan shellfish farming. Wiley, Hoboken, p 296

Soyez C, Huvet A, Gueguen Y, Lo C, Le Moullac G (2011) Determination of gender in the pearl oyster *Pinctada margaritifera*. J Shellfish Res 30(2):231–240

Staeger WH (1974) Cryobiological investigations of the gametes of the pacific oyster, *Crassostrea gigas*. Department of Fisheries and Wildlife. Oregon Stae University, Corvallis

Suquet M, Dreanno C, Fauvel C, Cosson J, Billard R (2000) Cryoconservation of sperm marine fish. Aquac Res 31:231–243

Suquet M, Labbé C, Brizard R, Donval A, Le Coz JR, Quéré C, Haffray P (2010) Changes in motility, ATP content, morphology and fertilization capacity during the movement phase of tetraploid Pacific oyster (*Crassostrea gigas*) sperm. Theriogenology 74:111–117

Suquet M, Cosson J, Donval A, Labbé C, Boulais M, Haffray P, Bernard I, Fauvel C (2012) Marathon versus sprint racers: an adaptation of sperm characteristics to the reproductive strategy of Pacific oyster, turbot and seabass. J Appl Ichthyol 28:956–960

Suquet M, Quéré C, Maingant C, Lebrun L, Ratiskol D, Miner P, Cosson J (2013) Effect of sampling location, release technique and time after activation on the movement characteristics of scallop (*Pecten maximus*) sperm. Aquat Living Resour 26:215–220. https://doi.org/10.1051/alr/2013048

Suquet M, Gourtay C, Donval A, Le Goïc N, Quéré C, Malo F, Le Grand J, Ratiskol D, Maingant C, Fauvel C (2016) The quality of great scallop (*Pecten maximus*) sperm after thawing. Gen Comp Endocrinol 229:127–131

Suquet M, Quéau I, Le Grand J, Ratiskol D, Pouvreau S (2017) Caractéristiques biologiques des gamètes et larves d'huître plate européenne (Ostrea edulis): données préliminaires. Rapport interne IFREMER/PFOM/PLI, pp 21

Suquet M, Pouvreau S, Queau I, Boulais M, Le Grand J, Ratiskol D, Cosson J (2018) Biological characteristics of sperm in European flat oyster (*Ostrea edulis*). Aquat Living Resour 31:20. https://doi.org/10.1051/alr/2018008

Tanabe T, Yuan Y, Nakamura S, Itoh N, Takahashi KG, Osada M (2010) The role in spawning of a putative serotonin receptor isolated from the germ and ciliary cells of the gonoduct in the gonad of the Japanese scallop, *Patinopecten yessoensis*. Gen Comp Endocrinol 166:620–627. https://doi.org/10.1016/j.ygcen.2010.01.014

Thielley M, Weppe M, Herbaut C (1993) Ultrastructural study of gametogenesis in the French Polynesian black pearl oyster *Pinctada margaritifera* (Mollusca Bivalvia) I-spermatogenesis. J Shellfish Res 12:41–47

Tiersch TR, Yang H, Jenkins JA, Dong Q (2007) Sperm cryopreservation in fish and shellfish. In: Roldan ERS, Gomendio M (eds) Spermatology. Nottingham University Press, Nottingham, SRF supplement 65. pp 493–508

Tranter DJ (1958) Reproduction in Australian pearl oysters (Lamellibranchia). IV. *Pinctada margaritifera* (Linnaeus). Aust J Mar Freshwater Res 9:509–525

Treen N, Itoh N, Miura H, Kikuchi I, Ueda T, Takahashi KG, Ubuka T, Yamamoto K, Sharp PJ, Tsutsui K, Osada M (2012) Mollusc gonadotropin-releasing hormone directly regulates gonadal functions: a primitive endocrine system controlling reproduction. Gen Comp Endocrinol 176:167–172. https://doi.org/10.1016/j.ygcen.2012.01.008

Ubertini M, Lagarde F, Mortreux S, Le Gall P, Chiantella C, Fiandrino A, Bernard I, Pouvreau S, d'Orbcastel ER (2017) Gametogenesis, spawning behavior and larval abundance of the Pacific oyster Crassostrea gigas in the Thau lagoon: evidence of an environment-dependent strategy. Aquaculture 473:51–61

Uddin J, Park K, Kang DH, Choi KSA (2007) Comparative reproductive biology of Yezo scallop, *Patinopecten yessoensis*, under two different culture systems on the east coast of Korea. Aquaculture 265(1):139–147. https://doi.org/10.1016/j.aquaculture.2007.01.047

Vanderhost G, Dott H, Samuels J, Genade A (1985) Short-term and long-term storage of viable oyster sperm. S Afr J Sci 81:404–405

Van Dyk M (2011) *Pinctada margaritifera* black lipped pearl oyster. Invertebrates of the Coral Sea. https://www.gbri.org.au. Accessed 20 July 2019

Van Houcke J, Medina I, Maehre HK, Cornet J, Cardinal M, Linssen J, Luten J (2017) The effect of algae diets (*Skeletonema costatum* and *Rhodomonas baltica*) on the biochemical composition and sensory characteristics of Pacific cupped oysters (*Crassostrea gigas*) during land-based refinement. Food Res Int 100(1):151–160. https://doi.org/10.1016/j.foodres.2017.06.041

Velasco LA, Barros J, Acosta E (2007) Spawning induction and early development of the Carribbean scallops Argopecten nucleus and Nodipecten nodosus. Aquaculture 266:153–165

Vélez A, Alifa E, Azuaje O (1990) Induction of spawning by temperature and serotonin in the hermaphroditic tropical scallop, *Pecten ziczac*. Aquaculture 84:307–213

Vitiello V, Carlino PA, Del Prete F, Langellotti AL, Sansone G (2011) Effects of cooling and freezing on the motility of *Ostrea edulis* (L., 1758) spermatozoa after thawing. Cryobiology 63:118–124

Wada KT, Temkin I (2008) Taxonomy and phylogeny. In: Southgate PC, Lucas JS (eds) The pearl oyster. Elsevier Press, Oxford, pp 37–76

Wang Q, He M (2014) Molecular characterization and analysis of a putative 5-HT receptor involved in reproduction process of the pearl oyster *Pinctada fucata*. Gen Comp Endocrinol 204:71–79. https://doi.org/10.1016/j.ygcen.2014.05.010

Wilson-Leedy JG, Ingermann RL (2007) Development of a novel CASA system based on open source software for characterization of zebrafish sperm motility parameters. Theriogenology 67:661–672. https://doi.org/10.1016/j.theriogenology.2006.10.003

Yamamoto G (1950) Ecological note of the spawning cycle of the scallop *Pecten yessoensis* (Jay). Sci Rep Tôhoku University (Ser 4 Biol) 18:477–481

Yamamoto G (1951a) Induction of spawning in the scallop *Pecten yessoensis* (Jay). Sci Rep Tôhoku University (Ser 4 Biol) 19(1):7–10

Yamamoto G (1951b) Ecological study on the spawning cycle of the scallop *Pecten yessoensis* in Mutsu Bay. Bull Japan Soc Fish 17:53–56

Yamamoto G (1952) Further study on the ecology of spawning in the scallop in relation to lunar phases, temperature and plankton. Sci Rep Tôhoku University (Ser 4 Biol) 19:247–254

Yamamoto G (1964) Studies on the propagation of the scallop, *Patinopecten yessoensis* (Jay) in Mutsu Bay. Suisan Zoyosko-ku Sosho 6:1–77

Yang PM, Yang AG, Liu ZH, Zhou LQ (2008) Ultrastructure and cryodamage studies of *Patinopecten yessoensis* sperm by electron microscopy. Marine Fisheries Research

Yang PM, Yang AG, Liu ZH, Zhou LQ (2007) Cryopreservation of *Patinopecten yessoensis* sperm and its application to hybridization. J Shanghai Fish Univ 16:351–356

Yankson K, Moyse J (1991) Cryopreservation of the spermatozoa of *Crassostrea tulipa* and three other oysters. Aquaculture 97:259–267

Young RJ, Laing JC (1990) Biogenic amine binding sites in rabbit spermatozoa. Biochem Int 21:781–787

Zhang G, Yan X (2006) A new three-phase culture method for Manila clam, *Ruditapes philippinarum*, farming in northern China. Aquaculture 258(1–4):452–461. https://doi.org/10.1016/j.aquaculture.2006.04.046

Cryopreservation of Germ Stem Cells in Fish 12

Roman Franěk and Martin Pšenička

Abstract

Germ cells have drawn much attention as gamete precursors because of their exceptional potential for fundamental biological studies, as well as for being a potential alternative to gene resource banking for breeding and restocking purposes via surrogacy. Research has developed markers for germ cell identification, shedding light on their origin and specification mechanism. Then, pioneering studies focused on germ cell transplantation into the surrogate host with donor-derived gamete production have followed suit. The advent of transgenic technologies allowed the development of lines expressing green fluorescent protein (GFP) in their germ cells exclusively or the use of artificially synthesised mRNA conjugated with GFP to label and trace germ cells in vivo and facilitate studies on transplantation. The power of germ cell manipulation techniques was significantly enhanced by the development of a wide range of cryopreservation methods; with these approaches, valuable genetic material can be cryopreserved even during the blastula stage, as well as harvested from an adult specimen. In turn, the germ cells in the cryogenic storage can be recovered by transplantation in the surrogate host, producing donor-derived gametes after maturation. The aim of this chapter is to provide basic knowledge about a relatively novel biotechnological approach in fish reproduction and summarise performed studies on germ cell cryopreservation.

Keywords

Germ cells · Fish · Surrogacy · Transplantation · Cryopreservation · Spermatogonia · Oogonia

R. Franěk (✉) · M. Pšenička
Faculty of Fisheries and Protection of Waters, South Bohemian Research Center of Aquaculture and Biodiversity of Hydrocenoses, University of South Bohemia in České Budějovice, Vodňany, Czech Republic
e-mail: franek@frov.jcu.cz; psenicka@frov.jcu.cz

J. Betsy, S. Kumar (eds.), *Cryopreservation of Fish Gametes*,
https://doi.org/10.1007/978-981-15-4025-7_12

12.1 Germline in Teleost Fish

All sexually reproducing organisms have two distinct cell lineages derived at early embryonic development (Kunwar and Lehmann 2006). The body is derived from the somatic lineage, and it is mortal because somatic cells are not capable of passing their genetic information to the progeny. In contrast to the somatic lineage, the germ cell lineage is capable of undergoing self-renewal and meiosis, resulting in the production of gametes. These cells are responsible for the preservation of future generations because they are the founder cells (precursors) of gametes (Starz-Gaiano and Lehmann 2001). Primordial germ cells (PGCs) transport genetic information from parents to offspring (McLaren 2003), and the genetic link between all generations is provided (Braat et al. 1999). Because of this unique feature, the germ cell lineage is considered immortal (Braat et al. 1999). Germ cells are considered to belong to the germline during all stages of life (Raz 2003). These cells are first formed during very early development as presumptive primordial germ cells. They undergo unequal division when one cell is incorporated into the soma and the second cell is responsible for germ stem line maintenance. The cell which has both daughter cells committed to the germline can be called a PGC. PGCs are morphologically distinct from other somatic cells; they are bigger (10–20 μm) with large nuclei (6–10 μm) (Braat et al. 1999).

12.1.1 Primordial Germ Cells—Specification

The mechanisms responsible for germ lineage formation in fish have been described extensively when PGCs are specified by the maternally deposited germplasm (or so-called nuage) (Dosch 2015). The germplasm contains an electro-dense structure with determinants responsible for PGC specification and maintenance, such as *vasa* (DEAD box RNA helicase) (Yoon et al. 1997), *nanos-1* (RNA binding zinc fin protein) (Dosch 2015), *dazl* (DAZ family RNA-binding protein), *tudors* (Tudor domain proteins) and *bucky ball* (Xu et al. 2010). After fertilisation, the germplasm is rapidly moved towards the animal pole by yolk streams and segregated between cleavage furrows during the two-cell stage (Dosch 2015). The germplasm is afterwards segregated into four clusters at the four- and eight-cell stages, still located at the margins of the cleavage furrows. The germplasm remains unequally segregated into four PGCs (by one daughter cell maintaining) until the 512-cell stage. Germplasm distribution during cell division is precluded by its local integration within the cell. At 30% epiboly, the germplasm is distributed symmetrically through the cell and is inherited by both daughter cells (Knaut et al. 2000).

12.1.2 PGC Migration

The PGCs in developing embryo are formed in four different random positions out of the future gonad, so they need to undergo migration towards the genital ridge from distinct positions (Yoon et al. 1997). The PGCs in four clusters in a square-like configuration are the basis for the formation of two bilateral groups. These groups subsequently set the basis for the germline (Braat et al. 1999). Early migration is additionally supported by the movement of surrounding somatic tissues, so the first phases of migration are considered passive (Weidinger et al. 1999). In an active migration event, PGCs undergo morphological changes before they reach the genital ridge (Raz and Reichman-Fried 2006). At the time of the four formed clusters, the PGCs possess a simple cell morphology attributed to the non-migratory phase. However, the onset of morphological changes is described 1.5 h after specification, when the PGCs become elongated and polarised, with developed pseudopodia. At 6 h post fertilisation (hpf), almost all PGCs undergo active migration towards the future genital ridge (Blaser et al. 2005).

The event of active migration towards the genital ridge is guided by chemoreceptors stimulated by chemokine signals when the migrating PGCs respond to attractive cues, which are produced alongside their migration path. Zebrafish PGCs have been shown to express chemokine receptor CXCR4, which responds to stromal-derived factor (SDF-1) expression. The necessity of chemotaxis for proper PGC migration into the genital ridge was demonstrated in CXCR4 and SDF-1 knockdown using antisense morpholino oligonucleotide, resulting in correct migration failure, when the PGCs were found in ectopic positions; moreover, alterations in PGC polarity were observed in morphants (Doitsidou et al. 2002). The migration process is further regulated by several genes, such as *staufen*, *dead end*, *Igf* and *Pik3* (responsible for migration), *hmgcr* and *quemao* (attraction to the mesoderm) and sdf1 (chemotaxis). The aforementioned genes, which are responsible for germline development, are suggested to be conserved amongst different species across animals (Xu et al. 2010).

The process of migration seems to be strongly controlled with a conserved pathway across fish. The guiding signals controlling the migration of PGCs are not related to phylogenetic distance. This was proven several times by xenogenous transplantation, followed by the observation of successful migration amongst cyprinids (Saito et al. 2008, 2010), sturgeon (donor) and goldfish (recipient) (Saito et al. 2014) or between Japanese eel (donor) and zebrafish (recipient) (Saito et al. 2011).

The PGCs settle in the region of the genital ridge after the migration event is completed (Fig. 12.1). Gonadal PGCs proliferate and later undergo sex differentiation into spermatogonia or oogonia with the support of gonadal somatic cells. The process of sex differentiation in fish is determined by genetic factors, but endocrine or ultimately exogenous factors can override primary sex fate (Strüssmann and Nakamura 2002). Sex differentiation is not always straightforward in fish, as a juvenile ovary phase common for all zebrafish, followed by meiotic oocyte apoptosis, is known to preclude testis development (Pradhan and Olsson 2014). The capacity of the gonads to produce gametes is maintained not only through self-renewal but also through differentiation abilities. A-type undifferentiated spermatogonia are capable of entering mitosis and producing again A-type spermatogonia or differentiating into B-type spermatogonia. Then, the differentiated

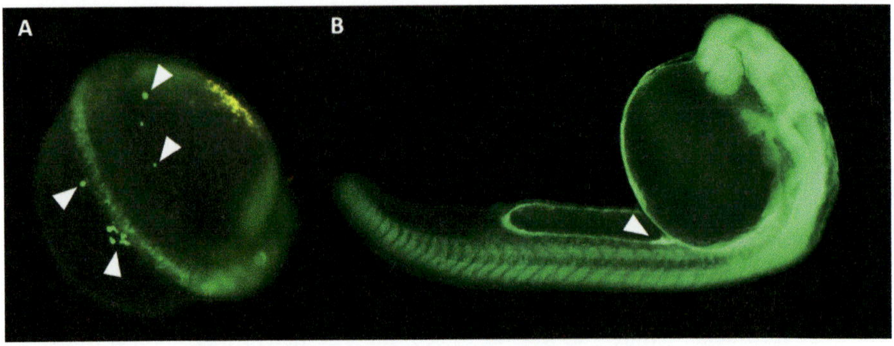

Fig. 12.1 PGC migration in transgenic vas:EGFP zebrafish embryos at 10-somite stage (**a**) and prim-6 stage with PGCs in the genital ridge (**b**). PGCs are indicated by white arrowheads

spermatogonia enter into meiosis, producing primary and secondary spermatocytes that differentiate into functional spermatozoa (Schulz et al. 2010). Female gonad differentiation is initiated after the proliferation of PGCs and their transformation into oogonia. The oocytes then enter meiosis, followed by vitellogenic growth. Before oocyte maturation and germinal vesicle breakdown, the development of inner and outer oocyte envelopes is completed. Prior to ovulation, the oocyte is extruded from the follicle, and a second polar body is extruded after activation, resulting in the fusion of the haploid oocyte nucleus with the haploid sperm and thus giving rise to the diploid zygote (Lubzens et al. 2010).

12.2 Germ Cell Technologies in Fish

A new era of biotechnologies started in fish after the identification of various germ cell markers, allowing investigations into the processes connected with germline development. In turn, this led to the development of techniques for germ cell transfer in various species in order to study the biology of PGCs and introduce the cells into different hosts and produce donor-derived progeny. Surrogate reproduction could be beneficial for species preservation, aquaculture and basic research. The techniques of germ cell transplantation rely on several consecutive steps, according to the chosen methodology and mainly the developmental stage of the host and the recipient (Goto and Saito 2019). PGC transfer can be performed as early as 3 h post fertilisation when zebrafish embryo reaches the 1 k cell stage (Lin et al. 1992), whereas differentiated germ stem cells (GSCs) can be transplanted even between adults (Lacerda et al. 2010). Together with cryopreservation possibilities, the use of germ cell technologies in fish can help in endangered species preservation. An International Union for Conservation of Nature report that assessed about 15,000 fish species indicates that up to one-third of fish species are already endangered, whereas dozens are extinct or extinct in the wild (IUCN 2016). Additionally, GSC cryopreservation in fish has been recently suggested to serve as an alternative to routine fixation in ethanol or formaldehyde (Hagedorn et al. 2018). Common fixatives can be problematic from the perspective of long-term DNA integrity, lesions and difficult DNA extraction because of tissue hardening (Zimmermann et al. 2008).

12.2.1 Germ Cell Transplantation in Fish

12.2.1.1 Embryonic Cell Transplantation

Blastomere Transplantation

As described above, germplasm containing PGC determinants is segregated into four random clusters at the 1 k cell stage located close to the margins of the bottom of the animal pole. The initial experiments were performed as an intraspecific transplantation between a zebrafish-pigmented donor and an albino recipient. The success of the transplantation was confirmed by the production of pigmented progeny from unpigmented parents. The basis of this transplantation is a collection of blastomeres (20–100 cells) from the presumed location of the PGCs and their transplantation into approximately the same position as the recipient's blastula (Fig. 12.1, upper panels). To trace the fate of the transplanted cells in the recipients, fluorescein dextran and phenol red were injected into the donor embryos after fertilisation. The mating of mature germline chimeras yielded a mixture of pigmented and non-pigmented progeny, suggesting that exogenous PGCs are capable of adopting migration patterns and proceeding regularly through gametogenesis (Lin et al. 1992). Further studies showed that the blastomere transplantation (BT) technique has its limitation when applied between species. The migration ability of PGCs decreases with increased phylogenetic distance, probably because of the aggregation of PGCs with co-transplanted somatic cells (Saito et al. 2010). One of the main prerequisites for BT transplantation is synchronisation of the recipient and the donor, which can be restrictive when applied in species with incompletely mastered artificial reproduction. Obviously, the number of created germline chimeras is limited not only by operator skills but also by the developmental rate of embryos because the optimal stage for transplantation is between the 1 and 4 k cell stages.

Blastoderm Transplantation

Similar to BT, blastoderm transplantation (BdT) is based on the localisation of PGCs during the blastula stage, when the lower part of the blastoderm containing PGCs is cut and placed into the central part of the previously cut host's blastodisc. This method can be called sandwich transplantation according to insertions of the donor's blastoderm into the recipient (Fig. 12.2). The difference between the BT and BdT techniques is that BT depends on a randomly chosen region of blastodiscs, which could contain PGCs. By contrast, in BdT, all PGCs are more likely transplanted within a single graft. As a result, the created germline chimeras between triploid crucian carp donors and diploid goldfish hosts are fertile. As sterilisation for the goldfish host was not performed, the germline chimeras produced eggs of both

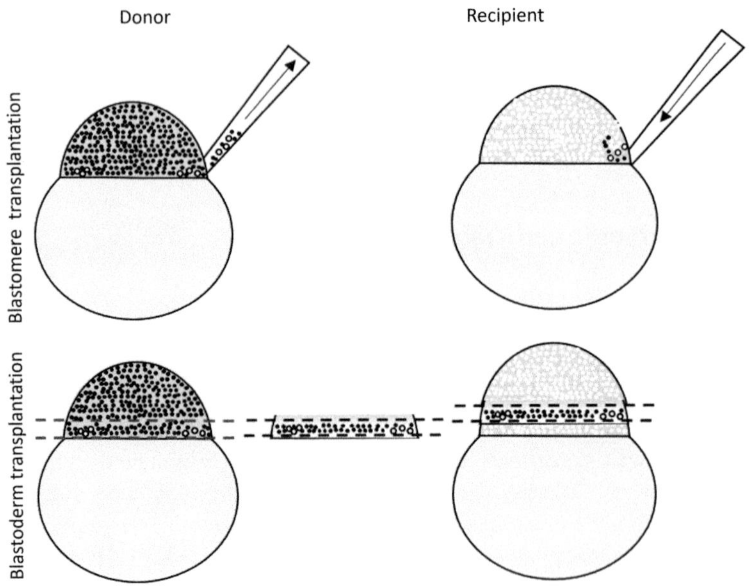

Fig. 12.2 Blastomere and blastoderm transplantation from the blastula stage host into the recipient in the same stage

species (Yamaha et al. 2001). BdT was also successfully applied when the donor goldfish's lower blastoderms were grafted into the hybrid embryos of female goldfish and supermale common carp, and a donor-derived sperm from XX goldfish PGCs was obtained after germline chimera maturation (Yamaha et al. 2003). Both BD and BdT do not require labelling of the donor cells to perform the transplantation, which is advantageous for application when an artificial label, such as GFP-nos 1 3′UTR mRNA (Saito et al. 2006) or transgenic lines allowing PGC identification, is unavailable; thus, it can be applied without limitations. Moreover, BD and BdT can be methods of choice when the recipient species are highly sensitive to the anaesthesia needed during intraperitoneal transplantation into larva, as is known for pike perch (Güralp et al. 2016, 2017).

Single PGC Transplantation

As the PGCs are undergoing migration, isolating them before they reach their destination in the genital ridge and transferring them into the host's blastula are possible. The visualisation of donor PGCs is a crucial prerequisite, so donors are injected with artificially synthesised mRNA green fluorescent protein (GFP) conjugated with zebrafish *nos 1* 3′UTR (GFP-*nos 1* 3′UTR mRNA) (Saito et al. 2006), or a transgenic donor strain with an expression of a reporter protein in its PGCs is utilised. According to Saito et al. (2010), donor embryos at the 10–15 somite stage are optimal for PGC isolation and transplantation, mainly because of the preservation of their migration capabilities. The use of more advanced developmental embryo stages resulted in decreased migration activity, which is naturally given by the period of regular migration event. PGCs can be isolated either through their simple preparation from the donor embryos with forceps or after treatment with citric acid, which helps with the loosening and makes the aspiration of single PGCs easier (Saito et al. 2010). Alternatively, PGCs from a simple culture prepared from dissociated blastomeres are left to differentiate for at least 1 day, or yolk-depleted embryoids that are allowed to develop for 1 day can be used (Kawakami et al. 2010). Both methods that use PGCs obtained as single cells have higher migration and colonisation rates, especially when this transplantation is applied between species. Although the PGC migration patterns in fish are very conserved, PGCs are always accompanied by somatic cells in interspecific BT, and they have low transplantation efficiency because cell aggregates are often formed (Saito et al. 2010).

Theoretically, single PGCs could be isolated even from blastula-stage embryos after their identification using the bucky ball GFP transgenic line, which enables early localisation of PGCs containing the germplasm (Riemer et al. 2015). Furthermore, using cell sorting approaches might be possible, as has been described in trout (Takeuchi et al. 2002; Kobayashi et al. 2004), but the ratio of the target PGCs appearing during a migration event to the number of somatic cells could be below the detection level for the cell sorter.

Intraperitoneal PGC Transplantation

The large size of transgenic salmonid embryos, together with their slow embryonic development, allows the dissection of their genital ridges containing undifferentiated PGCs and their use for intraperitoneal transplantation into hatched trout embryos (Fig. 12.3). However, this method has not been performed in other teleost model species, such as zebrafish, medaka or goldfish, because of their small body size, which does not allow the excision of the colonised gonad with the PGCs (Takeuchi et al. 2002). The collected gonads containing GFP-positive PGCs and also somatic cells are further treated enzymatically to obtain the cell suspension, and they can be optionally enriched using flow cytometry based on the large size of the PGCs (~20 μ) (Kobayashi et al. 2004). Afterwards, the PGCs were intraperitoneally transplanted into the hatched rainbow trout. The transplanted single PGCs were able to colonise the genital ridge and to differentiate in recipient testis or ovaries, resulting in donor-derived gamete production. Interestingly, the loss of their colonisation capability was similar to that reported by Saito et al. (2010), when the transplantation of PGCs from older donors/recipients (40 and 45 days post fertilisation [dpf]) was less successful in comparison with that of younger donors/recipients (35 dpf), indicating that the mechanisms responsible for the genital ridge colonisation are lost progressively (Takeuchi et al. 2003).

The disadvantage of species having small embryos for genital ridge dissection and PGC transplantation can be addressed by in vitro PGC induction. Riesco et al. developed an in vitro culture technique using dissociated blastodisc from vas:EGFP transgenic zebrafish with specific factors increasing the number of PGCs. The functionality of the in vitro induced PGCs was confirmed by intraperitoneal

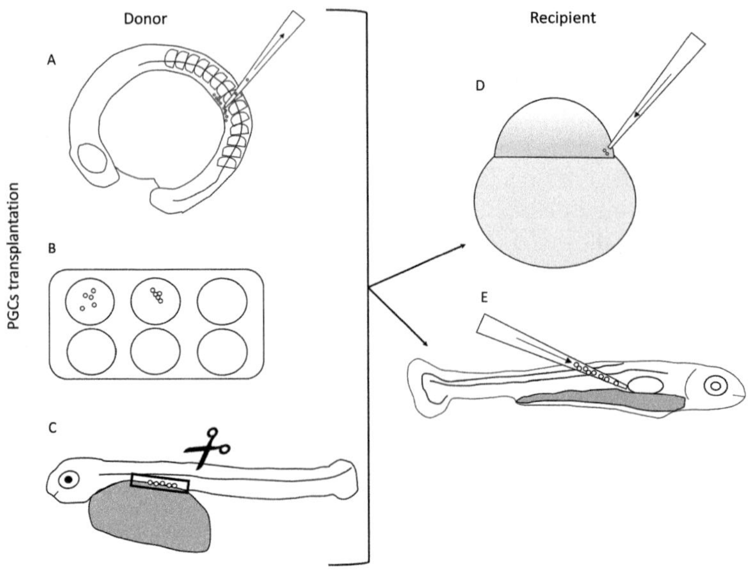

Fig. 12.3 Illustration of PGC transplantation in fish. PGCs can be isolated during (**a**) migration, (**b**) from cultured blastomeres or (**c**) from an undifferentiated genital ridge in the case of large embryos. PGCs can be transplanted into the recipient in (**d**) the blastula stage or (**e**) intraperitoneally into the hatched larva

transplantation into zebrafish 7 dpf and observation of active migration towards the genital ridge (Riesco et al. 2014), as well as their further differentiation in PGC-depleted recipient testis (Robles et al. 2017). This method could have extraordinary importance because PGCs are introduced into a more robust host, ensuring good post-transplantation survival. On the contrary, PGC transplantation into the blastula host is performed during the onset of the sensitive period of embryonic development, which could be further complicated by the injection of antisense morpholino targeting *dnd1* mRNA to sterilise the fish; this further decreases the survival rate.

Intraperitoneal Spermatogonia and Oogonia Transplantation

The first work on the transplantation of GSCs from differentiated gonads was performed using testicular cell suspension containing spermatogonia, which were transplanted into an allogenic rainbow trout recipient. The process of spermatogonia and oogonia transplantation is explained in Fig. 12.4. This study was the first report on the stemness of the partial population of spermatogonia, as they were capable of transdifferentiating into oogonia in the environment of the recipient's ovary and ultimately giving rise to donor-derived eggs; donor-derived sperm was also obtained. Remarkably, the colonisation rates for male and female recipients were comparable, and the developmental rate of oocytes derived from transdifferentiated spermatogonia was identical to that of endogenous oocytes. After maturation, 13 of

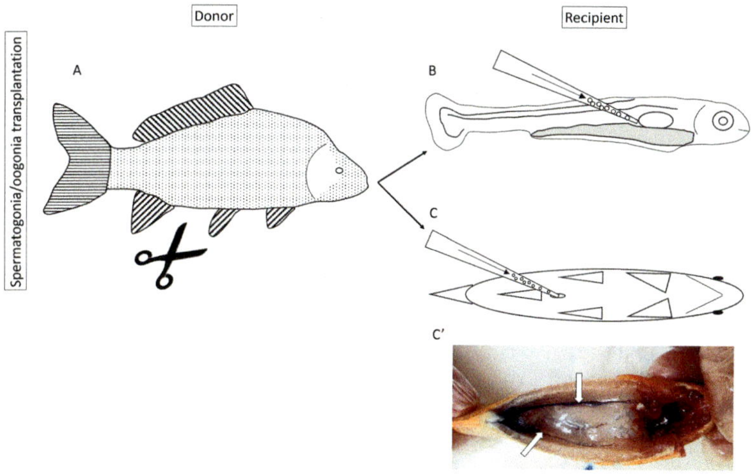

Fig. 12.4 Spermatogonia and oogonia transplantation. Isolated from a juvenile/adult donor (**a**) and transplanted into (**b**) the larva by intraperitoneal injection or (**c**) the adult using transplantation via the genital papilla, (**c′**) PGC-depleted adult goldfish injected with trypan blue trough the genital papilla. Arrows indicate gonads labelled by trypan blue

the 26 male recipients were identified to produce a fraction of the donor-derived sperm, and 16 of the 40 female recipients produced a fraction of the donor-derived egg (Okutsu et al. 2006). Afterwards, the stemness property of the female germ cells was confirmed by allogeneic transplantation using rainbow trout triploid, when a donor-derived sperm from transdifferentiated oogonia was obtained from the male recipients (Yoshizaki et al. 2010). These studies created a landmark for GSC transplantation obtained from differentiated gonads in fish, enabling the restoration of both sexes because whether undifferentiated PGC transplantation is the only way to restore both sexes was unknown.

The preservation of stemness capacity with a population of gonadal cells was further confirmed by allogenic spermatogonia transplantation in the marine species Nibe croaker (*Nibea mitsukurii*), when half of the transplanted recipients developed an ovary (Takeuchi et al. 2009), and later on, donor-derived progeny was obtained using triploid allogenic recipients (Yoshikawa et al. 2017). Similar results have been also obtained after transplantation zebrafish OSCs and subsequent production of donor-derived sperm (Wong et al. 2011; Franěk et al. 2019b). The possible acceptance of gonadal cells after xenogenic transplantation to improve the efficiency of aquaculture production and the preservation of endangered species was tested on Nibe croaker donors and chub mackerel (*Scomber japonicas*) recipients (Yazawa et al. 2010). The ultimate aim was to verify mackerel as a suitable host for Pacific bluefin tuna (*Thunnus orientalis*) germ cells in order to overcome problems associated with the holding and reproduction of enormously sized tuna broodstock (Yazawa et al. 2013). Besides mackerel, yellowtail kingfish (*Seriola lalandi*) was assessed as a surrogate for southern bluefin tuna (*Thunnus maccoyii*) germ cells;

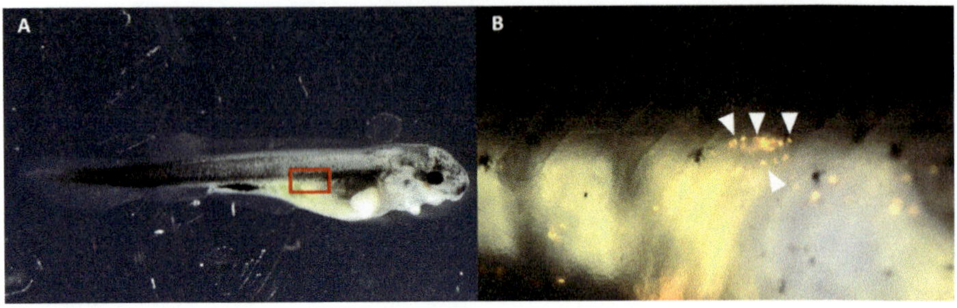

Fig. 12.5 Transplantation of PKH-26 labelled germ cells in sturgeons. Sterlet larvae transplanted by labelled oogonia from Russian sturgeon at 12 days post fertilisation (**a**). The red rectangle depicts the magnified caption on the PKH-26 labelled germ cells located in the genital ridge 1 week post transplantation (**b**)

despite the successful colonisation of the genital ridge after transplantation, tuna germ cells were not detected in the subsequent molecular analysis, indicating that the suitability of surrogates cannot be evaluated only through the detection of membrane labelled cells (Bar et al. 2016). Similar to work on tuna species, germ cell transplantation from large sturgeon species (Fig. 12.5) into smaller surrogates was performed to attempt to shorten the extremely long reproductive cycle and decrease the body size of broodstock (Pšenička et al. 2015, 2016; Ye et al. 2017). Intraperitoneal transplantation was further applied with other approaches, such as the use of PGC-depleted recipients with the method of gene knockdown (Yoshizaki et al. 2016) or knock out (Li et al. 2017) and the use of sterile hybrids (Wong et al. 2011; Yoshikawa et al. 2018).

Nowadays, it seems that intraperitoneal transplantation is mostly favoured in fish. Donor germ cells can be harvested at any time, which enables synchronisation according to the recipients' availability. There is no need to use transgenic lines, whereas the membranes of the cells can be labelled instantly with commercially available dyes prior to transplantation in order to validate the success rate in vivo and later trace the transplanted cells (Hamasaki et al. 2017). Compared with GSC delivery through the genital papilla, the injection of cells into the coelom cavity needs only a few thousand cells; a further increase in the number of introduced cells brings no benefits (Seki et al. 2017). The optimisation of intraperitoneal transplantation would be desirable. Usually, there is a concern on transplantation timing; the embryonic gonad has been suggested to remain *open* only shortly, which could negatively influence the colonisation rate in older recipients because the introduced cells probably cannot colonise gonads encompassed by somatic cells (Hamasaki et al. 2017; Octavera and Yoshizaki 2019). Contrary to the decreasing effectivity in older recipients, younger recipients are more likely to suffer from mortality associated with the transplantation procedure. A comparison of 11 and 14 dpf-old medaka recipients surprisingly showed a more than tenfold loss of colonisation rate, and the cells transplanted 19 dpf are incapable of colonising the gonad (Seki et al. 2017).

Intrapapillary Transplantation

In several aspects, this technology is similar to the previous type of transplantation in which a suspension of single cells obtained from the gonads is introduced into the recipient. Compared with the previous type of transplantation, interpapillary transplantation is based on the delivery of GSCs through the genital papilla of adult specimens into sexually mature gonads. The use of adult recipients and the minimal requirements for equipment prior to transplantation are the main advantages of this method. Despite these advantages, however, it is necessary to consider that a relatively large amount of donor gonadal tissues need to be used when 2×10^6–10^7 cells after sorting by Percoll gradient are transplanted per one individual of Nile tilapia (Lacerda et al. 2006, 2010). This technique was utilised to produce the donor-derived sperm and eggs of *Odontesthes bonariensis* from Patagonian pejerrey (*Odontesthes hatcheri*) recipients (Majhi et al. 2014) and the donor-derived sperm after allogeneic transplantation of zebrafish testicular cells (Nóbrega et al. 2010). After transplantation, donor-derived gametes are obtained rapidly because GSCs are introduced into sexually competent fish when donor-derived gametes were produced 9 weeks post-transplantation in tilapia (Lacerda et al. 2010), while 7–11 months were necessary to obtain the eggs and sperm of *O. bonariensis* from Patagonian pejerrey surrogates (Majhi et al. 2014). All the aforementioned intrapapillary transplantations utilised recipients that are previously sterilised with a cytostatic drug, busulfan, combined with heat treatment for several weeks, leading to the depletion of spermatogenic cysts in the testis. To achieve the maximum depletion of endogenous spermatogenesis, optimisation of the busulfan treatment (dose, repeated application) with temperature fluctuation was tested. However, only temperature treatment itself can be effective for spermatogenesis depletion compared with busulfan combined with temperature treatment in particular species. Furthermore, concerns regarding the safety of busulfan for staff and animal welfare during treatment have been raised (de Siqueira-Silva et al. 2015). Whether other sterilisation methods, such as PGC depletion using knock-down, triploidisation or hybridisation, are convenient prior to intrapapillary transplantation has not yet been proven. We found that a simple injection into PGC-depleted adult goldfish is relatively difficult, as empty gonads are very thin and can be ruptured easily during transplantation (unpublished data). Another question concerns the inner capacity of gonads and perhaps more importantly the low incidence of developed somatic gonadal cells forming a spermatogenic cyst (Goto et al. 2012). Thus, temporal spermatogenesis depletion seems to be most convenient for recipient preparation prior to intrapapillary transplantation.

12.2.2 Germ Cell Cryopreservation

The application of germ cell transplantation technologies in fish allowed the use of GSC cryopreservation procedures, which have immense potential for aquaculture, species preservation and line preservation in fish models. A very complex strategy can serve important benefits for end users, as cryopreservation protocols are being

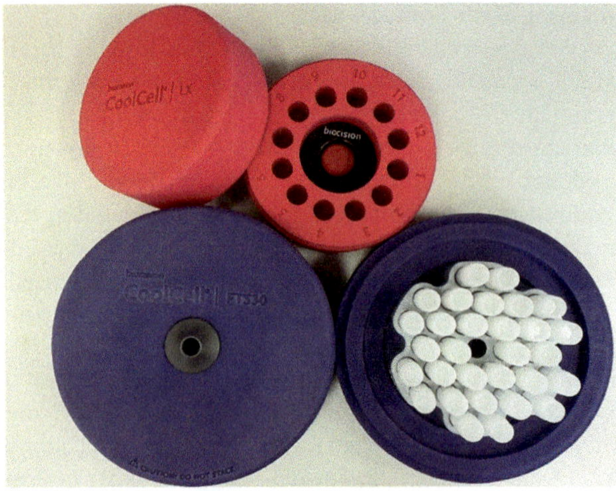

Fig. 12.6 Equipment for slow-rate freezing. Example of commercially available freezer boxes to ensure a cooling rate of −1 °C/min in a −80 °C freezer, allowing grams of the gonadal tissue to be cryopreserved

developed with satisfactory viability, ensuring a good outlook for the following germ cell transplantation. Moreover, the introduction of cryopreservation procedures of germ cells can significantly reduce the risk of losing valuable fish because of disease outbreaks, natural catastrophes or technology failure, especially in the case of closed recirculation systems. A further benefit is a reduction in perquisite space for keeping large quantities of broodstock when a part of genetic resources can be cryopreserved and stored for a virtually indefinite period.

The main aim of cryopreservation techniques in fish is to develop a protocol that ensures the preservation of intact tissue/cells in sub-zero temperatures for an extended period of time. Each cryopreservation procedure consists of steps including (1) the preparation of the cryomedia with optimised cryoprotectants and extenders, (2) the cryopreservation process itself using rapid or slow freezing, (3) cryogenic storage of the biological material and (4) recovery of the cryopreserved material by warming, with ultimate application in GC transplantation. Each of these steps should be optimised to keep cryoinjury at the lowest level possible and to maximise the output by obtaining a high percentage of viable cells in good health.

Two main methods are used for germ cell introduction into the frozen state. The first is ultra-rapid freezing and vitrification in which concentrated cryoprotectants are used; the material is frozen rapidly, and a glass-like state is formed, preventing the formation of ice crystals. The second method is slow-rate freezing (Fig. 12.6) in which a lower concentration of CPA is used, and the material is frozen at a rate in accordance with the protocol (usually 1 °C/min). The principle of slow-rate freezing is based on consecutive water efflux from the cells, minimising intracellular ice formation, whilst viscosity is increased further, preventing ice formation. Then, the biological material is cryopreserved and stored at −196 °C, in which the occurrence of any changes that could be detrimental is almost not possible (Mazur 1984).

Several peculiarities are well known in the preservation of fish germplasm. Only sperm cryopreservation procedures have been developed on the level enabling to facilitate the needs of aquaculture (Cabrita et al. 2010). Nowadays, breeding programs are applied in various fish species in aquaculture, and research when preservation of transgenic or mutant lines in model fish species has become recognised to be crucial for future needs (Carmichael et al. 2009; Robles et al. 2009). Besides sperm cryopreservation, much efforts have been invested into the cryopreservation of fish oocytes. The preservation of oocytes in fish is limited because of different biological properties in comparison with the spermatozoa. The structure of fish oocytes is very inconvenient for successful cryopreservation because of the low permeability of cryoprotectants, which is caused by chorion and the large yolk volume. Several studies testing the cryopreservation of vitellogenic oocytes (Godoy et al. 2013) and early-stage oocytes (Guan et al. 2008, 2010; Tsai et al. 2010) have been performed on zebrafish. However, ovarian follicles were severely damaged after cryopreservation, resulting in failure during their growth in an in vitro culture (Tsai et al. 2010; Anil et al. 2018).

Embryo cryopreservation research has been mainly focused on particular steps, including increasing the permeability for cryoprotectants or embryo handling rather than cryopreservation itself. One of the major obstacles is how to ensure a proper thawing procedure and avoid ice formation. To the best of our knowledge, these issues have been successfully addressed by the co-injection of cryoprotectant with nanoparticles into zebrafish embryo. The embryos were then thawed using a laser pulse responsible for thorough nanoparticle excitation, resulting in uniform warming (Khosla et al. 2017). However, such a procedure is obviously far from being applied in a large scale.

12.2.2.1 Primordial Germ Cell Cryopreservation

More notable progress has been achieved in PGC cryopreservation, as the use of undifferentiated cells for cryopreservation is advantageous from several points. Diploid and sexually bipotent cells allow the possibility for both sexes to be restored when cryopreservation is followed by transplantation into the surrogate host (Inoue et al. 2012), in which even a single PGC is sufficient to recover the fertility of an individual (Saito et al. 2008). Moreover, very early access to PGCs is also important in comparison to sperm, SSCs (spermatogonial stem cells) or OSCs (oogonial stem cells) when an adult or at least a juvenile individual is needed. PGC cryopreservation is conditioned by visualisation of PGCs achieved by GFP-nos1-3′UTR injection (Kawakami et al. 2010) or the use of transgenic lines (Kobayashi et al. 2007; Riesco et al. 2012), as PGCs need to be distinguished from other cells prior to cryopreservation and transplantation procedures. PGCs can be cryopreserved in several stages of embryonic development as soon as the blastula stage is reached (Higaki et al. 2009), during a migration event (Riesco et al. 2012) and until the post-migratory phase when the PGCs are already localised in the genital ridge (Kobayashi et al. 2007). However, being aware of the progressive loss of PGC migration capability when donors from later embryonic stages are used is necessary (Saito et al. 2010).

Loach embryos during somitogenesis with a depleted yolk previously injected with GFP-nos1–3′UTR were exposed to a cryoprotectant and then vitrified. The thawed embryos were dissociated in citric acid, and a single PGC transplantation was performed into a host in the blastula stage. The PGCs retained their viability and migrated towards the genital ridge actively (Inoue et al. 2012). PGC cryopreservation with the production of donor-derived progeny was achieved in rainbow trout genital ridges by slow-rate freezing (−1 °C/min). Subsequent transplantation was performed in rainbow trout hatchlings (Kobayashi et al. 2007). Alongside the complete procedure to generate live fish from vitrified PGCs described in zebrafish and rainbow trout (Kobayashi et al. 2007; Kawakami et al. 2010), several studies have been performed to optimise cryopreservation procedures and maximise post-thaw viability. When the cryopreservation of zebrafish whole embryos, dissected genital ridges and single PGCs was tested together with different cryopreservation methods, an almost 100% post-thaw viability was found (Riesco et al. 2012). Other fish species, such as common carp and Japanese eel, were also utilised for PGC cryopreservation and transplantation (Kawakami et al. 2012a, b).

Considering the potential for the practical application of PGC cryopreservation, the major advantage is very early access to the cells, whereas a pool of genetic diversity can be preserved when each embryo represents a unique genotype. On the other hand, the preparation of embryos prior to cryopreservation can be regarded as relatively difficult. Removing the chorion membrane, injecting GFP-nos1-3′UTR to label PGC and removing the yolk are necessary. Then, after cryopreservation and subsequent thawing, the PGCs are transplanted into the blastula-stage embryo or swim-up larvae.

12.2.2.2 Spermatogonia and Oogonia Cryopreservation

The cryopreservation of differentiated GSCs obtained from juvenile and adult individuals has been studied extensively mainly on salmonids and cyprinid fish species; various methodologies have been applied to develop efficient protocols, ensuring satisfactory post-thaw viability and then facilitating transplantation. Foremost, it is necessary to stress that OSC cryopreservation is, to the best of our knowledge, probably the only approach that effectively preserves maternal genetic resources, which can be recovered using transplantation into the surrogate recipient. The development of cryopreservation techniques could also have commercial potential when important aquaculture species are targeted. Moreover, whole procedures involving SSCs and OSCs are less demanding in terms of time and equipment use. The target cells are generally available at a higher amount according to the size of the gonad, the cells do not need to be labelled prior to the validation of cryopreservation success and subsequent transplantation does not need to be performed.

Overall, SSCs and OSCs have been efficiently cryopreserved via slow-rate freezing, which uses commercially available boxes made from various materials placed in a −80 °C freezer to ensure a cooling rate of about −1 °C/min or in programmable freezers (Franěk et al. 2019a, b; Lee and Yoshizaki 2016; Pšenička et al. 2016), or via ultra-rapid vitrification (Seki et al. 2017). The donor needs to be sacrificed prior to cryopreservation, but in species such as sturgeon, having a skin suitable for suture would enable small surgery and the nonlethal collection of a gonad fragment (M. Pšenička, unpublished). The fragment of gonadal tissue could be subsequently used for immediate transplantation or cryopreservation as a very efficient protocol that ensures 70% post-thaw survival (Pšenička et al. 2016). After collection of the gonads, the tissue can be immediately cryopreserved or digested into a single cell suspension and cryopreserved afterwards. Cryopreservation of whole tissue has been shown to be superior compared with cryopreservation of dissociated cells because higher survival rates are achieved; more importantly, thawed tissue, besides containing live cells, also has dead cells, which are digested afterwards during enzymatic dissociation. Thus, the cell suspension obtained from whole frozen tissue is almost free of dead cells, which could potentially hamper the success of the transplantation procedure (Pšenička et al. 2016).

Recently, new cryopreservation strategies have been developed in fish, and they have good potential for application in the field. Needle-immersed vitrification (NIV) uses pieces of gonadal tissue which are shortly exposed in an equilibration and vitrification solution and then are plunged directly into liquid nitrogen. This method is very rapid, as samples can be efficiently cryopreserved in minutes with minimal equipment (Lujić et al. 2017; Marinović et al. 2018). The simplest developed procedure for SSC cryopreservation is freezing of whole fish by placing it in a −80 °C freezer or on dry ice when the reported cooling rate inside the fish body is close or identical to the well-accepted optimum (−1 °C/min). Surprisingly, no statistical difference in the number of retrieved spermatogonia from rainbow trout males was reported during an observation period of 1–1113 days when whole fish were stored in −80 °C continuously or transferred into liquid nitrogen after freezing to −80 °C (Yoshizaki and Lee 2018). Thus, this method can have an exceptional advantage, as fish can be frozen immediately without any preparations.

Generally, cryoprotectants in a lower concentration (≤2 M), such as ethylene glycol, dimethyl sulfoxide and methanol, have often been identified as optimal for slow-rate freezing performed using freezing boxes and programmable freezers (Table 12.1). A further improvement in post-thaw viability was achieved by testing sugar and protein supplementation, yielding 40–80% viable germ cells post-thawing. Higher cryoprotectant concentrations are used in vitrification performed on dissociated cells in straws (Seki et al. 2017; Higaki et al. 2018), gonadal fragments or whole gonads (in case of zebrafish, medaka or juvenile trout) with the NIV method (Lujić et al. 2017; Marinović et al. 2018), or gonads placed on a metal mesh (e.g. copper mesh) plunged into liquid nitrogen (Seki et al. 2017). The post-thaw viability after the vitrification of fish germ cells was, in most cases, lower (~50%) than that with slow-rate freezing. Despite the lower post-thaw survival after vitrification, it is noteworthy to mention that this method is very rapid in terms of sample preparation, and it requires minimal equipment, such as a Styrofoam box with liquid nitrogen, so it can be very convenient for field sampling. On the other hand, not all species have been found to be suitable for vitrification, such as when we identified a fourfold lower survival rate after vitrification compared with that for slow-rate freezing for common carp testicular cells (Franěk et al. 2019a).

Table 12.1 List of selected studies on the cryopreservation of GSCs in fish

Species	Cryopreservation method	Cryopreserved material	Optimised cryoprotectant	Thawing temperature	Viability evaluation	Post-thaw viability	Reference
Brown trout *Salmo trutta*	NIV	OT	ES: 1.5 M ME, 1.5 DMSO VS: 3 M PG, 3 M DMSO	Room temperature	TB	40.34%	Lujić et al. (2017)
Common carp *Cyprinus carpio*	SRF	TT	2 M DMSO, 0.3 M trehalose	26 °C	TB	40.7 ± 9.2%	Franěk et al. (2019a)
		OT	1.5 M DMSO, 0.3 M glucose	26 °C	TB	66.0 ± 8.6%	Franěk et al. (2019c)
Goldfish *Carassius auratus*	SRF	TCS	2 M DMSO, 50 mM glucose	38 °C	Calcein-AM propidium iodide	~72%	Marinović et al. (2016)
Honmoroko *Gnathopogon caerulescens*	V	OCS	5 m DMSO	37 °C	Proliferation in cell culture	110%	Higaki et al. (2018)
		TCS	5 M PG			50%	
Manchurian trout *Brachymystax lenok*	SRF	TT	1.3 M ME, 0.2 M trehalose	30 °C	FC and TB	81.0 ± 1.3%	Lee and Yoshizaki (2016)
Medaka *Oryzias latipes*	V	TT	5 M EG, 21% Ficoll, 0.35 M sucrose	25 °C	GFP fluorescence	~43%	Seki et al. (2017)
Nile tilapia *Orechromis niloticus*	SRF	TCS	1.3 M DMSO, 10% FBS	25 °C	TB, cell culture	No data	Lacerda et al. (2010)

Rainbow trout *Oncorhynchus mykkis*	SRF	TT	1.3 M DMSO, 0.1 M trehalose	10 °C	GFP fluorescence	35.1 ± 5.3%	Lee et al. (2013)
		TT	1.3 M DMSO 0.1 M trehalose	10 °C		No data	Lee et al. (2016a)
		OT	1 M DMSO, 0.1 M trehalose	10 °C	FC and TB	72.9 ± 6.2%	Lee et al. (2016b)
Siberian sturgeon *Acipenser baerii*	SRF	TT	1.5 M EG, 50 mM glucose	38 °C	Calcein-AM propidium iodide	64.3 ± 6.1%	Pšenička et al. (2016)
		OT	1.5 M EG, 50 mM glucose			52.0 ± 7.3%	
Starry gobby *Asterropteryx semipunctata*	SRF		1.3 M DMSO, 0.1 M trehalose	30 °C	FC	~60%	Hagedorn et al. (2018)
Tench *Tinca tinca*	SRF	TCS	1.5 M GLY, 50 mM glucose	38 °C	Calcein-AM propidium iodide	57.7 ± 16.8%	Linhartová et al. (2014)
		TT	3 M DMSO, 50 mM glucose	38 °C	Calcein-AM propidium iodide	~55%	Marinović et al. (2016)
Tiger puffer *Takifugu rubripes*	SRF	TT	1.3 M DMSO, 0.1 M trehalose	20–22 °C	TB	61.2 ± 2.7%	Yoshikawa et al. (2018)
Zebrafish *Danio rerio*	NIV	TT	ES: 1.5 M ME and PG VS: 3 M DMSO and PG	25 °C	TB	50–72 ± 1–13%	Marinović et al. (2018)

NIV needle-immersed vitrification, *SRF* slow-rate freezing, *V* vitrification, *OT* ovarian tissue, *TT* testicular tissue, *TCS* testicular cell suspension, *OCS* ovarian cell suspension, *ME* methanol, *DMSO* dimethylsulfoxide, *PG* propylene glycol, *GLY* glycerol, *EG* ethylene glycol, *ES* equilibration solution, VS vitrification solution, *TB* trypan blue exclusion test, *FC* flow cytometry

12.2.3 Application of Cryopreserved PGCs and GSCs

With the use of GSC cryopreservation, the genetic material from putative donors can be sampled very early from juvenile specimens, thus decreasing the risk of their accidental mortality and saving space and cost. Similarly, the presented results achieved in oogonia and spermatogonia cryopreservation in commercially important species can be beneficial in gene resource banking for valuable line preservation. As a result, the importance of cryopreserved germ cells can be similar to that of cryopreserved sperm, whereas cryopreservation and the recovery of female genetic resources can have exceptional importance because current methods for female genetic resource preservation are not close enough to facilitating real application (Table 12.2). However, being aware that the use of robust technology for the recovery of cryopreserved cells in the surrogate host is crucial and that whole germ cell manipulations need to be considered in the long term is crucial. So far, germ cell technologies as complex strategies including the cryopreservation of oogonia and spermatogonia, transplantation and production of donor-derived egg and sperm have been applied successfully only in rainbow trout (Yoshikawa et al. 2018). For many other species, including endangered ones, such as bitterlings (Octavera and Yoshizaki 2019) and sturgeons (Pšenička et al. 2015, 2016; Ye et al. 2017), success has been achieved as cryopreservation protocols are developed, convenient recipients are identified or the production of sperm or egg is achieved. Research in this field is attracting much attention recently, so germ cell manipulation techniques will soon be applied in various fish species. Rainbow trout is probably the most prominent species with mastered germ cell manipulation techniques, with sole credit to the group of Professor Yoshizaki for this discovery. The inclusion of developed technologies for breeding work has been suggested (Yoshizaki and Yazawa 2019) when, for example, an all-female trout population is produced using sperm derived from transplanted oogonia (Lee et al., 2016b). A similar technology using GSC transplantation can be expected in other fish species, with interest in monosex stocks, such as tilapia.

A situation in which only a few superior individuals with favourable characteristics are alive can happen, whilst under normal conditions, an insufficient number of progenies for further work may be obtained from their reproduction. This can be precluded by GSC transplantation from those individuals into hundreds of hosts, resulting in promising expectations that a larger amount of progeny will be obtained. The power of surrogacy can be further improved with cryopreservation. Many unique inbred or isogenic lines have been developed previously, but unfortunately, most of them have been lost because of different reasons, such as space requirements, momentary uselessness or difficulties with their maintenance and breeding. The cryopreservation of GSCs seems to be a useful tool for such situations when efficient protocols could allow the conservation of both sexes whilst eggs and sperm can be recovered; therefore, the original genetic constitution is likely to remain unaltered.

Germ cell cryopreservation is obviously only the beginning. However, developed and optimised protocols have shown that even simple and cost-efficient methods for

Table 12.2 Comparison of cryobanking of different resources of germplasm

Germplasm		Collection is possible from	Repeated collection from the same donor	Feasibility of obtaining the material	Availability of cryopreservation protocols for target species	Method for cryopreservation	Potential for application in gene banking	Efficiency of cryopreservation	Application of thawed germ plasm
Sperm		Adult	Yes	Easy	~50 species	Slow rate freezing/ vitrification	High	High	Fertilization
Oocytes	Immature	Juvenile/ adult	Yes	Easy	[a]		Unlikely	None	Fertilization
	Mature	Juvenile/ adult	Yes/no	Moderately difficult	[a]		Unlikely	None	Cell culture
Embryos		Hpf/dpf	No	Easy	[b]		Unknown	Low	Rearing
PGCs		Hpf/dpf	No	Difficult	4 species		Low	Moderate	Transplantation/ differentiation in cell culture
Germ cells	Spermatogonia	Juvenile/ adult	Yes/no	Easy	11 species		High	High	
	Oogonia	Juvenile/ adult	Yes/no	Easy	5 species		High	High	

Hpf hours post fertilization, *dpf* days post fertilization

[a]Currently, no robust method for oocyte cryopreservation has been developed

[b]Developed method using vitrification and warming via nanoparticles injected into embryos is probably applicable for most of telost fish embryos. However, cryopreserved/thawed embryos were not reared further; therefore, their suitability for gene banking has not been proven

cryopreservation, such as freezing in a deep freezer, NIV or freezing and storage of whole fish with gonads in a −80 °C freezer, can yield satisfactory results. The further development of cryopreservation protocols can be expected in other important aquaculture and model fish species. However, from a realistic point of view, the application of GSC cryopreservation for breeding or restocking programs remains far from reality. Although sperm cryopreservation protocols have been developed and optimised extensively in dozens of species, including those of utmost importance for aquaculture, such as salmon, trout, carp or catfish, only a few reports are available regarding their use in breeding programs. A similar scenario can be found in conservation programs in which sperm cryopreservation procedures are usually recognised as a backup for worst-case scenarios. In this case, the cryopreserved sperm might not be sufficient to restore the original genetic pool. Furthermore, it is obvious that even hundreds or thousands of sperm samples from different individuals are useless when the number of females is limited. In order to address this issue, a more robust planning of breeding and conservation programs is necessary to obtain the full advantages of cryopreservation.

GSC cryopreservation can become a more realistic alternative for sperm cryopreservation, as nowadays, much efforts have been invested into the standardisation and cryopreservation of large volumes. To place GSC cryopreservation on the same level of importance as sperm cryopreservation, much work needs to be done on the development of the germ cell transplantation procedure, as it is considered as robust proof of their recovery after thawing (Robles et al. 2017). Traditional differential staining methods based on cell membrane integrity, such as trypan blue exclusion, are preferred, but molecular analysis monitoring the state of cryopreserved cells can also be involved. Approaches evaluating overall DNA integrity were suggested to serve as a thorough evaluation of the molecular changes caused by cryopreservation (Riesco and Robles 2013). The strong influence of cryoprotectant agents was identified to cause decreased methylation levels in cryopreserved spermatozoa, with further negative consequences for embryonic development and offspring viability (de Mello et al. 2017). Thus, similar changes can be expected for GSC cryopreservation. Similarly, the altered gene expression after PGC cryopreservation in zebrafish was attributed to cryopreservation and thawing procedures, in which several transcripts were found to be downregulated (Riesco and Robles 2013). All studies on GSC cryopreservation with subsequent transplantation utilised the optimised freezing protocol. This approach is logical, but there is no evidence of the future impact of cryopreservation on transplantation success. Only a few individual cells are capable of colonising the gonad primary, as it was documented by transplantation of GSCs after membrane labelling (Takeuchi et al. 2009) or GSCs from transgenic lines with reporter expression using GFP or red fluorescent protein (Kobayashi et al. 2004; Wong et al. 2011). Comparing the impact of cryopreservation protocols with different post-thaw viabilities on transplantation success would be very informative. In this way, even those cryopreservation protocols that are not very efficient can be reconsidered as suitable to use prior to transplantation.

A complete evaluation of cryopreservation consequences is difficult and time-consuming to perform. The introduction of a short-term culture could potentially improve the state of cryopreserved cells after freezing and thawing procedures. In rainbow trout, GSCs were cultured for several days prior to transplantation, and an improved colonisation rate was reported after transplantation in comparison with cells directly obtained from dissociated tissue. It was suggested that the recovery of cell-surface proteins responsible for adhesiveness could play a role after the introduction of the cells into the body cavity, where the cells need to attach in the genital ridge (Shikina et al. 2013). The enzymes used for obtaining a single-cell suspension prior to transplantation were also discussed, as trypsin is used very frequently, but it is well known to alter cell membranes because of its proteolytic activity. However, a direct effect on cell viability after dissociation was not observed when different concentrations of trypsin and collagenase were tested for sturgeon gonad dissociation (Pšenička et al. 2015). The cell culture has been shown to improve the overall state of the GSCs after enzymatic dissociation. However, being aware of the additional labour required when culture conditions should be optimised to promote GSCs instead of somatic cells, as well as the potential complications because of contamination and the poor state of cells after cryopreservation, is needed. Moreover, a culture of GSCs after cryopreservation, followed by transplantation, has not yet been tested. Thus, the direct harvesting of GSCs from a sacrificed donor with subsequent transplantation still seems to be the most effective method whilst eliminating potential risks because of failure during cryopreservation and cell culture.

One could suppose that other methods could alter GSC manipulation in fish with donor-derived gamete production. Specifically, in fish, the induction of uniparental inheritance is possible using gynogenesis or androgenesis. Androgenesis is performed with sperm from the target species and oocytes with inactivated nuclei by using irradiation either by UV or gamma rays. However, fish resulting from androgenesis are nucleocytoplasmic hybrids carrying nuclear DNA derived from the father, whereas most of the mitochondrial DNA is provided by the oocyte. Another challenge with this method is its extremely low yield when the produced fish are homozygous and have impaired viability and reproductive performance (Komen and Thorgaard 2007; Fujimoto et al. 2010); thus, the potential of androgenesis for genetic resource recovery is not realistic. The application of gynogenesis using cryopreserved oocytes is not possible, as methods for their cryopreservation are not available, whilst their potential consequences will be similar to those of androgenesis. Another theoretical option for gene banking is cryopreservation of somatic cells with application for nuclear transfer. However, the resulting embryos are similar to androgenesis when they are nucleocytoplasmic hybrids; the yield of clones is also very low (Zhu and Sun 2000).

12.3 Conclusion

GSCs cryopreservation itself is not an issue when the similarity of optimised cryomedia, the freezing methods and the yield of the obtained cells can be regarded as highly comparable across all species that have been cryopreserved so far. Thus, extending the list of species in which their GSCs have been cryopreserved successfully is a matter of research interest and willingness to devote time to it. A major issue, which is probably not so prominent yet, is the further utilisation and validation of cryopreserved GSCs. As stated previously, the GSCs of most of species with commercial or research interest can be cryopreserved, which makes such kinds of experiments interesting and relatively easy targets for investigation. However, there are only two potential applications of cryopreserved GSCs—transplantation and the production of donor-derived gametes from a surrogate host or their introduction into a cell culture with subsequent in vitro differentiation into gametes, or a combination of cell culture with transplantation, which is rather theoretical. The development of cryopreservation procedures without transplantation or the establishment of cell culture systems can be regarded as a work lacking in concept. If this issue will not be addressed, it can result in a contradictory situation in which many GSC cryopreservation protocols are available, but there will be no possibilities for further application because convenient recipients and methods for sterilisation are not identified for target species. Obviously, surrogate production or in vitro gamete production is a very laborious and time-consuming task, but it needs to be considered as a very integral part of cryopreservation studies because it is proof of their usability for breeding and their use in conservation programs in species of interest. Moreover, mastering the transplantation procedure is likely to be more essential for overall success in germ cell manipulation—the production of donor-derived gametes originating from cryopreserved GSCs. Thus, the transplantation of sexually bipotent germ cells is probably the only method to properly recover genetic resources. Both paternal and maternal genetic resources can be efficiently cryopreserved with satisfactory post-thaw viability, but several studies have shown that no difference exists in the transplantation of cryopreserved/thawed cells compared with the non-cryopreserved control in terms of colonisation rate (Franěk et al. 2019c, a) and gamete production when fertility was compared with that of intact fish (Lee et al. 2016b).

Acknowledgement The study was financially supported by the Ministry of Education, Youth and Sports of the Czech Republic, projects CENAKVA (LM2018099), Biodiversity (CZ.02.1.01/0.0/0.0/16_025/0007370), and NAZV QK1910428.

References

Anil S, Rawson D, Zhang T (2018) Development of molecular markers for zebrafish (*Danio rerio*) ovarian follicle growth assessment following in-vitro culture in cryopreservation studies. Cryobiology 83:75–83. https://doi.org/10.1016/j.cryobiol.2018.05.004

Bar I, Smith A, Bubner E, Yoshizaki G, Takeuchi Y, Yazawa R, Chen BN, Cummins S, Elizur A (2016) Assessment of yellowtail kingfish (*Seriola lalandi*) as a surrogate host for the production of southern bluefin tuna (*Thunnus maccoyii*) seed via spermatogonial germ cell transplantation. Reprod Fertil Dev 28:2051. https://doi.org/10.1071/RD15136

Blaser H, Eisenbeiss S, Neumann M, Reichman-Fried M, Thisse B, Thisse C, Raz E (2005) Transition from non-motile behaviour to directed migration during early PGC development in zebrafish. J Cell Sci 118:4027–4038. https://doi.org/10.1242/jcs.02522

Braat A, Speksnijder JE, Zivkovic D (1999) Germ line development in fishes. Int J Dev Biol 43:745–760

Cabrita E, Sarasquete C, Martínez-Páramo S, Robles V, Beirão J, Pérez-Cerezales S, Herráez MP (2010) Cryopreservation of fish sperm: applications and perspectives. J Appl Ichthyol 26:623–635. https://doi.org/10.1111/j.1439-0426.2010.01556.x

Carmichael C, Westerfield M, Varga ZM (2009) Cryopreservation and in vitro fertilization at the zebrafish international resource center. Methods Mol Biol (Clifton, NJ) 546:45–65. https://doi.org/10.1007/978-1-60327-977-2_4

de Mello F, Garcia JS, Godoy LC, Depincé A, Labbé C, Streit DP Jr (2017) The effect of cryoprotectant agents on DNA methylation patterns and progeny development in the spermatozoa of *Colossoma macropomum*. Gen Comp Endocrinol 245:94–101. https://doi.org/10.1016/j.ygcen.2016.06.003

de Siqueira-Silva DH, dos Santos Silva AP, Ninhaus-Silveira A, Veríssimo-Silveira R (2015) The effects of temperature and busulfan (Myleran) on the yellowtail tetra *Astyanax altiparanae* (Pisces, Characiformes) spermatogenesis. Theriogenology 84:1033–1042. https://doi.org/10.1016/j.theriogenology.2015.06.004

Doitsidou M, Reichman-Fried M, Stebler J, Köprunner M, Dörries J, Meyer D, Esguerra CV, Leung T, Raz E (2002) Guidance of primordial germ cell migration by the chemokine SDF-1. Cell 111:647–659. https://doi.org/10.1016/S0092-8674(02)01135-2

Dosch R (2015) Next generation mothers: maternal control of germline development in zebrafish. Crit Rev Biochem Mol Biol 50:54–68. https://doi.org/10.3109/10409238.2014.985816

Franěk R, Marinović Z, Lujić J, Urbányi B, Fučíková M, Kašpar V, Pšenička M, Horváth Á (2019a) Cryopreservation and transplantation of common carp spermatogonia. PLoS One 14:e0205481. https://doi.org/10.1371/journal.pone.0205481

Franěk R, Tichopád T, Fučíková M, Steinbach C, Pšenička M (2019b) Production and use of triploid zebrafish for surrogate reproduction. Theriogenology 140:33–43. https://doi.org/10.1016/j.theriogenology.2019.08.016

Franěk R, Tichopád T, Steinbach C, Xie X, Lujić J, Marinović Z, Horváth Á, Kašpar V, Pšenička M, Lujić J, Horváth Á, Pšenička M (2019c) Preservation of female genetic resources of common carp through oogonial stem cell manipulation. Cryobiology 87:78–85. https://doi.org/10.1016/j.cryobiol.2019.01.016

Fujimoto T, Saito T, Sakao S, Arai K, Yamaha E (2010) Developmental potential of embryonic cells in a nucleocytoplasmic hybrid formed using a goldfish haploid nucleus and loach egg cytoplasm. Int J Dev Biol 54:827–835. https://doi.org/10.1387/ijdb.092896tf

Godoy LC, Streit DP, Zampolla T, Bos-Mikich A, Zhang T (2013) A study on the vitrification of stage III zebrafish (*Danio rerio*) ovarian follicles. Cryobiology 67:347–354. https://doi.org/10.1016/j.cryobiol.2013.10.002

Goto R, Saito T (2019) A state-of-the-art review of surrogate propagation in fish. Theriogenology 133:216–227. https://doi.org/10.1016/j.theriogenology.2019.03.032

Goto R, Saito T, Takeda T, Fujimoto T, Takagi M, Arai K, Yamaha E (2012) Germ cells are not the primary factor for sexual fate determination in goldfish. Dev Biol 370:98–109. https://doi.org/10.1016/j.ydbio.2012.07.010

Guan M, Rawson DM, Zhang T (2008) Cryopreservation of zebrafish (*Danio rerio*) oocytes using improved controlled slow cooling protocols. Cryobiology 56:204–208. https://doi.org/10.1016/j.cryobiol.2008.03.001

Guan M, Rawson DM, Zhang T (2010) Cryopreservation of zebrafish (*Danio rerio*) oocytes by vitrification. Cryo-Letters 31:230–238

Güralp H, Pocherniaieva K, Blecha M, Policar T, Pšenička M, Saito T (2016) Early embryonic development in pikeperch (*Sander lucioperca*) related to micromanipulation. Czech J Anim Sci 61:273–280. https://doi.org/10.17221/35/2015-CJAS

Güralp H, Pocherniaieva K, Blecha M, Policar T, Pšenička M, Saito T (2017) Migration of primordial germ cells during late embryogenesis of pikeperch Sander lucioperca relative to blastomere transplantation. Czech J Anim Sci 62:121–129. https://doi.org/10.17221/40/2016-CJAS

Hagedorn MM, Daly JP, Carter VL, Cole KS, Jaafar Z, Lager CVA, Parenti LR (2018) Cryopreservation of fish spermatogonial cells: the future of natural history collections. Sci Rep 8:1–11. https://doi.org/10.1038/s41598-018-24269-3

Hamasaki M, Takeuchi Y, Yazawa R, Yoshikawa S, Kadomura K, Yamada T, Miyaki K, Kikuchi K, Yoshizaki G (2017) Production of Tiger puffer *Takifugu rubripes* offspring from triploid grass puffer *Takifugu niphobles* parents. Mar Biotechnol 19:579–591. https://doi.org/10.1007/s10126-017-9777-1

Higaki S, Mochizuki K, Baba H, Akashi Y, Yamaha E, Katagiri S, Takahashi Y (2009) Feasibility of cryopreservation of zebrafish (*Danio rerio*) primordial germ cells by whole embryo freezing. Jpn J Vet Res 57:119–128

Higaki S, Todo T, Teshima R, Tooyama I, Fujioka Y, Sakai N, Takada T (2018) Cryopreservation of male and female gonial cells by vitrification in the critically endangered cyprinid honmoroko *Gnathopogon caerulescens*. Fish Physiol Biochem 44:503–513. https://doi.org/10.1007/s10695-017-0449-x

Inoue D, Fujimoto T, Kawakami Y, Yasui GS, Yamaha E, Arai K (2012) Vitrification of primordial germ cells using whole embryos for gene-banking in loach, *Misgurnus anguillicaudatus*. J Appl Ichthyol 28:919–924. https://doi.org/10.1111/jai.12058

IUCN (2016) The IUCN red list of threatened species. http://www.iucnredlist.org. Accessed 31 Jan 2017

Kawakami Y, Goto-Kazeto R, Saito T, Fujimoto T, Higaki S, Takahashi Y, Arai K, Yamaha E (2010) Generation of germ-line chimera zebrafish using primordial germ cells isolated from cultured blastomeres and cryopreserved embryoids. Int J Dev Biol 54:1491–1499. https://doi.org/10.1387/ijdb.093059yk

Kawakami Y, Ishihara M, Saito T, Fujimoto T, Adachi S, Arai K, Yamaha E (2012a) Cryopreservation of green fluorescent protein (GFP)-labeled primordial germ cells with GFP fused to the 3′ untranslated region of the nanos gene by vitrification of Japanese eel (*Anguilla japonica*) somite stage embryos1. J Anim Sci 90:4256–4265. https://doi.org/10.2527/jas.2011-4884

Kawakami Y, Saito T, Fujimoto T, Goto-Kazeto R, Takahashi E, Adachi S, Arai K, Yamaha E (2012b) Technical note: viability and motility of vitrified/thawed primordial germ cell isolated from common carp (*Cyprinus carpio*) somite embryos. J Anim Sci 90:495–500. https://doi.org/10.2527/jas.2011-4329

Khosla K, Wang Y, Hagedorn M, Qin Z, Bischof J (2017) Gold nanorod induced warming of embryos from the cryogenic state enhances viability. ACS Nano 11:7869–7878. https://doi.org/10.1021/acsnano.7b02216

Knaut H, Pelegri F, Bohmann K, Schwarz H, Nüsslein-Volhard C (2000) Zebrafish vasa RNA but not its protein is a component of the germ plasm and segregates asymmetrically before germline specification. J Cell Biol 149:875–888. https://doi.org/10.1083/jcb.149.4.875

Kobayashi T, Yoshizaki G, Takeuchi Y, Takeuchi T (2004) Isolation of highly pure and viable primordial germ cells from rainbow trout by GFP-dependent flow cytometry. Mol Reprod Dev 67:91–100. https://doi.org/10.1002/mrd.20003

Kobayashi T, Takeuchi Y, Takeuchi T, Yoshizaki G (2007) Generation of viable fish from cryopreserved primordial germ cells. Mol Reprod Dev 74:207–213. https://doi.org/10.1002/mrd

Komen H, Thorgaard GH (2007) Androgenesis, gynogenesis and the production of clones in fishes: a review. Aquaculture 269:150–173. https://doi.org/10.1016/j.aquaculture.2007.05.009

Kunwar PS, Lehmann R (2006) Germ-cell attraction. Dev Biol 421:226–227

Lacerda S, Batlouni S, Silva S, Homem C, França L (2006) Germ cells transplantation in fish: the Nile-tilapia model. Anim Reprod 3:146–159

Lacerda SMSN, Batlouni SR, Costa GMJ, Segatelli TM, Quirino BR, Queiroz BM, Kalapothakis E, França LR (2010) A new and fast technique to generate offspring after germ cells transplantation in adult fish: the nile tilapia (*Oreochromis niloticus*) model. PLoS One 5:1–9. https://doi.org/10.1371/journal.pone.0010740

Lee S, Yoshizaki G (2016) Successful cryopreservation of spermatogonia in critically endangered Manchurian trout (*Brachymystax lenok*). Cryobiology 72:165–168. https://doi.org/10.1016/j.cryobiol.2016.01.004

Lee S, Iwasaki Y, Shikina S, Yoshizaki G (2013) Generation of functional eggs and sperm from cryopreserved whole testes. PNAS 110:1640–1645. https://doi.org/10.1073/pnas.1218468110

Lee S, Iwasaki Y, Yoshizaki G (2016a) Long-term (5 years) cryopreserved spermatogonia have high capacity to generate functional gametes via interspecies transplantation in salmonids. Cryobiology 73:5–9. https://doi.org/10.1016/j.cryobiol.2016.08.001

Lee S, Katayama N, Yoshizaki G (2016b) Generation of juvenile rainbow trout derived from cryopreserved whole ovaries by intraperitoneal transplantation of ovarian germ cells. Biochem Biophys Res Commun 478:1478–1483. https://doi.org/10.1016/j.bbrc.2016.08.156

Li Q, Fujii W, Naito K, Yoshizaki G (2017) Application of *dead end*-knockout zebrafish as recipients of germ cell transplantation. Mol Reprod Dev 84(10):1100–1111. https://doi.org/10.1002/mrd.22870

Lin S, Long W, Chen J, Hopkins N (1992) Production of germ-line chimeras in zebrafish by cell transplants from genetically pigmented to albino embryos. Proc Natl Acad Sci 89:4519–4523. https://doi.org/10.1073/pnas.89.10.4519

Linhartová Z, Rodina M, Guralp H, Gazo I, Saito T, Pšenička M (2014) Isolation and cryopreservation of early stages of germ cells of tench (*Tinca tinca*). CJAS 59:381–390

Lubzens E, Young G, Bobe J, Cerdà J (2010) Oogenesis in teleosts: how fish eggs are formed. Gen Comp Endocrinol 165:367–389. https://doi.org/10.1016/j.ygcen.2009.05.022

Lujić J, Marinović Z, Sušnik Bajec S, Djurdjevič I, Kása E, Urbányi B, Horváth Á (2017) First successful vitrification of salmonid ovarian tissue. Cryobiology 76:154–157. https://doi.org/10.1016/j.cryobiol.2017.04.005

Majhi SK, Hattori RS, Rahman SM, Strüssmann CA (2014) Surrogate production of eggs and sperm by intrapapillary transplantation of germ cells in cytoablated adult fish. PLoS One 9(4): e95294. https://doi.org/10.1371/journal.pone.0095294

Marinović Z, Lujić J, Kása E, Bernáth G, Urbányi B, Horváth Á (2016) Cryosurvival of isolated testicular cells and testicular tissue of tench *Tinca tinca* and goldfish *Carassius auratus* following slow-rate freezing. Gen Comp Endocrinol. https://doi.org/10.1016/j.ygcen.2016.07.005

Marinović Z, Lujić J, Kása E, Csenki Z, Urbányi B, Horváth Á (2018) Cryopreservation of zebrafish spermatogonia by whole testes needle immersed ultra-rapid cooling. J Vis Exp 2–7. https://doi.org/10.3791/56118

Mazur P (1984) Freezing of living cells: mechanisms and implications. Am J Phys 247(3 Pt 1): C125–C142. https://doi.org/10.1152/ajpcell.1984.247.3.C125

McLaren A (2003) Primordial germ cells in the mouse. Dev Biol 262:1–15. https://doi.org/10.1016/S0012-1606(03)00214-8

Nóbrega RH, Greebe CD, van de Kant H, Bogerd J, de França LR, Schulz RW (2010) Spermatogonial stem cell niche and spermatogonial stem cell transplantation in zebrafish. PLoS One 5:1–16. https://doi.org/10.1371/journal.pone.0012808

Octavera A, Yoshizaki G (2019) Production of donor-derived offspring by allogeneic transplantation of spermatogonia in Chinese rosy bitterling. Biol Reprod 100:1108–1117. https://doi.org/10.1093/biolre/ioy236

Okutsu T, Suzuki K, Takeuchi Y, Takeuchi T, Yoshizaki G (2006) Testicular germ cells can colonize sexually undifferentiated embryonic gonad and produce functional eggs in fish. Proc Natl Acad Sci U S A 103:2725–2729. https://doi.org/10.1073/pnas.0509218103

Pradhan A, Olsson P-E (2014) Juvenile ovary to testis transition in zebrafish involves inhibition of Ptges1. Biol Reprod 91:1–15. https://doi.org/10.1095/biolreprod.114.119016

Pšenička M, Saito T, Linhartová Z, Gazo I (2015) Isolation and transplantation of sturgeon early-stage germ cells. Theriogenology 83:1085–1092. https://doi.org/10.1016/j.theriogenology.2014.12.010

Pšenička M, Saito T, Rodina M, Dzyuba B (2016) Cryopreservation of early stage Siberian sturgeon Acipenser baerii germ cells, comparison of whole tissue and dissociated cells. Cryobiology 72:199–122. https://doi.org/10.1016/j.cryobiol.2016.02.005

Raz E (2003) Primordial germ-cell development: the zebrafish perspective. Nat Rev Genet 4:690–700. https://doi.org/10.1038/nrg1154

Raz E, Reichman-Fried M (2006) Attraction rules: germ cell migration in zebrafish. Curr Opin Genet Dev 16:355–359. https://doi.org/10.1016/j.gde.2006.06.007

Riemer S, Bontems F, Krishnakumar P, Gömann J, Dosch R (2015) A functional Bucky ball-GFP transgene visualizes germ plasm in living zebrafish. Gene Expr Patterns 18:44–52. https://doi.org/10.1016/j.gep.2015.05.003

Riesco MF, Robles V (2013) Cryopreservation causes genetic and epigenetic changes in zebrafish genital ridges. PLoS One 8:e67614. https://doi.org/10.1371/journal.pone.0067614

Riesco MF, Martínez-Pastor F, Chereguini O, Robles V (2012) Evaluation of zebrafish (*Danio rerio*) PGCs viability and DNA damage using different cryopreservation protocols. Theriogenology 77:122–130.e2. https://doi.org/10.1016/j.theriogenology.2011.07.024

Riesco MF, Valcarce DG, Alfonso J, Herráez MP, Robles V (2014) In vitro generation of zebrafish PGC-like cells. Biol Reprod 91(5):114. https://doi.org/10.1095/biolreprod.114.121491

Robles V, Cabrita E, Paz Herráez M (2009) Germplasm cryobanking in zebrafish and other aquarium model species. Zebrafish 6:281–293. https://doi.org/10.1089/zeb.2009.0592

Robles V, Riesco MF, Psenicka M, Saito T, Valcarce DG, Cabrita E, Herráez P (2017) Biology of teleost primordial germ cells (PGCs) and spermatogonia: biotechnological applications. Aquaculture 472:4–20. https://doi.org/10.1016/j.aquaculture.2016.03.004

Saito T, Fujimoto T, Maegawa S, Inoue K, Tanaka M, Arai K, Yamaha E (2006) Visualization of primordial germ cells in vivo using GFP-nos1 3′UTR mRNA. Int J Dev Biol 50:691–700. https://doi.org/10.1387/ijdb.062143ts

Saito T, Goto-Kazeto R, Arai K, Yamaha E (2008) Xenogenesis in teleost fish through generation of germ-line chimeras by single primordial germ cell transplantation. Biol Reprod 78:159–166. https://doi.org/10.1095/biolreprod.107.060038

Saito T, Goto-Kazeto R, Fujimoto T, Kawakami Y, Arai K, Yamaha E (2010) Inter-species transplantation and migration of primordial germ cells in cyprinid fish. Int J Dev Biol 54:1479–1484. https://doi.org/10.1387/ijdb.103111ts

Saito T, Goto-Kazeto R, Kawakami Y, Nomura K, Tanaka H, Adachi S, Arai K, Yamaha E (2011) The mechanism for primordial germ-cell migration is conserved between Japanese eel and zebrafish. PLoS One 6(9):e24460. https://doi.org/10.1371/journal.pone.0024460

Saito T, Pšeničkka M, Goto R, Adachi S, Inoue K, Arai K, Yamaha E (2014) The origin and migration of primordial germ cells in sturgeons. PLoS One 9(2):e86861. https://doi.org/10.1371/journal.pone.0086861

Schulz RW, Renato L, França D, Lareyre J, Legac F, Chiarini-garcia H, Henrique R, Miura T (2010) Spermatogenesis in fish. Gen Comp Endocrinol 165:390–411. https://doi.org/10.1016/j.ygcen.2009.02.013

Seki S, Kusano K, Lee S, Iwasaki Y, Yagisawa M, Ishida M, Hiratsuka T, Sasado T, Naruse K, Yoshizaki G (2017) Production of the medaka derived from vitrified whole testes by germ cell transplantation. Sci Rep 7:43185. https://doi.org/10.1038/srep43185

Shikina S, Nagasawa K, Hayashi M, Furuya M, Yoshizaki G (2013) Short-term in vitro culturing improves transplantability of type A spermatogonia in rainbow trout (*Oncorhynchus mykiss*). Mol Reprod Dev 773:n/a. https://doi.org/10.1002/mrd.22208

Starz-Gaiano M, Lehmann R (2001) Moving towards the next generation. Mech Dev 105:5–18. https://doi.org/10.1016/S0925-4773(01)00392-6

Strüssmann CA, Nakamura M (2002) Morphology, endocrinology, and environmental modulation of gonadal sex differentiation in teleost fishes. Fish Physiol Biochem 26:13–29. https://doi.org/10.1023/A:1023343023556

Takeuchi Y, Yoshizaki G, Kobayashi T, Takeuchi T (2002) Mass isolation of primordial germ cells from transgenic rainbow trout carrying the green fluorescent protein gene driven by the vasa gene promoter. Biol Reprod 67:1087–1092. https://doi.org/10.1095/biolreprod.102.005660

Takeuchi Y, Yoshizaki G, Takeuchi T (2003) Generation of live fry from intraperitoneally transplanted primordial germ cells in rainbow trout. Biol Reprod 69:1142–1149. https://doi.org/10.1095/biolreprod.103.017624

Takeuchi Y, Higuchi K, Yatabe T, Miwa M, Yoshizaki G (2009) Development of spermatogonial cell transplantation in Nibe croaker, *Nibea mitsukurii* (Perciformes, Sciaenidae). Biol Reprod 81:1055–1063. https://doi.org/10.1095/biolreprod.109.077701

Tsai S, Rawson DM, Zhang T (2010) Development of in vitro culture method for early stage zebrafish (*Danio rerio*) ovarian follicles for use in cryopreservation studies. Theriogenology 74:290–303. https://doi.org/10.1016/j.theriogenology.2010.02.013

Weidinger G, Wolke U, Köprunner M, Klinger M, Raz E (1999) Identification of tissues and patterning events required for distinct steps in early migration of zebrafish primordial germ cells. Development 126:5295–5307

Wong T-T, Saito T, Crodian J, Collodi P (2011) Zebrafish germline chimeras produced by transplantation of ovarian germ cells into sterile host larvae. Biol Reprod 84:1190–1197. https://doi.org/10.1095/biolreprod.110.088427

Xu HY, Li MY, Gui JF, Hong YH (2010) Fish germ cells. Sci China Life Sci 53:435–446. https://doi.org/10.1007/s11427-010-0058-8

Yamaha E, Kazama-Wakabayashi M, Otani S, Fujimoto T, Arai K (2001) Germ-line chimera by lower-part blastoderm transplantation between diploid goldfish and triploid crucian carp. Genetica 111:227–236. https://doi.org/10.1023/A:1013780423986

Yamaha E, Murakami M, Hada K, Otani S, Fujimoto T, Tanaka M, Sakao S, Kimura S, Sato S, Arai K (2003) Recovery of fertility in male hybrids of a cross between goldfish and common carp by transplantation of PGC (primordial germ cell)-containing graft. Genetica 119:121–131

Yazawa R, Takeuchi Y, Higuchi K, Yoshizaki G (2010) Chub mackerel gonads support colonization, survival, and proliferation of intraperitoneally transplanted xenogenic germ cells. Biol Reprod 82:896–904

Yazawa R, Takeuchi Y, Morita T, Ishida M, Yoshizaki G (2013) The Pacific bluefin tuna (*Thunnus orientalis*) dead end gene is suitable as a specific molecular marker of type A spermatogonia. Mol Reprod Dev 80:871–880. https://doi.org/10.1002/mrd.22224

Ye H, Li C-J, Yue H-M, Du H, Yang X-G, Yoshino T, Hayashida T, Takeuchi Y, Wei Q-W (2017) Establishment of intraperitoneal germ cell transplantation for critically endangered Chinese sturgeon *Acipenser sinensis*. Theriogenology 94:37–47. https://doi.org/10.1016/j.theriogenology.2017.02.009

Yoon C, Kawakami K, Hopkins N (1997) Zebrafish vasa homologue RNA is localized to the cleavage planes of 2- and 4-cell-stage embryos and is expressed in the primordial germ cells. Development (Cambridge, England) 124:3157–3165

Yoshikawa H, Takeuchi Y, Ino Y, Wang J, Iwata G, Kabeya N, Yazawa R, Yoshizaki G (2017) Efficient production of donor-derived gametes from triploid recipients following intraperitoneal germ cell transplantation into a marine teleost, Nibe croaker (*Nibea mitsukurii*). Aquaculture 478:35–47. https://doi.org/10.1016/j.aquaculture.2016.05.011

Yoshikawa H, Xu D, Ino Y, Yoshino T, Hayashida T, Wang J, Yazawa R, Yoshizaki G, Takeuchi Y (2018) Hybrid sterility in fish caused by mitotic arrest of primordial germ cells. Genetics 209:507–521. https://doi.org/10.1534/genetics.118.300777

Yoshizaki G, Lee S (2018) Production of live fish derived from frozen germ cells via germ cell transplantation. Stem Cell Res 29:103–110. https://doi.org/10.1016/j.scr.2018.03.015

Yoshizaki G, Ichikawa M, Hayashi M, Iwasaki Y, Miwa M, Shikina S, Okutsu T (2010) Sexual plasticity of ovarian germ cells in rainbow trout. Development 137:1227–1230. https://doi.org/10.1242/dev.044982

Yoshizaki G, Takashiba K, Shimamori S, Fujinuma K, Shikina S, Okutsu T, Kume S, Hayashi M (2016) Production of germ cell-deficient salmonids by dead end gene knockdown, and their use as recipients for germ cell transplantation. Mol Reprod Dev 83:298–311. https://doi.org/10.1002/mrd.22625

Yoshizaki G, Yazawa R (2019) Application of surrogate broodstock technology in aquaculture. Fish Sci 85:429–437. https://doi.org/10.1007/s12562-019-01299-y

Zhu ZY, Sun YH (2000) Embryonic and genetic manipulation in fish. Cell Res 10:17–27. https://doi.org/10.1038/sj.cr.7290032

Zimmermann J, Hajibabaei M, Blackburn DC, Hanken J, Cantin E, Posfai J, Evans TC (2008) DNA damage in preserved specimens and tissue samples: a molecular assessment. Front Zool 5:18. https://doi.org/10.1186/1742-9994-5-18

Cryopreservation and Storage of Oocytes, Embryos and Embryonic Cells of Fish

13

P. Routray

Abstract

Worldwide hatchery seed production has seen a manifold growth in terms of quantity of seed produced due to advancements in reproductive biotechnology and brood stock management. However, the quality of seed is an important parameter for profitable aquaculture. Both male and female gametes and stored embryos of animals play prominent roles in the genetic management of species and conservation. However, storage by cryopreservation of oocytes or embryos of fish like carps and other teleost is elusive and few claims are not reproducible in nature. Production of improved stock using cryopreserved milt is being practiced in many hatcheries. Male gametes of improved stock are cryopreserved and utilized for quality seed production as well as upgrading the brood stock in carps. However, cryogenic and non-cryogenic storage of oocytes and embryos has not been possible due to several reasons. Researchers have successfully cryopreserved and revived embryonic cells of fish in many laboratories around the world. Here an attempt has been made to describe the principles, practice, issues and way forward in the field of cryopreservation of fish egg (oocytes), embryo and embryonic cells both for long-term and short-term duration storage.

Keywords

Cryopreservation · Storage · Oocyte · Embryo · Embryonic cell · Fish

P. Routray (✉)
Cryobiology Laboratory, ICAR-Central Institute of Freshwater Aquaculture, Bhubaneswar, India

J. Betsy, S. Kumar (eds.), *Cryopreservation of Fish Gametes*,
https://doi.org/10.1007/978-981-15-4025-7_13

13.1 Introduction

Supplying quality fish seed at right time is essential for profitable fish farming. Fish production is hampered by the high mortality rate of fry during transfer from hatchery to rearing ponds (Islam and Hossain 2013), sometimes up to 70% of mortality recorded due to transportation stress (Husen and Sharma 2015) that is attributed to insufficient packaging ratio, depletion of dissolved oxygen, accumulation of CO_2, change in water quality, physical handling etc. Similarly, inbreeding also contributes to the problem of poor growth of carps and subsequently production of poor quality seed. To overcome this and make aquaculture forward looking and more pragmatic in general and hatchery seed production in particular, a possible option would be in vitro storage of gametes (both sperm and ova) for fertilization and seed production. Gametes of improved stock are cryopreserved and utilized for quality seed production as well as upgrading the brood stock of fish. Utilization of fresh and stored fish milt (spermatozoa) for seed production, progeny testing, genetic selection programmes and overcoming asynchronous spawning is in vogue in many countries. The quality of a gamete is determined by its ability to fertilize or to be fertilized and eventually develop into a normal embryo (Bobe and Labbe 2010). These gametes play prominent roles in the genetic management of species and conservation of aquatic resources. However, storage by cryopreservation of oocytes or embryos of fish like carps and other teleost is elusive and few claims are not reproducible in nature. Both sperm and embryo cryopreservation have become routine procedures in human-assisted reproduction and oocyte cryopreservation is being introduced into clinical practice and is getting more and more widely used. Embryo cryopreservation has decreased the number of fresh embryo transfers and maximized the effectiveness of the IVF cycle (Konc et al. 2014). The problems of embryo preservation in teleost fish are known to a larger extent and many researchers have reported that the thick chorion, membrane impermeability, large amount of yolk, cryoprotectant toxicity and intracellular ice formation are the major stumbling blocks in the path of successful cryopreservation of fish embryos. Similarly, fish oocytes are difficult to manage after ovulation because of its quick activation and cortical reactions when it comes in contact with any kind of solutions or for that matter for any change in the osmolality of solutions besides ionic imbalances and temperature of storage. The success of embryonic cell preservation in fish has shown some light towards species restoration/resurrection by advanced biotechnological means. However, it is pertinent to mention that all these biological entities (oocyte, embryo and embryonic cells) stored in laboratory conditions for varying duration of time and for different low temperature regimes could not convince the aquaculture industry to take up these and apply in field conditions. Here an attempt has been made to describe the principles, practice, issues and way forward in the field of cryopreservation of fish egg (oocytes), embryo and embryonic cells.

13.2 Storage of Oocytes, Embryos, Embryonic Cells

13.2.1 Storage

Storage of gametes (male and female) can be done by cryopreservation which has been considered as a practical tool (Betsy et al. 2014). Systematic germplasm conservation by cryopreserving fish spermatozoa, eggs and embryos can have a profound effect on aquaculture by allowing the maintenance of a large gene pool for genetic manipulation and conservation; ensure timely provision of fish seed and decreasing aquaculture costs by reducing the transport cost enabling long periods of storage of oocytes and embryos without changes in their native character. In vitro storage of gametes of fish (both sperm and ova) for fertilization and seed production has been attempted for a long time and success could be achieved in spermatozoa only. A schematic diagram describing the importance of storage of fish gametes is shown in Fig. 13.1. It is pertinent to mention that cryopreservation is a method to store or preserve live materials at ultra-low temperature without affecting the viability.

13.2.2 Cryopreservation

Cryopreservation of biological entities can be defined as the preservation or storage of living materials such as cells, tissues, oocytes, embryos, spermatozoa etc. at liquid nitrogen temperature (−196 °C) using specific extenders, embryo solution, vitrification media, cryoprotectants for a long period of time without affecting their original nature. This technique is widely used to preserve gametes, embryos and embryonic cells of many animal species. Polge et al. (1949) reported the first successful cryopreservation of mammalian sperm cells. Thereafter, sperm cell preservation of mammals and other animal species including fish has become a regular practice around the world. The general protocol of cryopreservation of small cells and spermatozoa is as follows: (1) loading, which is the equilibrium of the specimen in cryoprotective agents (CPA); (2) dehydration in a solution (extender + CPA); (3) plunging the specimen in liquid nitrogen; (4) warming the specimen and (5) unloading of CPAs, by progressive dilution from the specimens. Wilmut (1972) and Whittingham et al. (1972) reported firstly that mouse embryos could be successfully frozen at −196 °C. Although the general protocols were similar with that of spermatozoa, the slow cooling and thawing rates were preferred for better survival in embryos. Thus, animal embryos and many biological materials could be successfully cryopreserved worldwide.

Cryopreserved embryos are widely used in assisted reproduction of domestic animals and in human in vitro fertilization (IVF). However, the technique was empirical and depended largely on the kind of materials employed. For example, it varies greatly according to species, the appropriate cryoprotectant and cooling speed. It is known that one CPA like dimethyl sulfoxide (DMSO) penetrates into the cell and protects proteins and membranes, whereas other CPAs which do not penetrate

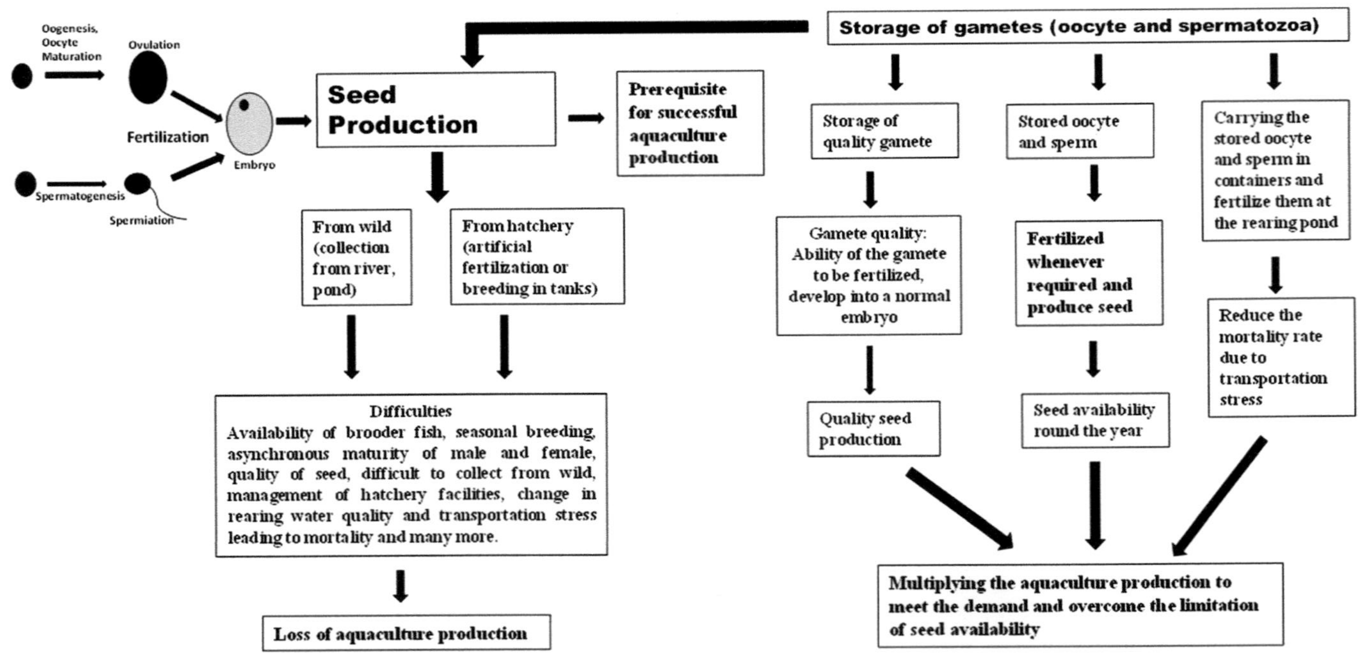

Fig. 13.1 Schematic representation of formation of fish seed, factors affecting seed production and importance of gamete storage for improvement of aquaculture

the cells, and thus remains in the extracellular medium promote dehydration. The intracellular and extracellular environments associated with water transport and membrane permeability form the basis behind the cryopreservation of cells. It is generally thought that larger cells that are less permeable should be frozen slowly and the smaller cells should be frozen quickly (Mazur 1965). The reason behind this concept deals with cellular damage during the freezing process by osmotic shock, intracellular ice formation, increased intracellular concentration of solutes and solution effects (changes resulting from dehydration of cells) (Mazur 1977). Despite such empirical successes, the principle of cryopreservation technique remained unclear until 1985.

Rall and Fahy (1985) showed that glass transition of the materials during the freezing process is the main principle behind the success of cryopreservation of embryonic cells. After that, on the basis of the glassification principle, several highly effective cryopreservation protocols have been developed for mammalian oocytes and embryos (Hurst et al. 1997; Karlsson et al. 1996; O'Neil et al. 1997). The most common protocols include impregnation of the cells with a permeating cryoprotectant such as DMSO prior to cryopreservation. This protocol is apparently not so different from empirically established methods; however, any technique has to be justified by scientific reasoning. The presence of this chemical, like other permeating cryoprotectants, reduces the gap between freezing and vitrification temperatures and thereby prevents fatal, widespread intra- and extracellular ice formation.

13.2.3 Principles of Cryopreservation (Glassification)

The aim of most cryopreservation protocols is to vitrify the materials without ice formation or, in other words to achieve a glass state. The solution forms an amorphous glass as a result of rapid cooling by direct immersion into liquid nitrogen. The resultant glass retains the normal molecular/ionic distribution but remains in an extremely viscous state. The glass is devoid of any ice crystals and the embryos are not subjected to the physical damage that is associated with ice crystal formation during the freezing process, which is common in conventional cryopreservation method.

Freezing implies two processes, namely the lowering of the temperature and the change of phase from liquid to solid. Both phases are accompanied by a reduction in heat content of the material (Reid 1993). The freezing process, which includes supercooling, nucleation, ice crystal propagation and maturation, is strongly influenced by thermodynamic, kinetics and material variables (Sahagian and Goff 1996). The freezing point of biological materials is the temperature at which a minute crystal of ice exists in equilibrium with the surrounding water or when the chemical potential between the solids and the liquid phase is at equilibrium (Sahagian and Goff 1996).

According to the above physico-chemical process, the freezing behaviour of a solution could be explained as follows: although the melting point or equilibrium freezing point of pure water is 0 °C, those of water-containing ions and other substances are at lower temperatures depending on the solute concentration. In a usual solution, as pure water freezes out of the solution, it leaves behind a solution of increased solute concentration. The equilibrium freezing point of the remaining solution will therefore decline. This process is called freezing point depression. If the temperature is decreased further and more water is removed in the form of ice, the concentration of solutes in the remaining liquid increases, a process called freeze concentration, which results in a continuous drop in the equilibrium freezing point of the solution (Fig. 13.2). It was believed that as long as the ambient temperature continues to decline, these changes occur until the entire solvent and its contained solutes (and other substances) freeze completely at the eutectic point. This is true for some salt solutions such as NaCl. However, it has been found that many solution systems do not have an eutectic point, but the increasingly concentrated solution by freezing concentration process do turn into glassy state at Tg′ passing through an equilibrium eutectic point, which is called a glass transition temperature of maximum freeze concentrate solution (Franks 1982). Furthermore, an important finding relating to this was that the glass transition temperature for a binary system of water and solute is hardly depended on the water content as shown in Fig. 13.2. In case of low moisture materials, the transition temperature is above zero. This entire behaviour is shown in Fig. 13.1, which is called a state diagram. When a conventional cryopreservation process is considered, many researchers aim to realize the pass A in Fig. 13.2. To attend this, it is required to cool down at ultimate high speed, which is often called a quench operation, because the temperature range between equilibrium freezing point and the glass transition temperature is under unstable super cooled state for the solution, so that ice crystallization from the solution easily occurs. An application of this quick cool technique is useful for small cell such as spermatozoa because the resistance of the heat transfer to the centre is negligible.

However, for large organ or cell such as fish embryo, the heat transfer resistance from surface to centre is not negligible to derive a significant delay on the temperature depression around the cell centre even if the surrounding is cooled by liquid nitrogen. Therefore, in case of large cells, alternative method should be considered. As described, it has been known better for embryos to be cooled down ‘slowly’.

Now this reason could be understood as follows; when the solution including cells and CPA is cooled to subzero temperatures, ice crystallization from the surrounding solution with the cell occurs causing to freeze concentration on the solution around the cell. Then, the cell is dehydrated and the impregnation of CPA is promoted, as a resultant the CPA concentration inside the cell becomes high during slow cooling. Again, as shown in Fig. 13.1, it is found that the solution with high CPA or solute in the concentration has a high glass transition temperature, which means the solution forms glass easily. It has been known for glassification of the concentrated system do not require quick cooling as in case of low concentration solution. Therefore, dehydrated and concentrated cell by slow cooling is understood to turn into glassy state easily despite slow cooling.

According to above principle, the important things to be focused is how to increase the concentration of inner cell as well as to know the glass transition

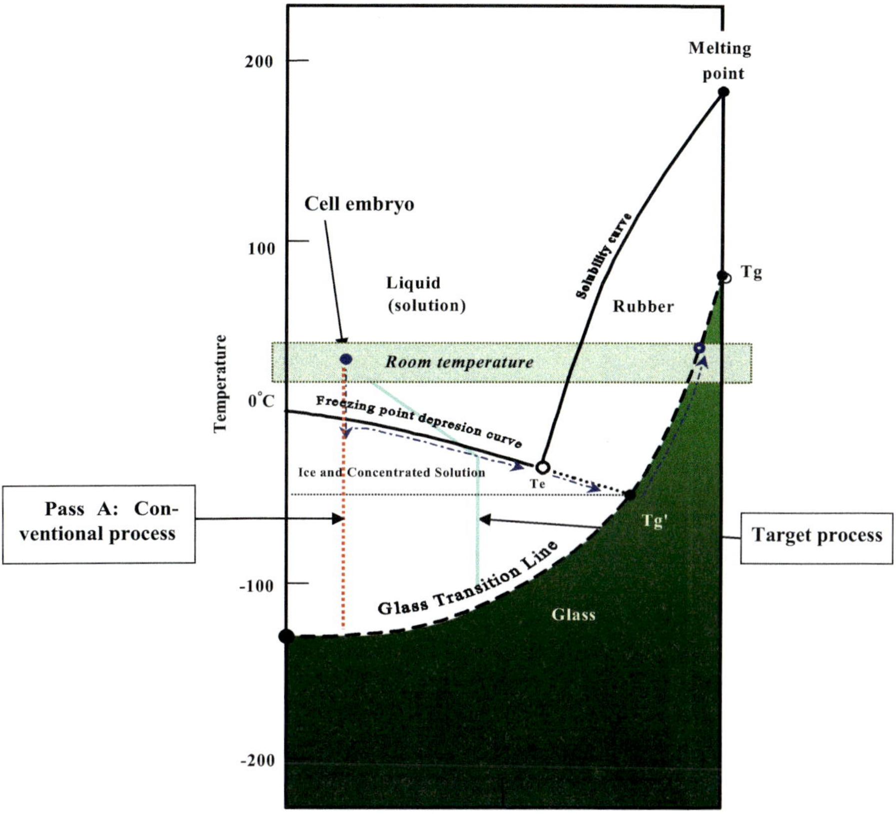

Fig. 13.2 Schematic state diagram of water showing the relation between glass transition temperature and solute content of solution

temperature–concentration of various CPAs solutions, i.e., a glass transition line as shown in Fig. 13.2.

13.3 Present Status and Problems of Fish Embryo and Gamete Cryopreservation

Male gamete cryopreservation first became possible in 1949 when Polge et al. (1949) successfully frozen and thawed human and avian spermatozoa using glycerol as a cryoprotectant.

Shortly thereafter, Blaxter (1953) applied a similar approach to teleost gametes and reported success with Atlantic herring spermatozoa, achieving approximately 80% cellular motility upon thawing. Since then spermatozoa of many species of fish has been successfully cryopreserved for a host of teleosts like carps, salmonids, catfish, cichilids, medaka, pejerrey, grouper, cod, silver barb and zebrafish (Arii et al.

1987; Harvey et al. 1982; Leung and Jamieson 1991; Rana 1995; Routray et al. 2003; Routray et al. 2008; Sahu et al. 2011). Presently, the cryopreserved spermatozoa have been used for artificial breeding, cryo-banking and to offset the variation of spawning and stock enhancement programs in different parts of the world. However, the real issue of oocyte and embryo cryopreservation in fish could not be resolved due to several unsolved and unanswered scientific questions. It may be noted that the whole embryo cryopreservation has been a failure. Some old researches have reported the survival of embryos after lowering of temperature to −50 °C. But these results are not reproducible or consistent as they were mostly based on empirical hit and trial methods. Probably, it can be supposed that the keeping time was short, the temperature of embryo might not truly attain −50 °C and the inner embryo were still under super cooled condition. A comparative status of oocyte and embryo cryopreservation is presented in Table 13.1.

Many researches have overlooked the principle of glassification process though it is very important for success of cryopreservation. In recent studies, cryopreservation protocols based on data for mammalian embryos considering glass transition principle, have been tested in isolated cells such as blastomeres (Leveroni et al. 1998; Nilsson and Cloud 1993; Strüssmann et al. 1999; Routray et al. 2009) and whole eggs and embryos (Chao and Liao 2001; Gwo 2000; Hagedorn and Kleinhans 2000; Routray et al. 2003) of fish, but successful cryopreservation was obtained only with the isolated cells. It must be noted also that isolated cell cryopreservation, although technically feasible, is not practical in large-scale hatchery activities (Urbányi et al. 2000). The difficulty with whole fish eggs and embryos apparently lies in the peculiar characteristics of these biological materials. For instance, fish egg and embryos are several times larger than those of mammals, have a thick chorion, a large volume of yolk segregated from the active cytoplasm, and show considerable structural complexity early in development (Hagedorn and Kleinhans 2000; Hagedorn et al. 1997; Harvey 1983; Harvey and Ashwood-Smith 1982; Zhang and Rawson 1996). Therefore, it is very difficult to achieve a swift and uniform permeation of the cryoprotectant into the various compartments of fish eggs and embryos, resulting in toxicity to some cells while in others the concentration of cryoprotectant is not sufficient to prevent cryoinjuries.

Researchers have tried to circumvent these problems in several ways including the impregnation of eggs and embryos with cryoprotectant after dechorionation (Hagedorn et al. 1997), under negative pressure (Leung and Jamieson 1991), using hydrostatic pressure (Routray et al. 2002a) or by direct delivery of cryoprotectant into the yolk by microinjection (Janik et al. 2000a) but as observed in earlier studies, failed to achieve successful cryopreservation. These facts underscore the need for fundamental information on the factors that affect the uptake of cryoprotectants in order to develop successful protocols for intact fish egg and embryo cryopreservation (Hagedorn and Kleinhans 2000).

Recently successful cryopreservation of zebrafish embryos using gold nanorods (GNRs) to assist in the warming process has been reported (Khosla et al. 2017) where propylene glycol was microinjected into zebrafish embryos along with GNRs, and the samples were cooled at a rate of 90,000 °C/min in liquid nitrogen and thawed

Table 13.1 Present status of cryopreservation in embryos and embryonic cells of fishes and other aquatic species

Name of the species	Stage studied	Extender/EM used	CPA used	Max. temp. °C	Freezing/thawing rate	Success/remark	References
Rainbow trout (*Salmo gairdneri*)	Unfertilized egg	Hank's salt solution (HSS)	DMSO	−20	Cooling rate not mentioned. Thawing in 50–60 °C water	Hatched 66%	Zell (1978)
Rainbow trout (*S. gairdneri*)	Zygote	HSS	DMSO	−20	do	Hatched 68%	Zell (1978)
Rainbow trout (*S. gairdneri*)	Eyed	HSS	DMSO	−20	do	Hatched 39%	Zell (1978)
Brook trout (*Salvelinus fontinalis*)	Zygote	HSS	DMSO	−50	do	Hatched 20–70%	Zell (1978)
Zebrafish (*Bracydanio rerio*)	Fertilized 100% epiboly	Ionic buffer 0.040 mOsm)	Propylene glycol and DMSO	−196	3 min over LN_2 vapour	0%	Janik et al. (2000)
Zebrafish (*B. rerio*)	10 stages (epiboly to eye stage)	–	Sucrose/trehalose methanol	0 to −15	Supercooled	30.2–4.2% survival	Zhang and Rawson (1995)
Medaka (*Oryzias latipes*)	Embryo	Yamamoto ringer	2 M DMSO (5 h)	−40	−1 °C/min to 10 °C then at −2 °C/min to −6 °C then −10 °C/min to −40	57.1% survival	Arii et al. (1987)
Goldfish (*Carassius auratus*)	Heart formation stage	–	12% DMSO stepwise 5, 8, 10, and 12%	−35 (60 min stored)	−1 °C to 4 °C, −0.35 °C to −35 °C. Thawing in 24 °C water	60% survival and 35% hatching	Liu et al. (1993)
Rotifer (*Brachionus plicatilis*)	Stage III (symmetrical embryo)	–	10% DMSO 30 min	−196	Two-step freezing	53–63% survived	Toldeo and Kurokura (1990)

(continued)

Table 13.1 (continued)

Name of the species	Stage studied	Extender/EM used	CPA used	Max. temp. °C	Freezing/thawing rate	Success/remark	References
Oyster (*Crossostrea gigas*)	Early larvae	Sea water	2 M DMSO + 0.06 M Trehalose	−196	−1 °C/min to −12 °C, −2 °C/min to −35 °C	62–75.1% survival	Chao et al. (1997)
Hard clam (*Meretrix lusoria*)	Early larvae	Sea water	2 M DMSO + 0.06 M Trehalose	−196	−1 °C/min to −12 °C, −2 °C/min to −35 °C then quenching in LN_2	73.3–84.2% survival	Chao et al. (1997)
Teleost fish *Prochilodus marggravii*	Oocyte	Nil, only air filled in inflated bag	Nil	18 for 2 h	Cooled from 26 to 18 °C	Less than 30% fertilized	Rizzo et al. (2003)
Flounder, *Paralichthys olivaceus*	Embryo (14-somite to the pre-hatching stage)	Four vitrifying solution	Propylene glycol and methanol	−196 for 1–7 h	Vitrification	20 viable embryos recovered from 292 cryopreserved embryos	Chen and Tian (2005)
Zebrafish, *Danio rerio*	Stage-III oocyte	KCl buffer, L-15 medium	Methanol and glucose	−196	Plunging in liquid nitrogen	88% survival assessed by trypan blue assay	Guan et al. (2008a)
Zebrafish, *D. rerio*	Embryo	–	Propylene glycol along with gold nanorods microinjected	−196	Cooling at 90,000 °C/min, thawing (unfreeze) at (1.4×10^7 °C/min) by irradiation with a 1064 nm laser pulse for 1 ms	10% after 24 h post taw and −31% after 1-h	Khosla et al. (2017)

Gorgonian (*Junceella juncea*)	Oocytes	PG-based ES (1 M PG + 0.5 M EG + 0.5 M methanol) or (2 M PG + 1 M EG + 1 M methanol)	VS was composed of propylene glycol (PG), ethylene glycol (EG) and methanol with concentrations of 3.5 M, 1.5 M and 2 M	−196	Vitrification	The success of the vitrification determined by ATP levels in cooled thaw oocytes and the highest viability was 76.6%.	Tsai et al. (2015)
Zebrafish, *D. rerio*	Ovarian tissue including immature oocytes	2 M methanol + 0.1 M trehalose + 10% egg yolk solution	2 M methanol	−196	Vitrification	63.5% membrane integrity in primary growth oocytes	Marques et al. (2019)
Rainbow trout, *Oncorhynchus mykiss*	Isolated blastomeres	Niu twitty's solution	8.7% DMSO	−196	At 1 °C/min to −4 °C, at 10 °C/min to −90 °C then plunged in LN_2	19.4–35.9% survival	Nilsson and Cloud (1993)
Rainbow trout, *O. mykiss*	Isolated blastomeres	Ca^{2+} and Mg^{2+} free	1.4 M 1,2-propanediol	−196	At −1 °C/min to −6.6 °C hold 15 min, at −0.3 °C/min to −40 °C, at −2 °C/min to −80 °C hold 10 min then plunged in LN_2	53–88% survival	Laveroni Calvi and Maisse (1998)
Medaka, *O. latipes*; whiting, *Shillago shiama* and pejerrey, *Odontesthes bonariensis*	Isolated blastomeres	Ca^{2+} free Waymouth's culture medium	9–18% DMSO	−196	At −14.2 °C/min to −128.7 °C for whiting, at −1.0 °C/min to −14.2 °C for medaka and pejerrey	19.9% whiting, 34.15% medaka and 67.4% pejerrey	Strüssmann et al. (1999)

(continued)

Table 13.1 (continued)

Name of the species	Stage studied	Extender/EM used	CPA used	Max. temp. °C	Freezing/thawing rate	Success/remark	References
Leopard danio, *Brachydanio frankei*	Isolated blastomeres and embryonic stem-like cells	LDF comprised Leibowitz-15 (L-15), Dulbecco's modified Eagles medium (DMEM) and Ham's F12 (Sigma, St Louis, MO, USA) in 50:35:15 supplemented with sodium bicarbonate	0.6–1.8 M ethylene glycol (EG), propylene glycol (PG)	−196	At −1.0 °C/min to −4 °C/min up to −80 °C, and then plunged into liquid nitrogen (−196 °C)	70% for blastomeres and 65% for embryonic stem-like cells	Routray et al. (2009)

at a rate of 1.4 × 107 °C/min using 1064 nm laser pulse for 1 ms that resulted in revival of 10% embryos after 24 h post-thawing. This rapid warming process led to the outrunning of ice formation, which could have damaged the embryos. This finding is highly encouraging, but the technology-intensive results need further simplification and the cost will be a prohibitive factor for further implementation by hatchery owners. The role of GNRs and their biocompatibility or toxicity needs further studies.

However, on consideration of CPA uptake, it should not be neglected that all CPAs has toxicity for biological systems. This information would give a limitation of using CPA. In a study of DMSO permeation into intact fish embryos, Suzuki et al. (1995) measured the concentration of the cryoprotectant in the embryos and showed the toxicity. However, their study did not take into account the stage of development of the embryos and the duration of exposure to the cryoprotectant. Uptake dynamics of the cryoprotectant DMSO by intact unfertilized eggs (stage 0), 8-cell (stage 5) and eyed embryos (stage 30) of medaka, *O. latipes*, the relation of the internal concentration (Cin) of DMSO with fertilization and survival rates, and the effects of several factors on these processes have been reported (Routray et al. 2002b). Cryoprotectant permeation, estimated from the initial rates of DMSO uptake, was higher in embryos than in eggs and increased with embryonic development; however, the DMSO concentration inside eyed embryos reached a plateau at 1–5 min and could not be increased by prolonging the duration. This signifies the importance of developmental stage in the cryopreservation process of fish embryo. The stage of development is also an important aspect encountered during the fish embryo cryopreservation. There is a wide pattern of development in fish embryos. Some fishes hatch very fast within 18–24 h after spawning (carps) whereas some hatch after 10–11 days after spawning (medaka). So generally researchers find difficulty in choosing the proper species and stage of embryos for cryopreservation and the literature on this is very few. Nonetheless, the potential relevance of these factors for fish egg and embryo cryopreservation has been largely neglected. The study of hydrostatic pressure, for instance, is limited to a report by Leung and Jamieson (1991) on the effects of mild vacuum during impregnation.

13.3.1 Temperature of Intracellular Ice Formation (TIIF)

The last hurdle generally faced by many researchers is to find a proper cooling and thawing rate in order to make glass i.e. vitrification. Formation of ice crystals and their subsequent growth by recrystallization during warming leads to injury in most of the times (Mazur 1965; 1977). Intracellular ice formation in embryos of different animal species including human embryos has been reported (Ashwood-Smith 1986; Cohen et al. 1986; Veron et al. 1997; Schreuders et al. 1996). A critical parameter in understanding the freezing of cytoplasm or ice formation is the temperature at which the super cooled cytoplasm is assumed to freeze (Mazur 1977).

One of the reasons for failure of fish embryo cryopreservation is believed to be the high sensitivity to low temperatures or chilling injury (Zhang and Rawson 1995).

Chilling range of temperature for embryonic studies is generally 0 °C to −10 °C. Recent studies on *Drosophila melanogaster* embryos have suggested the importance of cold tolerance for the embryos in designing the cooling protocol (Steponkus et al. 1990). Similar to *Drosophila* embryos, embryos of fish also show chilling injury (Rall 1993). Arii et al. (1987) have reported that the medaka embryos survived at temperatures as low as −40 °C for a short duration, but they did not make clear the low temperature injury and nucleation temperatures of embryos, which is important information to confirm the ice nucleation state in fish embryos. However, the exact mechanism of damage to fish embryos during cooling or freezing has not yet been identified due to lack of systematic research. Intra- and extra-cellular ice formation during freezing is not desirable to reduce the extent of extreme hypothermic, mechanical and osmotic stress on fish embryos (Lovelock 1953; Meryman et al. 1977; Janik et al. 2000). Liu et al. (2000) showed that the ice formation within the egg, i.e. the intra-embryonic freezing, was the main factor affecting the survival of zebrafish (*B. rerio*) embryos. Recently, Zhang and Rawson (1995) also reported the differences in cold tolerance of zebrafish embryos treated with different cryoprotectants. Cryoprotectant like DMSO is believed to help the embryos in tolerating the chilling sensitivity (Arii et al. 1987). However, there are few reports on the IIF or cold tolerance and equilibrium melting temperature of fish (Routray et al. 2002b). Furthermore, though DMSO (Chao and Liao 2001) and trehalose (Kawai et al. 1992) have been used extensively as cryoprotectants, their effect on the nucleation behaviour of embryos has received little attention.

Information related to intracellular ice formation (IIF) in fish embryos is very scanty and moreover, this information is of prime importance, based on which cooling protocols for cryopreservation can be designed. Thawing rate is also an important factor, which can cause lethal damage to embryos by reversible ice formation during warming. One of the major requirements for the successful vitrification is to increase the viscosity of the solution either by increasing the concentration of the solution or by controlling the speed of cooling.

After all, attempts to cryopreserve fish eggs and embryos have been futile till now. The formidable constraints faced by researchers around the world could be divided into three major parts namely, (1) low impregnation of cryoprotective agents (CPAs) with proper concentration into the highly impermeable embryonic cells through the embryonic membrane, (2) finding out the proper stage of embryo for storage and their sensitivity to chilling temperature and (3) their IIF temperature and freezing by vitrification or making a perfect glass transition with proper cooling and thawing rates.

The cold tolerance and ice nucleation temperature of medaka embryos in three embryonic stages were investigated with different cryoprotectant treatments. The ice nucleation temperatures of the embryos in every stage showed a decreasing tendency by cryoprotectant treatment, in the order of trehalose, DMSO, and a combination of trehalose and DMSO. Temperature of intracellular ice formation (TIIF) has been studied in detail by Routray et al. (2001) in pejerrey and medaka oocytes and embryos by using diffraction scanning calorimetry (DSC). A typical DSC cooling and warming pattern of fish embryos (Pejerrey, *O. bonariensis*) with the TIIF and

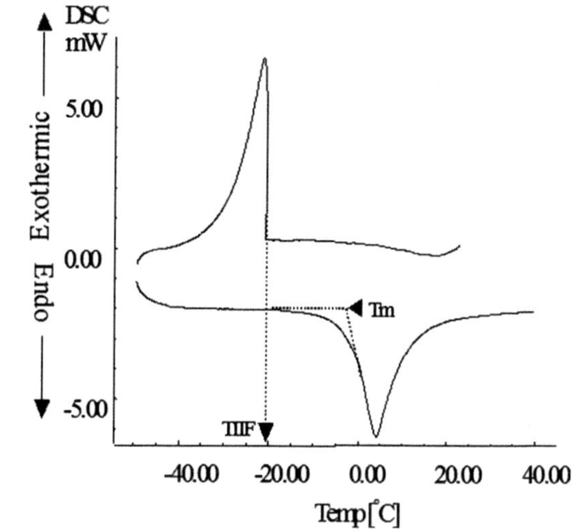

Fig. 13.3 Typical temperature and heat flow of eight-cell embryos of pejerrey. *O. bonariensis* obtained by DSC analysis. Arrow shows the temperature of intracellular ice formation (TIIF) and equilibrium melting point (Tm). Cooling and warming rate were −2 °C and 2 °C/min respectively. (Source: Routray et al. 2001)

equilibrium melting point (Tm) with a cooling and warming rate of −2°C and 2 °C/min respectively is shown in Fig. 13.3. From the onset temperature of the peak, the TIIF and Tm were estimated and the DSC graph showed a similar pattern of peaks as shown in Fig. 13.3. It is reported that slow cooling rates of −2 °C and −5 °C/min substantially reduced the TIIF to −25 °C in embryos that is desired for cryopreservation.

Relation between cooling rates and TIIF in pejerrey embryos treated with 2 M DMSO is shown in Fig. 13.4 (Source: Routray et al. 2001). Similarly, increase in cryoprotectant concentration (DMSO) also decreased the TIIF and equilibrium melting point of fish embryos (Fig. 13.5). Rall et al. (1983) reported depression of ice nucleation temperature by DMSO and glycerol in mouse embryos. However, TIIF studies by Routray et al. (2001) could not find out the fixed reasons for the above phenomena.

13.4 Cryopreservation of Fish Oocytes

Cryopreservation of fish oocytes is challenging because oocytes have low membrane permeability to water and cryoprotectant and are highly chilling sensitive. Therefore, there has been no successful protocol reported for cryopreservation of fish oocytes (Routray et al. 2002a; Isayeva et al. 2004; Plachinta et al. 2004; Guan et al. 2008b; Guan et al. 2010). The keeping quality and storage duration without compromising the oocyte viability (after stripping) differs from species to species and depends largely on temperature of storage and biochemical degradation.

The loss of oocyte viability after ovulation is one of the limiting factors in controlled reproduction of several fish species. This is more evident in tropical fish where the ovulated (stripped) oocytes cannot be stored for longer periods as they

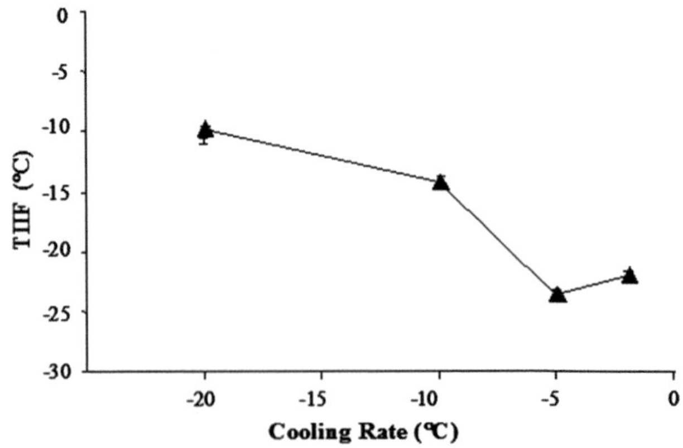

Fig. 13.4 Relation between cooling rates and TIIF in pejerrey embryos treated with 2 M DMSO. Data shown as mean ± SEM. (Source: Routray et al. 2001)

quickly lose their viability and fertilization capacity (Espinach Ros et al. 1984; Rizzo et al. 2003). Unlike salmon and trout, oocytes of the several fish examined to date lose their viability within a few hours of storage in the ovarian cavity (Espinach Ros et al. 1984; Harvey and Kelley 1985; Legendre et al. 2000). Similarly, if the ovulated (unfertilized) eggs are not released from the body of fish at a proper time, then also it loses its ability to be fertilized due to gradual morphological and biochemical changes, sometimes referred to as an over ripening condition in brood fish (Nomura et al. 1974; Bromage and Roberts 1995; Lahnsteiner 2000). Preservation of gametes is aimed at increasing post spawning longevity of gametes that may improve hatchery management, minimize inbreeding and asynchronous brooder maturation (Bromage and Roberts 1995). Storage of matured oocytes can be an alternative to embryo preservation as it has several advantages like smaller size, lower water content and absence of a fully developed chorion (Zhang et al. 2007).

Although a great deal of work has been done examining the cryobiology of zebrafish oocytes at various stages of maturation, the successful cryopreservation, maturation and subsequent fertilization of these oocytes has not been proven. Similarly, teleost embryo cryopreservation has been extensively attempted for the past three decades, and several issues have made it a challenging system to develop successful protocols. Successful short-term storage of ovulated oocytes of Indian major carp, rohu, *Labeo rohita* for a comparatively longer time without losing its fertilization ability (till 3 h) under different temperature regimes has been reported (Mishra et al. 2017). Present status of oocyte cryopreservation in fish and other aquatic species is depicted in Table 13.1. It is envisaged that even short-term storage of ovulated oocyte of carps may open new vistas in aquaculture in the Indian sub-continent as the stored oocytes can be transported between hatcheries and farms similar to that of milt for artificial insemination.

Fig. 13.5 DSC cooling curves showing freeze onset temperatures of the medaka embryos in three stages, (**a**) eight cell, (**b**) optic bud, (**c**) eyed stage, with different treatments. Scan rate were −2 °C/min for all. (Source: Routray et al. 2001)

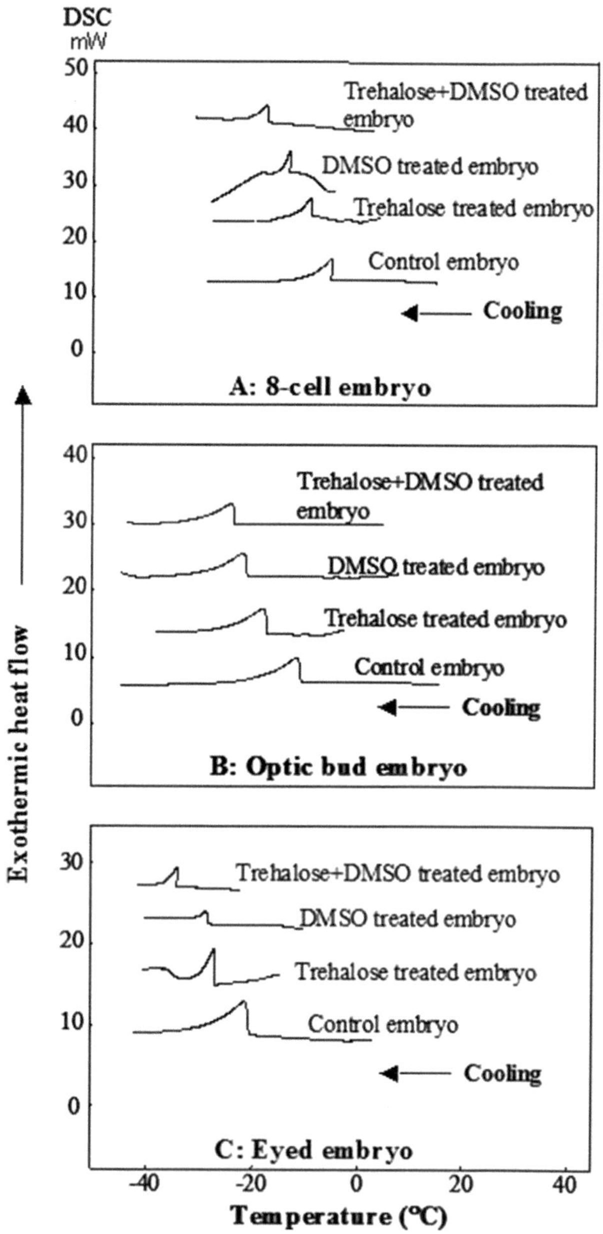

Table 13.2 Chronological history in cryopreservation of ES cells in different species

Name of the species	Methodology adopted	References
Human	Vitrification method	Reubinoff et al. (2001)
Human	Non-toxic cryoprotectants	Lin et al. (2004)
Human	Different vitrification solution	Zhou et al. (2004)
Primate	A first report of primate ES cell cryopreservation	Fujioka et al. (2004)
Human	Vitrification and slow cooling process	Heng et al. (2005)
Human	Study of ES cells on various temperature regimes	Heng et al. (2006)
Mouse	First report on embryoid body preservation and their post-thawed assessment was done	Ichikawa et al. (2007)
Fish		
Zebrafish	First report on fish ES cells cryopreservation	Fan et al. (2004)
Rohu	First report of fish ES cells cryopreservation using both penetrating and non-penetrating cryoprotetant	Dash et al. (2008)
Leopard danio	Cryopreservation of blastomeres and ES cells using slow freezing protocol	Routray et al. (2009)

13.5 Cryopreservation of Fish Embryonic Cells

Cryopreservation of fish embryonic cells in a group or in isolated condition is relatively easy compared to oocytes and embryos. Cryopreservation of embryonic cells/blastomeres was reported in few teleost species viz. zebrafish, and leopard danio (Harvey 1983; Routray et al. 2009), rainbow trout (*O. mykiss*) (Nilsson and Cloud 1993), Chum salmon (*Oncorhynchus keta*) (Kusuda et al. 2002), Whiting (*Shillago japonica*), pejerrey (*O. bonariensis*) and medaka (*O. latipes*) (Strüssmann et al. 1999) and mrigal (*Cirrhinus mrigala*) (Tripathy et al. 2012) (Table 13.2). A schematic representation of the process of embryonic cell isolation, cryopreservation and derivation of embryonic stem like cells is depicted in Fig. 13.6.

The success of embryonic cell cryopreservation and further in vitro culture depends on the stage of development from where it is harvested. Because the developmental time for different fish is variable. For instance the embryonic development of IMC (rohu, catla and mrigal) is dynamic in nature. Embryonic cells (blastomeres) isolated from early stages of developing embryos are considered to be best and most suitable for this. Embryonic cells isolated from (32-cell, 64-cell, 128-cell, 256-cell, 512-cell, 1024-cell, 2048-cell and gastrula) stage embryos of IMC (*C. mrigala*) were successfully cryopreserved using three cryoprotectants (CPAs).

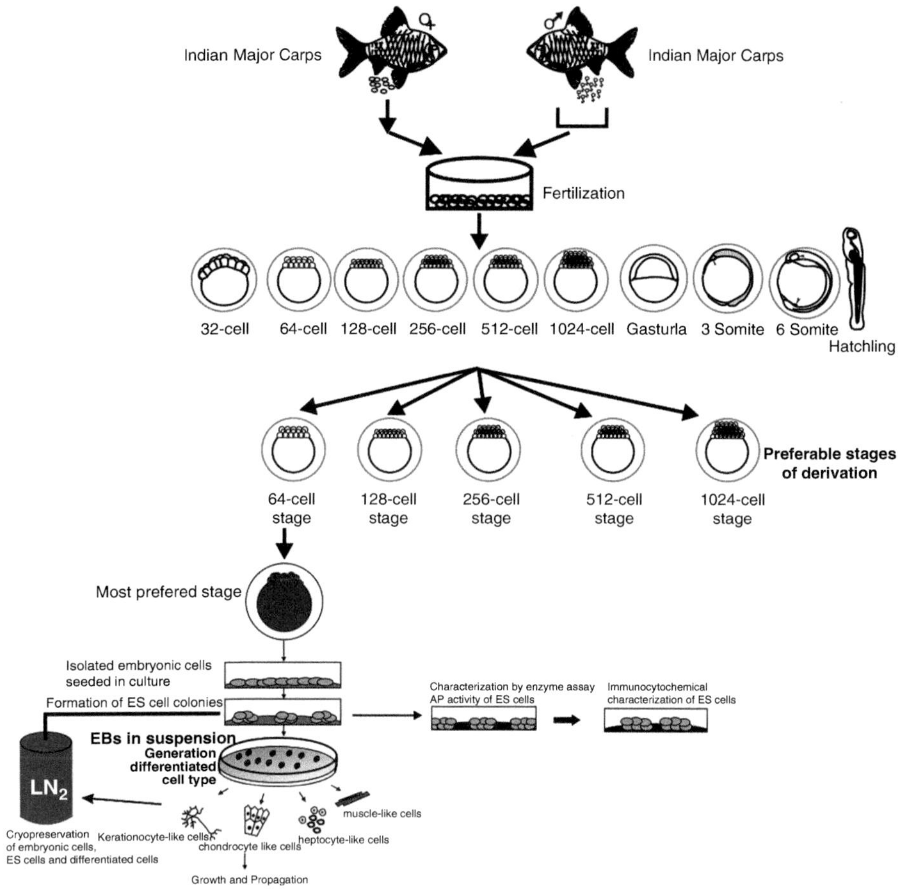

Fig. 13.6 Schematic representation of isolation, cryopreservation, proliferation and differentiation of embryonic cells of Indian major carps

Viability assessment of embryonic cells showed that the viability of embryonic cells was largely dependent on temperature, concentration of CPAs and exposure time. The viability of embryonic cells was maximum (>45%) with 10% DMSO concentration at 4 °C at 12 h of exposure. The isolated embryonic cells were successfully cryopreserved following slow freezing protocol at a cooling rate of −1 °C/min along with pre-cooled freezing medium (80% DMEM + 10% DMSO + 10% FBS).

A survival of more than 70% isolated blastomeres was observed in *Brachydanio frankei* following the slow cooling rate and using EG and PG as CPAs (Routray et al. 2009). In a similar study of isolated blastomeres, a viability rate of 19%, 34% and 67% was reported for whiting, *S. japonica*, medaka, *O. latipes* and pejerrey, *O. bonariensis* respectively (Harvey 1983). A repeatable technique for cryopreservation of rainbow trout blastomeres has been reported with 53% viability using1,2-Propanediol as CPAs (Laveroni Calvi and Maisse 1998). Successful cryopreservation of embryonic cells (blastomeres) of fish has also been reported in other aquatic species such as common carp (Kusuda et al. 2004).

13.6 Conclusion

The field of cryobiology has progressed tremendously during the last four decades. Successful cryopreservation of spermatozoa of aquatic species has paved the way for its application in fish hatcheries; however, the storage of oocytes and embryos could not be achieved for longer duration. Taking cue from oocyte and embryo of human and other mammals, it can be further studied to make this technology viable and industry ready. A silver lining is that the oocytes and embryos of some fish are stored for short duration and becoming a useful tool for hatchery managers and breeders.

References

Arii N, Namai K, Fujiya G, Nakazawa T (1987) Cryoprotection of medaka embryos during development. Zool Sci 4:813–818

Ashwood-Smith MJ (1986) The cryopreservation of human embryos. Hum Reprod 1(5):319–332

Betsy CJ, Stephen J, Kumar S (2014) Potential research leads in cryopreservation of fish gametes. Int J Res Fish Aquac 4(3):108–111

Blaxter JHS (1953) Sperm storage and cross-fertilization of spring and autumn spawning herring. Nature 172:1189–1190

Bobe J, Labbe C (2010) Egg and sperm quality in fish. Gen Comp Endocrinol 165:535–548

Bromage NR, Roberts RJ (1995) Preservation of gametes. In: Bromage NR, Roberts RJ (eds) Brodstock management and egg and larval quality. Blackwell, Oxford, pp 53–75

Chao NH, Liao IC (2001) Cryopreservation of finfish and shellfish gametes and embryos. Aquaculture 197:161–189

Chao NH, Lin TT, Chen YJ, Hsu HW, Liao IC (1997) Cryopreservation of late embryos and early larvae in the oyster and hard clam. Aquaculture 155:31–44

Chen SL, Tian YS (2005) Cryopreservation of flounder (*Paralichthys olivaceus*) embryos by vitrification. Theriogenology 63(4):1207–1219. https://doi.org/10.1016/j.theriogenology.2004.06.007

Cohen J, Simpsons RF, Fehilly CB, Edwards RG (1986) Factors affecting survival and implementation of cryopreserved human embryos. J In Vitro Fert Embryo Transf 3:46–52

Dash SN, Routray P, Dash C, Guru BC, Swain P, Sarangi N (2008) Use of the non-toxic cryoprotectant trehalose enhances recovery and function of embryonic stem cells following cryogenic storage. Curr Stem Cell Res Ther 3:277–287

Espinach Ros A, Amutio VG, Arceredillo JM, Orti G, Nani A (1984) Induced breeding of the South American catfish, *Rhamdia sapo* (C. & V.). Aquaculture 37(2):141–146

Fan L, Crodian J, Collodi P (2004) Production of zebrafish germline chimeras by using cultured embryonic stem (ES) cells. Methods Cell Biol 77:113–119

Franks F (1982) Water and aqueous solutions at subzero temperatures. In: Franks F (ed) Water a comprehensive treatise, vol 7. Plenum Press, New York, pp 1–334

Fujioka T, Yasuchika K, Nakamura Y, Nakatsuji N, Suemori H (2004) A simple and efficient cryopreservation method for primate embryonic stem cells. Int J Dev Biol 48(10):1149–1154

Guan M, Rawson DM, Zhang T (2008a) Cryopreservation of zebrafish (*Danio rerio*) oocytes using improved controlled slow cooling protocols. Cryobiology 56(3):204–208. https://doi.org/10.1016/j.cryobiol.2008.03.001

Guan M, Rawson DM, Zhang T (2008b) Development of a new method for isolating zebrafish oocytes (*Danio rerio*) from ovary tissue masses. Theriogenology 69(3):269–275

Guan M, Rawson DM, Zhang T (2010) Cryopreservation of zebrafish (*Danio rerio*) oocytes by vitrification. CryoLetters 31(3):230–238

Gwo JC (2000) Cryopreservation of eggs and embryos from aquatic organisms. In: Tiersch TR, Mazik PM (eds) Cryopreservation in aquatic species. World Aquaculture Society, Baton Rouge, pp 211–229

Hagedorn M, Kleinhans FW (2000) Problems and prospects in cryopreservation of fish embryos. In: Tiersch TR, Mazik PM (eds) Cryopreservation in aquatic species. World Aquaculture Society, Baton Rouge, pp 161–178

Hagedorn M, Kleinhans FW, Freitas R, Liu J, Hsu EW, Wildt DE, Rall WF (1997) Water distribution and permeability of zebrafish embryos, *Brachydanio rerio*. J Exp Zool 278:356–371

Harvey B (1983) Cooling of embryonic cells, isolated blastoderms, and intact embryos of the zebra fish, *Brachydanio rerio* to −196°C. Cryobiology 20:440–447

Harvey B, Ashwood-Smith MJ (1982) Cryoprotectant penetration and supercooling in the eggs of salmonid fishes. Cryobiology 19:29–40

Harvey B, Kelley RN (1985) Short-term storage of *Sarotherodon mossambicus* ova. Aquaculture 37:391–395

Harvey B, Kelley RN, Ashwood-Smith MJ (1982) Cryopreservation of zebrafish spermatozoa using methanol. Can J Zool 60:1867–1870

Heng BC, Kuleshova LL, Bested SM, Liu H, Cao T (2005) The cryopreservation of human embryonic stem cells. Biotechnol Appl Biochem 41(2):97

Heng BC, Ye CP, Liu H, Toh WS, Rufaihah AJ, Cao T (2006) Kinetics of cell death of frozen-thawed human embryonic stem cell colonies is reversibly slowed down by exposure to low temperature. Zygote 14(4):341–348

Hurst T, Shafir E, Lancaster P (1997) Assisted conception in Australia and New Zealand 1996. Assisted conception series number 3. AIHW National Perinatal Statistics Unit, Sydney, pp 1–71. ISSN 1038-7234

Husen MA, Sharma S (2015) Immersion of rohu fingerlings in clove oil reduced handling and confinement stress and mortality. Int J Fish Aquatic Stud 2(6):299–305

Ichikawa H, No H, Takei S, Takashimizu I, Yue F, Cui L, Ogiwara N, Johkura K, Nishimoto Y, Sasaki K (2007) Cryopreservation of mouse embryoid bodies. Cryobiology 54(3):290–293

Isayeva A, Zhang T, Rawson DM (2004) Studies on chilling sensitivity of zebrafish (*Danio rerio*) oocytes. Cryobiology 49(2):114–122. https://doi.org/10.1016/j.cryobiol.2004.05.005

Islam MN, Hossain MA (2013) Mortality rate of fish seeds (*Labeo rohita*) during traditional transportations system in the Northwest Bangladesh. J Sci Res 5(2):383–392

Janik M, Kleinhans FW, Hagedorn M (2000) Overcoming a permeability barrier by microinjecting cryoprotectants into zebrafish embryos (*Brachydanio rerio*). Cryobiology 41(1):25–34

Janik M, Kleinhans FW, Hagedorn M (2000a) Microinjection of cryoprotectants into the yolk of zebrafish embryos (*Brachydanio rerio*). Biol Reprod 62:146–146

Karlsson JO, Eroglu A, Toth TL, Cravalho EG, Toner M (1996) Fertilization and development of mouse oocytes cryopreserved using a theoretically optimized protocol. Hum Reprod 11:1296–1305

Kawai H, Sakurai M, Inoue Y, Chujo R, Kobayashi S (1992) Hydration of oligosaccharides: anomalous hydration ability of trehalose. Cryobiology 29(5):599–606

Khosla K, Wang Y, Hagedorn M, Qin Z, Bischof J (2017) Gold nanorod induced warming of embryos from the cryogenic state enhances viability. ACS Nano 11(8):7869–7878. https://doi.org/10.1021/acsnano.7b02216

Konc J, Kanyó K, Kriston R, Somoskői B, Cseh S. (2014). Cryopreservation of embryos and oocytes in human assisted reproduction. New Adv Rep Biomed. (Special Issue), Biomed Res Int. 2014;2014:307268. https://doi.org/10.1155/2014/307268

Kusuda S, Teranishi T, Koide N (2002) Cryopreservation of chum salmon blastomeres by the straw method. Cryobiology 45(1):60–67. https://doi.org/10.1016/S0011-2240(02)00106-2

Kusuda S, Teranishi T, Koide N, Nagai T, Arai K, Yamaha E (2004) Pluripotency of cryopreserved blastomeres of the goldfish. J Exp Zool A Comp Exp Biol 301:131–138. https://doi.org/10.1002/jez.a.20017

Lahnsteiner F (2000) Morphological, physiological and biochemical parameters characterizing the overripening of rainbow trout eggs. Fish Physiol Biochem 23:107–118

Legendre M, Slembrouck J, Subagja J, Kristanto AH (2000) Ovulation rate, latency period and ova viability after GnRH- or hCG-induced breeding in the Asian catfish *Pangasius hypophthalmus* (Siluriforme, Pangasiidae). Aquat Living Resour 13:145–151

Leung LKP, Jamieson BGM (1991) Live preservation of fish gametes. In: Jamieson BGM (ed) Fish evolution and systematics: evidence from spermatozoa. Cambridge University Press, Cambridge, pp 245–269

Leveroni Calvi S, Maisse G (1998) Cryopreservation of rainbow trout (Oncorhynchus mykiss) blastomeres: influence of embryo stage on postthaw survival rate. Cryobiology 36(4):255–262

Lin J, de Pablo JJ, Palecek SP (2004) Cryopreservation of adherent human embryonic stem cells. Biotechnol Bioeng 88(3):299–312

Liu KC, Chou TC, Lin HD (1993) Cryosurvival of goldfish embryo after subzero freezing. Aquat Living Resour 6:63–66

Liu XH, Zhang T, Rawson DM (2000) DSC studies on characterization of intraembryonic freezing and cryoprotectant penetration in zebrafish embryos. Cryobiology, (Abstract-71, presented in 37th annual meeting), p 355

Lovelock JE (1953) The mechanisms of the protective action of glycerol against haemolysis by freezing and thawing. Biochim Biophys Acta 11:28–36

Marques LS, Fossati AA, Rodrigues RB, Da Rosa HT, Izaguirry AP, Ramalho JB, Moreira JC, Santos FW, Zhang T, Streit DP (2019) Slow freezing versus vitrification for the cryopreservation of zebrafish (*Danio rerio*) ovarian tissue. Sci Rep 9(1):1–1. https://doi.org/10.1038/s41598-019-51696-7

Mazur P (1965) Causes of injury in frozen and thawed cells. Fed Proc 24(Suppl):S175–S182

Mazur P (1977) The role of intracellular freezing in the death of cells cooled at supraoptimal rates. Cryobiology 14:251–272

Meryman HT, Williams RJ, Douglas MS (1977) Freezing injury from 'solution effect' and its prevention by natural or artificial cryoprotection. Cryobiology 14:287–302

Mishra G, Patra S, Dash SK, Verma DK, Routray P (2017) In vitro storage of fish oocytes: effect of storage temperature, media conditions and storage duration on fertilization and larval hatchability of Indian major carp, rohu (*Labeo rohita*). Aquac Res 48(5):2486–2494

Nilsson E, Cloud JG (1993) Cryopreservation of rainbow trout (*Oncorynchus mykiss*) blastomeres. Aquat Living Resour 6:77–80

Nomura M, Sakai K, Takashima F (1974) The over-ripening phenomenon of rainbow trout. I. Temporal morphological changes of eggs retained in the body cavity after ovulation. Bull Jpn Soc Sci Fish 40:977–984

O'Neil L, Paynter SJ, Fuller BJ, Shaw RW (1997) Vitrification of mature mouse oocytes: improved results following addition of polyethylene glycol to a dimethyl sulphoxide solution. Cryobiology 34:295–301

Plachinta M, Zhang T, Rawson DM (2004) Studies on cryoprotectant toxicity to zebrafish (*Danio rerio*) oocytes. CryoLetters 25(6):415–424

Polge C, Smith AU, Parkes AS (1949) Revival of sprmatozoa after vitrification and dehydration at low temperatures. Nature 164:1–666

Rall WF (1993) In: Cloud JG, Thorgaard GH (eds) Genetic conservation of salmonid fishes. Plenum Press, NewYork, p 137

Rall WF, Fahy GM (1985) Ice free cryopreservation of mouse embryos at −196 °C by vitrification. Nature 313:573–575

Rall WF, Mazur P, McGrath JJ (1983) Depression of the ice nucleation temperature of rapidly cooled mouse embryos by glycerol and dimethyl sulfoxide. Biophys J 41:1–12

Rana K (1995) Preservation of gametes. In: Bromage NR, Roberts RJ (eds) Broodstock management and egg and larval quality. Blackwell Science, Oxford, pp 25–52

Reid DS (1993) Basic physical phenomena in the freezing and thawing of plant and animal tissues. In: Mallett CP (ed) Frozen food technology. Blackie Academic and Professional, Glasgow, pp 1–19

Reubinoff BE, Pera MF, Vajta G, Trounson AO (2001) Effective cryopreservation of human embryonic stem cells by the open pulled straw vitrification method. Hum Reprod 16 (10):2187–2194

Rizzo E, Godinho HP, Sato Y (2003) Short-term storage of oocytes from the neotropical teleost fish *Prochilodus marggravii*. Theriogenology 60(6):1059–1070. https://doi.org/10.1016/S0093-691X(03)00108-0

Routray P, Suzuki T, Kimizuka N, Kawai K, Strüssmann CA, Takai R (2001) Cold tolerance and ice nucleation temperature of medaka (*Oryzias latipes*) embryos with different cryoprotectant treatments. Cryobiol Cryotechnol 47(2):69–74

Routray P, Suzuki T, Strüssmann CA, Takai R (2002a) Factors affecting the uptake of cryoprotective agents by fish eggs and embryos. Fish Sci 68:965–966. https://doi.org/10.2331/fishsci.68.sup1_965

Routray P, Suzuki T, Strüssmann CA, Takai R (2002b) Factors affecting the uptake of DMSO by the eggs and embryos of medaka, *Oryzias latipes*. Theriogenology 58:1483–1496. https://doi.org/10.1016/S0093-691X(02)01076-2

Routray P, Gupta SD, Behera MK (2003) Cryogenics in fish hatchery technology. Fish Chimes 23:7–9

Routray P, Dash SN, Dash C, Swain P, Sarkar SK, Sarangi N (2008) Cryopreservation of silver barb *Puntius gonionotus* (Bleeker) spermatozoa: effect of extender composition, cryoprotective agents and freezing rate on their post thawing fertilization ability. Aquac Res 39:1597–1605

Routray P, Dash C, Dash SN, Tripathy S, Verma DK, Swain SK, Swain P, Guru BC (2009) Cryopreservation of isolated blastomeres and embryonic stem-like cells of Leopard danio, *Brachydanio frankei*. Aquac Res 41(4):579–589. https://doi.org/10.1111/j.1365-2109.2009.02456.x

Sahagian ME, Goff HD (1996) Fundamental aspects of the freezing process. In: Jeremiah LE (ed) Freezing effects on food quality. Marcel Dekker Inc., New York, pp 1–50

Sahu AD, Bhol C, Routray P, Sundaray JK (2011) Cryopreservation of male gonads of carps, *Labeo calbasu* and *Labeo rohita* for long term storage and its utilization in aquaculture. Indian J Sci Technol. 9th ISRPF issue;4:308–309

Schreuders PD, Smith ED, Cole KW, Laughinghouse A, Mazur P (1996) Characterization of intraembryonic freezing in *Anopheles gambiae* embryos. Cryobiology 33(5):487–501. https://doi.org/10.1006/cryo.1996.9999

Steponkus PL, Myers SP, Lynch DV, Gardner L, Bronshteyn V, Leibo SP, Rall WF, Pitt RE, Lin TT, Maclntyre RJ (1990) Cryopreservation of Drosophila melanogaster embryos. Nature 345 (6271):170–172

Strüssmann CA, Nakatsugawa H, Takashima F, Hasobe M, Suzuki T, Takai R (1999) Cryopreservation of isolated fish blastomeres: effects of cell stage, cryoprotectant concentration, and cooling rate on postthawing survival. Cryobiology 39:252–261

Suzuki T, Komada H, Takai R, Arii K, Kozima TT (1995) Relation between toxicity of cryoprotectant DMSO and its concentration in several fish embryos. Fish Sci 61:193–197

Toldeo JD, Kurokura H (1990) Cryopreservation of the euryhaline rotifer *Brachionus plicatilis* embryos. Aquaculture 91:385–394

Tripathy S, Dash C, Routray P (2012) In vitro differentiation ability of post-thawed blastomeres derived from early blastula stage embryos of Indian major carp, *Cirrhinus mrigala* (Ham.). e-planet 10(2):25–29

Tsai S, Yen W, Chavanich S, Viyakarn V, Lin C (2015) Development of cryopreservation techniques for gorgonian (*Junceella juncea*) oocytes through vitrification. PLoS One 10: e0123409. https://doi.org/10.1371/journal.pone.0123409

Urbányi B, Magyary I, Horváth A, Baranyai B, Dinnyés A (2000) Cryopreservation of sperm and eggs of sharptooth catfish. In: Tiersch TR, Mazik PM (eds) Cryopreservation in aquatic species. World Aquaculture Society, Baton Rouge, pp 88–198

Veron P, Vannier G, Luce JM (1997) Decrease of supercooling capacity during embryogenesis and larval growth in coleoptera. C R Acad Sci III 320(5):359–366

Whittingham DG, Leibo SP, Mazur P (1972) Survival of mouse embryos frozen to −196 °C to-296 °C. Science 178:411–414

Wilmut I (1972) The effects of cooling rate, warming rate, cryoprotective agent and stage of development on survival of mouse embryos during freezing and thawing. Life Sci 11:1071–1079

Zell RPS (1978) Cryopreservation of gametes and embryos of salmonid fishes. Ann Biol Anim Biophys 18:1089–1099

Zhang T, Rawson DM (1995) Studies on chilling sensitivity of zebrafish (*Brachydanio rerio*) embryos. Cryobiology 32:239–246

Zhang T, Rawson DM (1996) Feasibility studies on vitrification of intact zebrafish (*Brachydanio rerio*) embryos. Cryobiology 33:1–13

Zhang T, Rawson DM, Pekarsky I, Blais I, Esther L (2007) Low-temperature preservation of fish gonad cells and oocytes. In: Babin PJ, Cerda J, Lubzens E (eds) The fish oocyte from basic studies to biotechnological applications. Springer, New York, pp 411–436

Zhou CQ, Mai QY, Li T, Zhuang GL (2004) Cryopreservation of human embryonic stem cells by vitrification. Chin Med J 117(7):1050–1055

14 Potential of Fish Gamete Cryopreservation in Conservation Programs in Bangladesh

Md. Rafiqul Islam Sarder

Abstract

Bangladesh is enriched with aquatic biodiversity and ranked third in Asia after China and India. It has highly potential fisheries resources comprising 260 freshwater fish, 475 marine fish, 24 freshwater prawn, and 36 marine shrimp species. The country produced 4.134 million metric tons fish in the year 2017, in which open water capture fishery contributed 28.14% and aquaculture contributed 56.44%. Presently, Bangladesh ranked fifth in aquaculture production and third in open water capture fisheries in the world. Earlier, the lion share of the production came from open water capture fisheries, but the production has been reduced over the years due to environmental and human-created problems. As a result, natural recruitment of many fish species has been hampered and they became threatened. According to IUCN Bangladesh (2015), among the 260 fresh water fish species, 9 have been categorized as critically endangered, 30 as endangered, and 25 as vulnerable. Aquaculture production of the country is being increased day by day due to availability of seeds produced through hypophysation in the private and government hatcheries, but the quality of seeds has been deteriorated due to some genetics causes such as inbreeding, interspecific hybridization, and negative selection. Therefore, it is urgent to protect the endangered fish as well as to improve the quality of seeds of commercially important fish species. This can be done by applying cryopreservation techniques associated with genetic improvement program. Considering the potential of fish cryopreservation, research on sperm cryopreservation began in the Department of Fisheries Biology and Genetics, Bangladesh Agricultural University in 2002, and protocols have been developed for a number of fish species such as Indian major carps, exotic carps, catfishes, indigenous endangered fish species, barb, and tilapia. Breeding of Indian major carps using cryopreserved sperm in private hatcheries

M. R. I. Sarder (✉)
Department of Fisheries Biology and Genetics, Bangladesh Agricultural University, Mymensingh, Bangladesh

J. Betsy, S. Kumar (eds.), *Cryopreservation of Fish Gametes*,
https://doi.org/10.1007/978-981-15-4025-7_14

demonstrated excellent performance of cryopreserved sperm-originated seeds and broods that urged to establish cryogenic sperm bank of commercially important fish as well as endangered fish species in Bangladesh.

Keywords

Fish production · Decline of biodiversity · Gamete cryopreservation · Cryogenic sperm bank

14.1 Introduction

Bangladesh is a Southeast Asian country (Lat 20°34′ and 26°39′ N and Long 80°00′ and 90°41′ East) covering an area of 1,47,570 km^2. It is one of the world largest inland aquatic habitats and the third largest aquatic biodiversity in Asia after China and India (Hussain and Mazid 2001; Hossain 2014; Shamsuzzaman et al. 2017). It has diverse inland water resources of 4.76 million ha (DoF 2017) comprised of rivers, canals, depressions, ponds, lakes, and floodplains. The total area of the inland water body covering 47, 60,894 ha of which 82.49% is open water and 17.51% is closed water. The total production of fish of the country is estimated as 4.134 million metric tons in which open water capture fishery contributed 28.14% and aquaculture contributed 56.44%. The remaining 15.42% came from marine fisheries sector (DoF 2017). Bangladesh is recognized as a leading fish-producing country and ranked fifth in aquaculture production and third in open water capture fisheries in the world (FAO 2018). Fisheries sector has an enormous contribution to the development of Bangladesh. Almost 18 million people (11% of the total population) are directly and indirectly involved in fisheries sector for their livelihoods. This sector contributed 3.61% to the national gross domestic product (GDP) and almost one-fourth (24.41%) to the agricultural returns and 1.51% to export earnings. It also supplies 60% of the total animal protein in the diet of the people of Bangladesh (DoF 2017).

Bangladesh has highly potential fisheries resources having 260 freshwater fish, 475 marine fish, 24 freshwater prawn, and 36 marine shrimp species (DoF 2016). Earlier, the inland open water capture fisheries contributed a lion share of the total fish production but the production of this sector has been reduced over the years. For example, during 1960–1961, the inland capture fisheries contributed approximately 90% of the country's total fish production which declined to 62.59% in 1983–1984, 52.57% in 1993–1994, 34.83% in 2003–2004, and lastly 28.14% in 2016–2017 (DoF 2017). Over the past four decades, inland open water fish production and their biodiversity reduced increasingly owing to the combined effect of a number of environmental and anthropogenic causes, for instance, massive siltation in natural habitats and elevation of river beds, loss of breeding and nursery grounds, overexploitation, alteration of habitats, and ecological modifications such as blocking of migratory channels by construction of flood control dams, roads and highways, townships and other developmental infrastructures, irrigation schemes, destructive fishing pressure (IUCN Bangladesh 2015), aquatic pollution through pesticides and dumping of industrial wastes, eutrophication (Dudgeon et al. 2006), use of illegal fishing gears and methods, mismanagement in leasing system of *haor, baor, beels,* etc. (Parveen and Faisal 2002; World Bank 2005). Consequently, stocks of some fish species have been depleted to below replaceable levels and they are in the verge of extinction. According to IUCN Bangladesh (2015), among the 260 freshwater fish species, 9 have been categorized as critically endangered, 30 as endangered, and 25 as vulnerable, whereas only 15 years ago 54 freshwater fish species were recorded as endangered and threatened (IUCN Bangladesh 2000). The situation is really critical if this information is given due weightage. It is obvious that if this situation is continued to be unattended, we have to get prepared to face the serious consequences in fisheries sector in near future. Although some measures like establishing sanctuary in particular rivers, restricting discharge of industrial wastes to rivers without treatment etc. are taken by the Government against the above-mentioned factors to slow down the environmental degradation processes it is almost impossible to have a complete stop of the matter in the present situation of Bangladesh. It is therefore, logical that steps are taken to improve the aquatic environment in one hand and research is conducted to conserve the population of the commercial and threatened species by taking ex-situ conservation programs like sperm cryopreservation in another.

14.2 Fish Gamete (Sperm) Cryopreservation and its Progress in Bangladesh

Considering the importance and potential of fish cryopreservation, research on fish sperm cryopreservation was started in the Department of Fisheries Biology and Genetics, Bangladesh Agricultural University (BAU) in 2002 under a research project funded by DFID (Department for International Development), UK. After that, sperm cryopreservation work has been continued and protocols for a number of fish species have been developed which are shown in Table 14.1.

Recently, experiment on seed production using 30 days old cryopreserved sperm of Rohu (sperm motility 85%) was conducted in government and private hatcheries in Mymensingh district and 67% and 38% fertilization and hatching, respectively (compared to fresh sperm, 82% fertilization and 64% hatching) were obtained (Rashed 2016). Similarly, 60 days old cryopreserved sperm of Mrigal showed 72% motility and it produced 64% fertilization and 53% hatching against 67% and 56% fertilization and hatching by fresh sperm (Hossain et al. 2018). In both cases, cryopreserved sperm of Rohu and Mrigal of Halda river origin (genetically superior stock) were used while the fresh sperm (control) were used from respective hatcheries where the experiments were conducted. Seeds produced by cryopreserved sperm of both species showed higher growth than those produced by fresh sperm of hatchery-origin males. More interestingly, 30 days old cryopreserved sperm collected from cryopreserved sperm-originated F1 males of Rohu showed 75% fertilization and 57% hatching compared to fresh sperm of hatchery-origin males (84%

Table 14.1 Sperm cryopreservation protocols developed for fishes in Bangladesh

	Local name	Common name	Scientific name	References
1.	Bighead carp	Bighead carp	*Aristichthys nobilis*	Khan et al. (2004)
2.	Thai Sarputi	Silver barb	*Barbonymus gonionotus*	Uddowla et al. (2011), Rahman et al. (2009), Sarder et al. (2005)
3.	Carpio	Common carp	*Cyprinus carpio*	Sarder et al. (2005)
4.	Tilapia	Nile tilapia	*Oreochromis niloticus*	Sarder et al. (2006), Islam (2017)
5.	Rohu	Indian major carp	*Labeo rohita*	Rafiquzzaman et al. (2007), Sarder et al. (2011), Khan et al. (2015)
6.	Mrigal	Indian major carp	*Cirrhinus cirrhosus*	Sarder et al. (2009)
7.	Silver carp	Silver carp	*Hypophthalmicthys molitrix*	Hossain and Sarder (2009)
8.	Grass carp	Grass carp	*Ctenopharyngodon idella*	Sarder et al. (2010)
9.	Sarputi	Olive barb	*Puntius sarana*	Nahiduzzaman et al. (2011)
10.	Kalibaush	Indian major carp	*Labeo calbasu*	Nahiduzzaman et al. (2012)
11.	Bheda	Mud perch	*Nandus nandus*	Sarder et al. (2012)
12.	Pabda	Pabda catfish	*Ompok pabda*	Sarder et al. (2013)
13.	Koi	Climbing perch	*Anabas testudineus*	Nazrul et al. (2013)
14.	Gulsha	Gangetic mystus	*Mystus cavasius*	Islam et al. (2016)
15.	Shal Baim	Spiny eel	*Mastacembelus armatus*	Rahman et al. (2016)
16.	Thai Pangas	Striped catfish	*Pangasianodon hypophthalmus*	Hossain et al. (2016)
17.	Tangra	Striped dwarf catfish	*Mystus vittatus*	Sarder et al. (2017)
18.	Bhangna	Reba carp	*Cirrhinus reba*	Sultana et al. (2017)
19.	Bata	Bata labeo	*Labeo bata*	Noor et al. (2018)
20.	Shing	Stinging catfish	*Heteropneustes fossilis*	Azad (2011)
21.	Magur	Walking catfish	*Clarias batrachus*	Moniruzzaman (2010)
22.	Catla	Indian major carp	*Catla catla*	Sultana (2016)

fertilization and 68% hatching) and their progeny showed higher growth than those produced by fresh sperm (Ullah 2018).

14.3 Necessity of Establishing Sperm Bank

In the beginning of fish culture, i.e. in early 1970s, aquaculture practice was started with Indian major carps (IMCs) (*Catla catla, Labeo rohita, Cirrhinus cirrhosus*) and *Labeo calbasu* using naturally collected seeds. But along with the indigenous carps, a number of other fish species, such as exotic carps, catfishes, pangas, barbs, tilapia, prawn, and shrimp are contributing to aquaculture production. Aquaculture production is increasing due to availability of hatchery-produced seeds and adoption of new technologies, but has not achieved maximum production level (at present aquaculture production is around 4765 kg/ha) compared to other major fish-producing countries like China and India (DoF 2018). Sustainable production has not been ensured, and environmental and anthropogenic threats continue to cause reduced production. Limited production of quality seeds of indigenous carps, exotic carps, and other species and inadequate supply to the farmers is a major problem leading to decreased production. The country produced 6,63,462 kg of seeds of Indian major carps and exotic carps from 139 government (12,826 kg) and 818 private hatcheries (6,50,636 kg) (DoF 2018) but quality has been deteriorated due to inbreeding, interspecific hybridization, negative selection, and improper broodstock management (Simonsen et al. 2004, 2005; Hansen et al. 2006; Alam and Islam 2005). As a result fry shows slow growth, high mortality, deformities, and disease susceptibility. To overcome these problems, the Government has initiated broodstock improvement program for Indian major carps through establishing live brood banks in government hatcheries and distribution of broods to other hatcheries but it is difficult to supply broods to the large number of public and private hatcheries. Current brood banks are developed by rearing naturally collected seeds of Indian major carps (from the Halda, Padma, and Jamuna rivers) but availability and quality of wild seeds are a big question. However, seed production in natural environment is drastically reduced and contributes less than 1% of the total seed production. Seed quality of exotic carps is also diminishing and improvement of their broodstock is quite difficult as collection of new stock from abroad is not easy and expensive. During seed production through induced breeding in the hatcheries, hybridization between silver carp and bighead carp, grass carp with other carps, and inter-specific introgression among the IMCs is taking place. In this situation, setting up a cryogenic sperm repository of indigenous carps, exotic carps, and catfishes and use of cryopreserved sperm in seed production could resolve the existing genetic problems. It will also help to develop brood banks in the government and private sectors by providing a reliable source of quality germplasm. In fact, sperm from the cryogenic sperm bank can be used to produce seeds for brood production. This high-quality broods can eventually be used to propagate quality seeds in public and private hatcheries in commercial scale.

14.4 Conclusion

Aquaculture production has remarkably contributed to total fish production of the country but maximum and sustainable production has not been obtained. Among the many factors, deterioration of seed quality is the main obstacle for achieving the sustainable production. On the other hand, open water capture fish production has been decreasing due to environmental and human-created problems, and a considerable number of fish species became threatened. To overcome the problems, all the available conservation techniques should be applied and cryopreservation of genetically quality sperm and use in seed production could play a vital role in this regard. Thus, cryogenic sperm bank can improve the quality of seeds of commercially important fish species in one hand and conserve the germ plasm and restore the population of the threatened species at a sustainable level in another.

References

Alam MS, Islam MS (2005) Population genetic structure of *Catla catla* (Hamilton) revealed by microsatellite DNA markers. Aquaculture 246:151–160

Azad MAK (2011) Cryopreservation of spermatozoa of indigenous stinging catfish (*Heteropneustes fossilis*, Bloch) for *ex-situ* conservation. MS Thesis, Department of Fisheries Biology and Genetics, Bangladesh Agricultural University, Mymensingh, p 66

DoF (2016) National Fish Week 2016 Compendium (In Bengali). Department of Fisheries, Ministry of Fisheries and Livestock, Bangladesh, p 148

DoF (2017) Yearbook of Fisheries Statistics of Bangladesh 2016–17. Fisheries Resources Survey System (FRSS), Department of Fisheries, Director General, DoF, Bangladesh, p 129

DoF (2018) National Fish Week 2018 Compendium (In Bengali). Department of Fisheries, Ministry of Fisheries and Livestock, Bangladesh, p 160

Dudgeon D, Arthington AH, Gessner MO, Kawabata Z, Knowler DJ, Lévêque C, Naiman RJ, Prieur-Richard A, Soto D, Stiassny MLJ, Sullivan CA (2006) Freshwater biodiversity: importance, threats, status and conservation challenges. Biol Rev 81:163–182

FAO (2018) FAO year book. In: Fishery and aquaculture statistics. FAO, Rome, p 104

Hansen MM, Simonsen V, Mensberg KLD, Sarder MRI, Alam MS (2006) Loss of genetic variation in hatchery-reared Indian major carps, *Catla catla*. J Fish Biol 69:229–241

Hossain MAR (2014) Habitat and fish diversity: Bangladesh perspective. In: Wahab MA, Shah MS, Hossain MAR, Barman BK and Hoq ME (eds) Advances in Fisheries Research in Bangladesh: I. Proc. of 5th Fisheries Conference & Research Fair 2012, Bangladesh Agricultural Research Council, Dhaka, 18–19 January 2012, pp 1–26

Hossain MS, Sarder MRI (2009) Cryogenic freezing of silver carp spermatozoa for conservation of gene pool. Progress Agric 20(1&2):99–106

Hossain MR, Rahman MK, Akhter JN, Sultana S (2016) Short term cryo-milt preservation of *Pangasianodon hypophthalmus* at the freshwater station of BFRI in Bangladesh. Asian Australas J Biosci Biotechnol 1(2):398–401

Hossain MA, Rouf MA, Akter N, Sarder MRI, Rahman MM, Mollah MFA (2018) Quality assessment of cryopreserved sperm of mrigal, *cirrhinus cirrhosus* and its use in commercial seed production. J Aquac Tropics 33(3–4):187–200

Hussain MG, Mazid MA (2001) Genetic improvement and conservation of carp species in Bangladesh. BFRI and ICLARM publication, Dhaka

Islam MZ (2017) Production of YY supermale of Nile tilapia (*Oreochromis niloticus*) and cryopreservation of sperm. PhD Dissertation, Department of Fisheries Biology and Genetics, Bangladesh Agricultural University, Mymensingh

Islam MM, Noor MN, Islam AA, Sarder MRI, Islam MZ, Jahan DA (2016) Development of a sperm cryopreservation protocol of Gangetic Mystus (*Mystus cavasius*). Progress Agric 27 (4):517–529

IUCN Bangladesh (2000) International union for conservation of nature, Bangladesh, Bangladesher Bipanno Prani. IUCN, The World Conservation Union. Bangladesh, p 294

IUCN Bangladesh (2015) Red List of Bangladesh 5: Freshwater Fishes. International Union for Conservation of Nature, Bangladesh Country Office, Dhaka, Bangladesh, pp xvi+360

Khan MGQ, Mollah MFA, Ahmed AU, Hasanat MA, Uddowla MH (2004) Cryopreservation of bighead carp (*Aristichthys nobilis*) spermatozoa for gene resource conservation. Bangladesh J Fish 28(1–2):23–37

Khan NS, Sarder MRI, Faroque MAA, Mollah MFA (2015) Standardization of sperm cryopreservation techniques of Indian Major Carp Rohu (*Labeo rohita*, Hamilton 1822). Int J Fish Aquat Stud 2(6):175–181

Moniruzzaman (2010) Cryogenic preservation of sperm of indigenous catfish Magur, *Clarias batrachus* for gene pool conservation. MS Thesis, Department of Fisheries Biology and Genetics, Bangladesh Agricultural University, Mymensingh, p 69

Nahiduzzaman M, Hassan MM, Khanam UH, Mamun SNA, Hossain MAR, Tiersch TR (2011) Sperm cryopreservation of the critically endangered olive barb (Sarpunti) *Puntius sarana* (Hamilton, 1822). Cryobiology 62:62–67

Nahiduzzaman M, Hassan MM, Roy PK, Hossain MA, Hossain MAR, Tiersch TR (2012) Sperm cryopreservation of the Indian major carp, *Labeo calbasu*: effects of cryoprotectants, cooling rates and thawing rates on egg fertilization. Anim Reprod Sci 136:133–138

Nazrul KMN, Sarder MRI, Tonny US (2013) Sperm cryopreservation protocol of *Anabas testudineus* (Bloch 1792). LAP Lambert Academic Publishing, USA

Noor MN, Sarder MRI, Islam AA, Jahan DA, Islam MM (2018) Development of sperm cryopreservation technique of minor carp, bata (*Labeo bata* Ham. 1822) for *ex-situ* conservation. Bangladesh J Fish 30(2):121–133

Parveen S, Faisal IM (2002) People versus power: the geopolitics of Kaptai Dam in Bangladesh. Int J Water Resour Dev 18:197–208

Rafiquzzaman SM, Sarder MRI, Islam MS, Sultana R (2007) Sperm cryopreservation of rohu, *Labeo rohita*: cryodiluents, sperm: cryodiluents ratio and cryoprotectant concentration. Bangladesh J Fish Res 11(2):141–152

Rahman F, Sarder MRI, Rouf MA (2009) Comparison of growth performance between cryopreserved and fresh sperm-originated fry of *Barbonymus gonionotus*. J Bangladesh Agric Univ 7(1):145–149

Rahman MM, Ali MR, Sarder MRI, Mollah MFA, Khan NS (2016) Development of sperm cryopreservation protocol of endangered spiny eel, *Mastacembelus armatus* (Lacepede 1800) for *ex-situ* conservation. Cryobiology 73:316–323

Rashed M (2016) Production of seed of *Labeo rohita* (Hamilton, 1822) using cryopreserved seed and determination of their growth performance. MS Thesis, Department of Fisheries Biology and Genetics, Bangladesh Agricultural University, Mymensingh

Sarder MRI, Salam MA, Shafin SM, Hussain MG, Islam MS (2005) Cryopreservation of sperm of common carp, *Cyprinus carpio* and silver barb, *Barbonymus gonionotus* for genetically improved seed production. Bangladesh J Fish Res (Special Issue) 9(1):1–3

Sarder MRI, Rafiquzzaman SM, Islam MS, Sultana R (2006) Cryopreservation of spermatozoa of tilapia, *Oreochromis niloticus* L. by optimizing different factors. J Asiat Soc Bangladesh Sci 32 (1):89–96

Sarder MRI, Rafiquzzaman SM, Islam MF (2009) Cryopreservation of spermatozoa of Mrigal, *Cirrhinus cirrhosus* with a view to minimize inbreeding and hybridization. J Bangladesh Agric Univ 7(1):211–218

Sarder MRI, Hossain MA, Hossain MS (2010) Cryofreezing of grass carp (*Ctenopharyngodon idella*) spermatozoa for ex-situ conservation. Progress Agric 21(1 & 2):141–150

Sarder MRI, Rahman AKMA, Samad MS, Nazrul KMS, Bhuiyan MK (2011) Cryopreservation of sperm of *Labeo rohita* (Hamilton, 1822) and its use in fertilization and hatching of eggs. Progress Agric 22(1 & 2):137

Sarder MRI, Sarker MFM, Saha SK (2012) Cryopreservation of sperm of an indigenous endangered fish species *Nandus nandus* (Hamilton, 1822) for ex-situ conservation. Cryobiology 65:202–209

Sarder MRI, Saha SK, Sarker MFM (2013) Cryopreservation of sperm of an indigenous endangered fish, Pabda *(Ompok pabda,* Hamilton-Buchanan, 1822). N Am J Aquac 75:114–123

Sarder MRI, Islam AA, Islam MM, Noor MN, Jahan DA (2017) Cryopreservation of spermatozoa of threatened Asian striped dwarf catfish *Mystus vittatus* (Bloch 1794) for *ex-situ* conservation: an approach to poverty alleviation. Asian J Poverty Stud 3(2):157–169

Shamsuzzaman MM, Islam MM, Tania NJ, Al-Mamun MA, Barman PP, Xu X (2017) Fisheries resources of Bangladesh: present status and future direction. Aquac Fish 2:145–156

Simonsen V, Hansen MM, Sarder MRI, Alam MS (2004) High level of hybridization in three species of Indian major carps. NAGA WorldFish Center Q 27(1 & 2):65–69

Simonsen V, Hansen MM, Mensberg KLD, Sarder MRI, Alam MS (2005) Widespread hybridization among species of Indian major carps in hatcheries, but not in the wild. J Fish Biol 67:794–808

Sultana N (2016) Standardization of cryopreservation protocol of *Catla catla* spermatozoa for its gene pool conservation. MS Thesis, Department of Fisheries Biology and Genetics, Bangladesh Agricultural University, Mymensingh

Sultana S, Hasan MM, Hossain MS, Islam MR, Salimullah M, Sarder MRI, Islam MS, Moniruzzaman M, Alam J (2017) Cryogenic preservation of critically endangered *Cirrhinus reba* fish sperm. Int J Fish Aquat Stud 5(5):334–339

Uddowla MH, Salma U, Rahman MM, Hossain MS, Mollah MFA (2011) Effects of diluents, dilution ratios and cryoprotectants for the cryopreservation of Thai Sharpunti (*Barbodes gonionotus*) spermatozoa. Int J Nat Sci 1(2):44–51

Ullah AM (2018) Production of seeds using cryopreserved sperm-originated males (F1) of *Labeo rohita* and their growth performance. MS Thesis, Department of Fisheries Biology and Genetics, Bangladesh Agricultural University, Mymensingh

World Bank (2005) The International bank for reconstruction and development saving fishes and fishers. Towards sustainable and equitable governance of the global fishing sector 1818 H. Street. Washington DC. www.worldBank.org/fish

Gamete Manipulation and Conservation for Genetic Improvement in Penaeid Shrimp

15

Hugo H. Montaldo

Abstract

The development of gamete manipulation and conservation is penaeid shrimp may have important effects in breeding programs. Artificial insemination makes it possible to obtain unbiased estimates of genetic parameters under current hierarchical mating designs. The use of factorial mating designs coupled with external or in vitro fertilization may allow more accurate estimates of genetic parameters and the presence of maternal effects, as well as design selection programs with higher genetic responses, decreased levels of loss in genetic variability and lower inbreeding levels. In turn, the conservation of semen may be useful in reducing loss of genetic variation in conservation programs, measuring genetic gains without control populations, and facilitating the use of foreign germplasm and the design of crossbreeding programs.

Keywords

Cryopreservation · Genetic improvement · Shrimp · Conservation · Inbreeding

15.1 Introduction

Gamete manipulation and conservation in penaeid shrimp involve several techniques along the lines of that already more developed in terrestrial mammalian species. Some of them, namely artificial insemination and semen extension, have had a tremendous impact on breeding program design and genetic studies in cattle (Wilmut et al. 1992; van Vleck 1999). Several authors have reviewed the achieved and

H. H. Montaldo (✉)
Facultad de Medicina Veterinaria y Zootecnia, Universidad Nacional Autónoma de México, Ciudad Universitaria s/n, Coyoacán, Mexico
e-mail: montaldo@unam.mx

J. Betsy, S. Kumar (eds.), *Cryopreservation of Fish Gametes*,
https://doi.org/10.1007/978-981-15-4025-7_15

potential impact of these by now well-established reproductive technologies on the expected genetic gains in cattle and other terrestrial species (Wilmut et al. 1992). A general conclusion is that the potential contributions of reproductive biotechnologies for the genetic improvement of animals are vast and that include aquaculture animals.

In the case of shrimp, the developments obtained in reproductive techniques are more recent (Beirão et al. 2019); nonetheless, a number of the general principles to integrate reproductive technologies in breeding programs have common elements with those of terrestrial animals, or other aquaculture species, even bearing in mind the substantial biological differences between them. Penaeids are the most important shrimp species for aquaculture production. However there is a lack of a systematic examination on the potential impact of artificial reproductive technologies on the design of breeding programs. Therefore, this review will be oriented toward the analysis of current and possible future developments on several aspects of the genetic improvement of penaeid shrimp populations based on the integration of artificial reproductive techniques. We will review the potential impact of several of the principal possible reproductive techniques that are already developed, namely, artificial insemination (AI), currently used extensively by industry in some penaeid shrimp species such as *Litopenaeus vannamei* to create hierarchical sire/dam structures that are very useful in selection programs and genetic studies (Castillo-Juárez et al. 2015). We will also explore techniques that are still under development such as the extension of spermatophore gamete content to produce several fertilizing doses, thereby creating the possibility of increasing the female: male mating ratios, and the conservation of male gametes (including cryopreservation) to transport and select the time of using the gametes of particular males. These techniques can be used jointly to open the possibility of using in-vitro fertilization (IVF) to obtain an increased control of mating designs. These technologies could create a number of practical applications for the genetic management of penaeid shrimp breeding populations.

The primary uses of gamete manipulation for reproductive purposes in breeding programs in shrimp may be classified into five main approaches: (a) Design of studies to estimate genetic parameters, (b) Design of breeding programs (c) Diffusion of genetic superiority (d) Estimation of genetic trends, (e) Inbreeding control and conservation of genetic diversity, and (e) Migration and crossbreeding.

15.2 General Aspects of Gamete Management and Conservation in Penaeid Shrimp

While AI is a relatively mature technique in some penaeid shrimp species such as *L. vannamei*, some limitations remain. Using current AI technique based on spermatophore extrusion, the sire: dam ratio is approximately 1.3–1.6 from inseminations of two females per male, each with a single spermatophore of the two available (Campos-Montes et al. 2012; Montaldo et al. 2013). This relatively low sire: dam ratio is caused by a high proportion of fertilization failures. AI in other closed thelycum penaeids is probably more difficult. The consequences of this low hatching proportion on bias of genetic parameter estimation were examined by Montaldo et al. (2013) in *L. vannamei* and will be discussed further here in relation to the estimation accuracy of genetic parameters and selection program design.

The other techniques outlined above are still in their infancy, and the conservation of spermatophores in penaeid shrimp is only possible for a very short period and with low hatching rates (Beirão et al. 2019). Long-term efficient conservation of gametes would facilitate the migration of genes from populations geographically distant and design crossbreeding trials to efficiently estimate the associated parameters in purebreds and crossbreeds. Moreover, long-term gamete conservation can be used in the estimation of genetic progress without using control populations (Henderson 1973) and in the maintenance of genetic variability in small conservation populations (Sonesson et al. 2002).

The extension of sperm in further doses above the two spermatophore limit and use of IVF may allow the use of mating designs not limited to hierarchical mating arrangements, also accommodating factorial mating arrangements involving several male mates per female, similar to that possible in fish with external fertilization such as salmonids. This may be important in the estimation of quantitative genetic parameters, evaluation of breeding values and maximizing selection responses (Dupont-Nivet et al. 2002, 2006; D'Agaro et al. 2007; Beirão et al. 2019). The use of IVF in shrimp implied the development of several techniques to manipulate not only semen but also the female gametes. Organizing factorial mating using AI without external fertilization would be very difficult for logistical reasons, and may involve the need to develop spawning synchronization methods. To our knowledge there is no description in shrimp breeding literature about the use of factorial mating designs to date.

15.3 Estimation of Genetic Parameters

Genetic parameter estimation is a required step in the development and planning of selection programs. Traditionally, estimation of genetic parameters in penaeid shrimp has been made using a hierarchical design, characterized by the use of two females mated per male using AI. Although the advent of mixed model methodology and Restricted Likelihood methods (REML) with multigenerational pedigreed data (Thompson 2008), has somewhat relaxed the requirements to separate genetic and environmental variances efficiently (De los Ríos-Pérez et al. 2015), basic design aspects are key to the statistical separation of genetic from environmental effects. In aquaculture, single pair mating designs are only appropriate to estimate heritability (additive genetic variance/phenotypic variance) under very restricted assumptions of no maternal effects, usually considered jointly with genetic dominant and other environmental effects common to all progeny of one dam (such as, e.g., the spawning tank, etc.) as the common full-sibs environmental effects (c-effects). These nongenetic effects are capable of inflating the estimates of heritability (h^2) to a great extent. With simulated data from a single generation (Montaldo et al. 2013), determined that when using an animal model the heritability estimates would be biased by an amount equivalent to 2 × Vc/Vp, where Vc is the variance of common full-sib environmental effects, and Vp is the phenotypic variance, suppose true h^2 is 0.2 and c^2 = Vc/Vp is 0.1. In this case, the heritability estimate obtained in models without considering the c-effects would be $0.2 + 2 \times 0.1 = 0.4$. Therefore, h^2 inflation can be considerable, even if the c^2 value is not very high. Without AI, the separation of additive genetic variance from common full-sib environmental variance would not be possible.

A possible improvement on data used for genetic parameter estimation is to increase the average number of dams per sire on final data from current 1.3–1.6 to a value closer to 2, or even higher. For a trait with a h^2 of 0.2 and a c^2 of 0.1, changing from 1.3 to 1.7 dams per sire, and from 1.7 to 2.0 implies a more efficient use of data where a lower number of families and less total organisms need to be measured in order to obtain similar statistical accuracy (standard error of h^2/true $h^2 \approx 2$ estimated with methods by Kempthorne (1957)). A further reduction both in the number of families and in the numbers of organisms measured to get the same accuracy is obtained using a hierarchical design with four dams per sire. Similar to results by Blanc (2003), in our calculations, a complete factorial design had advantages over a hierarchical design. Results from the literature also indicate advantages in terms of the accuracy of estimated parameters of using partial and incomplete factorial designs, particularly designs in which males are mated with the same two females or another incomplete factorial design (Dupont-Nivet et al. 2002; Blanc 2003). In principle, at least, a complete factorial design can be used to estimate variances due to additive genetics, dominant genetics, and maternal effects (Blanc 2003), unlike nested models that could confound maternal with dominant genetic effects. Single pair design which only allow for the separation of additive genetics from the environmental effect, so that any dominant or common environmental effect would inflate the heritability estimates.

In summary, improvement in hatchability from AI in shrimp using nested designs and development of procedures to control external or in vitro fertilization to create factorial designs may be advantageous in terms of the accuracy of estimated genetic parameters.

15.4 Breeding Programs Design

Generally speaking, the evaluation of breeding programs should be ideally made based on biological and economic considerations related to the value of the expected selection responses, the associated costs for a particular testing, and selection strategy and the costs associated with the increase of inbreeding rate/loss of genetic variation for a given time horizon (Gjedrem 2005; Ponzoni et al. 2007; Lind et al. 2012; Nguyen 2016; Janssen et al. 2018). Given the complexity of the actual programs, and the difficulties in accurately determining the costs and the correct time horizon, an optimization of all factors affecting the program is seldom made, and many decisions in breeding operation in practice are ad hoc, and based on approximate considerations of efficiency of the different components of the entire program taken separately.

Several methods may be implemented when using selection for shrimp genetic improvement, ranging from individual selection with an undefined population structure, individual selection with single pair mating structures, within-family selection, family selection, and combined selection (Bentsen and Olesen 2002; Gjedrem 2005). Not all methods are suitable for the improvement of all traits, and in many instances combinations of methods can be used in programs that may use multistage selection. An important part of the testing stage in breeding programs is to determine the best mating structure design. So far, most family-based selection programs in shrimp are based on hierarchical sire/dam structures obtained with AI or single pair-mating schemes where sires and dams are used to create a number of full-sib families (Gjedrem 2005). Additionally, as mentioned above, factorial design, which can be useful to separate the effects of additive genetics, dominant genetics, and maternal effects, is another option to increase accuracy.

Research regarding genetic responses using either complete or incomplete factorial mating in fish breeding found advantages of using these designs over nested and single-pair designs, either from the point of view of increasing genetic gains, and reducing the loss of genetic variability and inbreeding increase, indicating an advantage also associated with programs with a larger number of families (Engström et al. 1996; Dupont-Nivet et al. 2006; D'Agaro et al. 2007; Busack and Knudsen 2007). In summary, nested designs could help improve the separation of c-effects from additive genetic effects compared to single-pair mating. Factorial mating designs would allow for increased rates of genetic response and reduced loss of genetic variability along the same line of results obtained for genetic parameter estimation.

15.5 Diffusion of Genetic Superiority

In terrestrial animals, one of the main uses of AI is to increase the relative use of genetically superior males and shorten the lag between the genetic level of the nucleus and the commercial population. In aquaculture animals including shrimp, this objective seems to be largely unnecessary because of the high fecundity of females. Therefore, there is a possibility of shortening any lag created by the presence of a multiplier layer by directly producing all commercial larvae directly from the nucleus. We can conclude that looking for extension of semen for the diffusion of genetic superiority seem unnecessary, therefore a relatively small increase in the female: male ratio may be sufficient to optimize the breeding program or genetic parameter estimation experiments at the nucleus level.

15.6 Control of Loss in Genetic Diversity and Inbreeding Increase

Strategic use of mating and relationship information are key to controlling the loss of diversity and inbreeding increase implied in genetic selection programs in any animal population (Gjedrem 2005; De los Ríos-Pérez et al. 2015). Therefore, flexibility in mating is a positive feature associated with the option of increasing the number of families. That management could begin with reducing the possibility of co-selection of related parents as much as possible by design (Bentsen and Olesen 2002). Moreover, mating could be made between selected animals, or mating could even be obtained avoiding as much as possible future relationships between animals in the population. The estimation of relationships may be based on either pedigree or genomic information (Yáñez et al. 2014). If a more complete control of reproduction is possible, optimization algorithms may be used to determine mating needed to maximize genetic response at fixed levels of inbreeding/loss of genetic variability (D'Agaro et al. 2007).

The use of frozen conserved semen from the first and second generations could be used in genetic conservation programs as a tool to avoid loss of diversity by reducing genetic drift in the population (Sonesson et al. 2002). In this program, inbreeding levels (F) will reach an asymptotic value of 1/(3 N) where N is the population size.

15.7 Migration and Crossbreeding Design

Whether used to replace current population or to create a new *synthetic* population based on a stable mixture of genes from two or more populations or to create a crossed commercial population by mating pure lines, migration is an important tool in changing populations (Gjedrem 2005). That possibility is even more significant in situations where the traits to be improved lack sufficient genetic variation and where substantial genetic differences between populations exist. An example of this situation is found in disease resistance between long exposed and recently exposed populations to particular diseases (Castillo-Juárez et al. 2018). Therefore, gamete conservation such as semen freezing may be a very useful tool to accelerate and facilitate the migration of genetic material between different places.

An additional area where gamete conservation may impact breeding programs is when implementing diallel experiments involving several lines to optimize crossbreeding programs. In this case, the availability of IVF or external fertilization would help obtain faster results, allocating the crosses.

15.8 Conclusion

Artificial insemination has been very useful to estimate reliable and accurate genetic parameters and to serve as the basis for genetic evaluation between families. The future development of artificial reproductive techniques in penaeid shrimp may have further implications on the efficiency of breeding studies and practical improvement programs. Most of these developments can derive from relatively modest increases in hatching rates in AI and in extending semen to increase the number of dams mated to a sire. Developments conducive to using factorial mating designs in shrimp may have positive effects on genetic gains and the conservation of genetic variability.

References

Beirão J, Boulais M, Gallego V et al (2019) Sperm handling in aquatic animals for artificial reproduction. Theriogenology 133:161–178. https://doi.org/10.1016/j.theriogenology.2019.05.004

Bentsen HB, Olesen I (2002) Designing aquaculture mass selection programs to avoid high inbreeding rates. Aquaculture 204:349–359. https://doi.org/10.1016/S0044-8486(01)00846-8

Blanc JM (2003) Comparison of experimental designs for estimating quantitative genetic parameters in fish. Aquac Res 34:1099–1105. https://doi.org/10.1046/j.1365-2109.2003.00919.x

Busack C, Knudsen CM (2007) Using factorial mating designs to increase the effective number of breeders in fish hatcheries. Aquaculture 273:24–32. https://doi.org/10.1016/j.aquaculture.2007.09.010

Campos-Montes GR, Montaldo HH, Martínez-Ortega A et al (2012) Genetic parameters for growth and survival traits in Pacific white shrimp Penaeus (Litopenaeus) vannamei from a nucleus population undergoing a two-stage selection program. Aquac Int 21(2):299–310

Castillo-Juárez H, Campos-Montes GR, Caballero-Zamora A, Montaldo HH (2015) Genetic improvement of Pacific white shrimp (Penaeus (Litopenaeus)vannamei): perspectives for genomic selection. Front Genet 5:1–4. https://doi.org/10.3389/fgene.2015.00093

Castillo-Juárez H, Montaldo HH, Campos-Montes GR et al (2018) Heritability, genetic line and inbreeding effects on resistance of Whiteleg shrimp Penaeus vannamei Boone 1931 to acute hepatopancreatic necrosis disease (AHPND) in Mexico. Asian Fish Sci 31S:88–101. (in press)

D'Agaro E, Woolliams JA, Haley CS, Lanari D (2007) Optimizing mating schemes in fish breeding. Ital J Anim Sci 6:795–796. https://doi.org/10.4081/ijas.2007.1s.795

De los Ríos-Pérez L, Campos-Montes GR, Martínez-Ortega A et al (2015) Inbreeding effects on body weight at harvest size and grow-out survival rate in a genetic selected population of pacific white shrimp penaeus (Litopenaeus) vannamei. J World Aquac Soc 46:53–60. https://doi.org/10.1111/jwas.12169

Dupont-Nivet M, Vandeputte M, Chevassus B (2002) Optimization of factorial mating designs for inference on heritability in fish species. Aquaculture 204:361–370. https://doi.org/10.1016/S0044-8486(01)00839-0

Dupont-Nivet M, Vandeputte M, Haffray P, Chevassus B (2006) Effect of different mating designs on inbreeding, genetic variance and response to selection when applying individual selection in fish breeding programs. Aquaculture 252:161–170. https://doi.org/10.1016/j.aquaculture.2005.07.005

Engström G, McMillan I, McKay L, Quinton M (1996) Comparison of mating systems in a fish population: a simulation study. J Anim Breed Genet 113:559–566. https://doi.org/10.1111/j.1439-0388.1996.tb00645.x

Gjedrem T (2005) Selection and breeding programs in aquaculture. Springer Dordrecht, Berlin

Henderson CR (1973) Sire evaluation and genetic trends. J Anim Sci 1973(Symposium):10–41. https://doi.org/10.1093/ansci/1973.Symposium.10

Janssen K, Saatkamp H, Komen H (2018) Cost-benefit analysis of aquaculture breeding programs. Genet Sel Evol 50:1–16. https://doi.org/10.1186/s12711-018-0372-3

Kempthorne O (1957) An introduction to genetic statistics. Whiley, New York

Lind CE, Ponzoni RW, Nguyen NH, Khaw HL (2012) Selective breeding in fish and conservation of genetic resources for aquaculture. Reprod Domest Anim 47:255–263. https://doi.org/10.1111/j.1439-0531.2012.02084.x

Montaldo HH, Castillo-Juárez H, Campos-Montes G, Pérez-Enciso M (2013) Effect of the data family structure, tank replication and the statistical model, on the estimation of genetic parameters for body weight at 28 days of age in the Pacific white shrimp (Penaeus (Litopenaeus) vannamei Boone, 1931). Aquac Res 44:1715–1723. https://doi.org/10.1111/j.1365-2109.2012.03176.x

Nguyen NH (2016) Genetic improvement for important farmed aquaculture species with a reference to carp, tilapia and prawns in Asia: achievements, lessons and challenges. Fish Fish 17:483–506. https://doi.org/10.1111/faf.12122

Ponzoni RW, Nguyen NH, Khaw HL (2007) Investment appraisal of genetic improvement programs in Nile tilapia (Oreochromis niloticus) 269:187–199. https://doi.org/10.1016/j.aquaculture.2007.04.054

Sonesson AK, Goddard ME, Meuwissen THE (2002) The use of frozen semen to minimize inbreeding in small populations. Genet Res 80:27–30. https://doi.org/10.1017/S0016672302005712

Thompson R (2008) Estimation of quantitative genetic parameters. Proc Biol Sci 275:679–686. https://doi.org/10.1098/rspb.2007.1417

van Vleck LD (1999) Implications of cloning for breed improvement strategies: are traditional methods of animal improvement obsolete? J Anim Sci 77(Suppl 2):111–121

Wilmut I, Haley CS, Woolliams JA (1992) Impact of biotechnology on animal breeding. Anim Reprod Sci 28:149–162. https://doi.org/10.1016/0378-4320(92)90101-I

Yáñez JM, Houston RD, Newman S, Palti Y (2014) Genetics and genomics of disease resistance in salmonid species 5:1–13. https://doi.org/10.3389/fgene.2014.00415